Dustjacket:
Photograph by M. Pope
Design by Ian Allan Studio

Back cover (top):
London, Brighton & South Coast Railway 'Atlantic' No 40 passing Tooting Bec Common in 1921 with the 'Southern Belle'.
From a classic coloured postcard of the day, issued by the Locomotive Publishing Company.

Back cover (middle):
Rebuilt 'West Country' Pacific No 34017 *Ilfracombe* speeds past Raynes Park with the down 'Bournemouth Belle' in soft December sunlight; 1961. *J. P. Mullett/Colour-Rail*

Back cover (bottom):
British Rail Mk 3b Pullman No 11088 *Sir John Brunner* at Liverpool Lime Street. Note the outward opening doors, unfortunately breaking with a time-honoured Pullman tradition. *J. R. Dungate*

PULLMAN

1

1
The interior of a Pullman Car in the 'Bournemouth Belle' rake. *BR*

2
The 'Golden Arrow' resumed running in April 1946 and was usually hauled by a Bulleid Pacific, decorated with large golden arrows and the flags of Britain and France. 'West Country' class No 21C140 is seen at Victoria.

GOLDEN ARROW
4
21C140

PULLMAN

Travelling in Style

The story of George Mortimer Pullman's classic American invention and how it came to Britain, by

BRIAN HARESNAPE

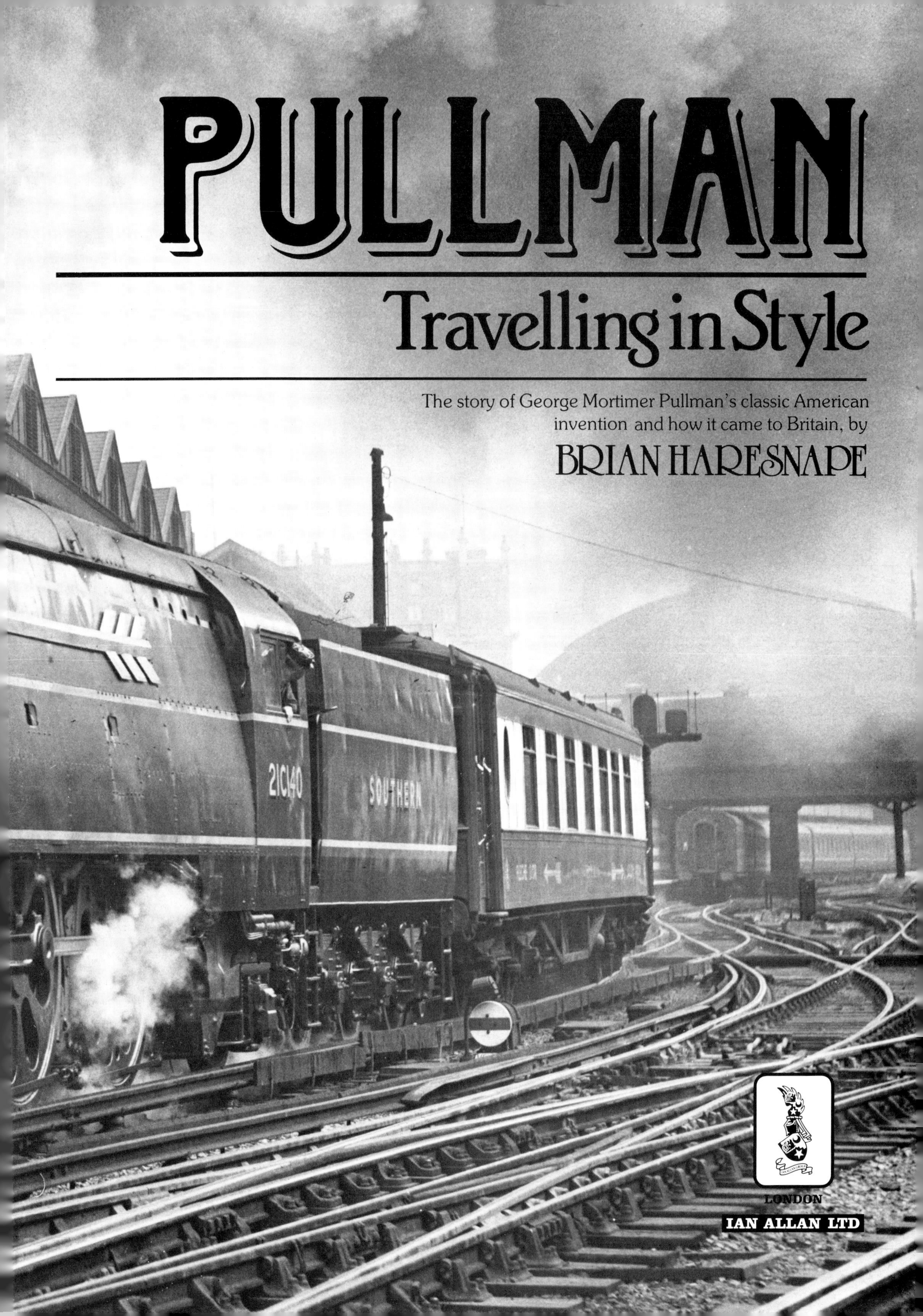

LONDON

IAN ALLAN LTD

For Michèle; who has never travelled in a Pullman car.

First published 1987

ISBN 0 7110 1648 8

Published by Ian Allan Ltd, Shepperton, Surrey; and printed by Ian Allan Printing Ltd at their works at Coombelands in Runnymede, England

'Improvement is our Watchword' —
George Mortimer Pullman

Photographs from the Locomotive & General Railway Photographs collection appear courtesy of David & Charles Ltd.

Contents

3

3
The down 'Sheffield Pullman' is seen at Welwyn Tunnel North headed by an almost new Type 4 diesel (later Class 40) No D248. *Derek Cross*

4
Copley Hill MPD has turned out Peppercorn Class A1 No 60120 *Kittiwake* to work the up 'Yorkshire Pullman' express from Leeds on 3 November 1951. The train is seen at Langley with No 60120 picking up water.
E. D. Bruton

YORKSHIRE PULLMAN
60120

Brian Haresnape: An Appreciation

The tragic and untimely death of Brian Haresnape coinciding with his completion of this book has been a sad loss to his friends. It was also a great loss to the railway community – particularly to British Rail and to its Design Panel whose Boswell he had been in his objective and independent assessment of the Board's design policy since the Panel's inception in 1956.

He was perhaps the only writer who consistently devoted his talent over the years to a continuous and critical appraisal of environmental design on the railways, giving a fair and constructive appreciation of its history and the potential of new developments. In this, his experience and research of his subject was unique.

As an artist he had been associated with British Rail in the development of potential new livery applications for rolling stock in anticipation of the introduction of the new corporate identity programme in 1965. He was generous in his encouragement and support of the Board's Industrial Design department and assisted in the implementation of this policy.

His books such as *Journey by Design* were based on a detailed personal assessment and experience of the history that he so ably unfolded to the reader.

This last book *Pullman* has a wider transatlantic and international connotation. George Pullman's original concept of rail travel created a standard of passenger comfort that led to the name Pullman passing into the language as synonymous with a quality of service and luxury previously associated with the great hotels of the world and brought romance to the railways.

The Pullman concept was indeed to provide a hotel on wheels – the operational and financial problems this involved are well documented in this book. The long journey times of several days on American railroads with no affective competition provided a ready market for the new and superior designs of sleeping and restaurant cars. This grand scale of American rolling stock could not of course be brought across the Atlantic to support the European interpretation of the original Pullman concept. British railway companies of the day with inherent gauge restrictions and much shorter journey times could only offer a somewhat condensed version of the

American Pullman experience. The subsequent decline of this service was due to economic and operating constraints rather than any rejection of the concept on the part of the passenger.

There has recently been a revival of the Pullman-style service by SNCF and BR InterCity. SNCF took the step of commissioning new interior designs of a high quality to provide a unique restaurant with a catering system that translates the epoch of the great chef to wheels in an operational method that promises economy with personalised service and gastronomic excellence.

The revival and relaunch of BR InterCity Pullman service has relied on the development of a 'modular' catering system akin to the airline methods, to provide economy in service and quality and choice of food. The relaunch relied for new rolling stock on the introduction of the Mk 3b first-class open saloon, which carried the BR version of the Pullman crest and continued the tradition of named coaches.

The basic Mk 3 carriage design had in the Mk 3b been updated with a new extruded alloy component system developed in-house by the BR industrial design team providing indirect ceiling lighting, new luggage racks, interior trim and seating. The reclining seats were the production version of the original APT prototype first-class seat. This interior concept, with pleated curtains and new moquette upholstery and end-wall carpet murals, was well received by passengers in market research.

With the introduction of new carriage design on the electrified East Coast main line, InterCity may be in the position to provide yet further improvements in the quality of the passenger Pullman environment.

Whether the marketing of the Pullman name in the context of contemporary rail travel is a reasonable translation of George Pullman's original concept can perhaps best be judged by the reading of this book. Certainly the true essence of the Edwardian grandeur of the experience can perhaps only be achieved by the attention to accommodation and service offered by such special trains as the revived 'Orient Express'.

The packaging of Pullman to meet the requirements of current accounting does not perhaps do justice to the memory of George Pullman.

The research that Brian Haresnape has undertaken in the preparation of this book and his personal interpretation of the story he unfolds is a further tribute to his memory as a railway journalist. Perhaps British Rail could be persuaded to mark the publication of this book by naming a Pullman Car after him. He would have enjoyed that. If there is a celestial railway, be sure that Brian Haresnape will be travelling Pullman.

James S. Cousins
July 1987

5

6

5
The 'Thanet Belle' near Chislehurst, Kent. This train was inaugurated in 1948 and is seen here headed by Bulleid Pacific No 21C157 *Biggin Hill*. *E. R. Wethersett*

6
The up 'Queen of Scots' Pullman headed by a Type 4 diesel-electric passing the down 'Queen of Scots' headed by an 'A3' Pacific on 30 July 1960. *L. H. Callon*

Special Note

Brian Haresnape died before he had had an opportunity to check the page proofs of this book. Since I had read his original draft manuscript in 1985 and had subsequently been in regular contact with him about the project, the publishers asked me if I would undertaken this task – which I was happy to do.

In the two years that the book was in preparation, many correspondents volunteered new information to illuminate some of the more shadowy corners of the Pullman story. With access to Brian's papers, I have been able to incorporate a number of well-attested emendations, both in the text and in the various stocklists. But it would be premature to claim that the tables constitute a definitive record of British Pullman cars. I would be pleased to hear from anyone (c/o Malaga Books) who has any additional or corrective information – particularly if it is supported by contemporary written records or photographs.

Charles Long
Leatherhead, Surrey
July 1987

Foreword by George Behrend MA FRGS MCIT

The decision to introduce new Pullman car services on British Rail's busy InterCity route to Yorkshire, and also to Merseyside from London in May 1985, has opened a new chapter in the long history of Pullman services in Great Britain, and, if run properly, will ensure that the qualities associated with the Pullman name in rail travel will become recognised by a whole new generation of travellers.

Like steam trains on BR, Pullmans simply refuse to die, and in the late 1970s, when BR had only one such train — the 'Manchester Pullman' — still in service, the preservationists were busy getting Pullman cars into service on private tourist railways, and lobbying to operate private trains on BR. In fact the Pullman revival really began in 1977 with the Kent & East Sussex Railway's restored cars in the 'Wealden Pullman' which carried 2,500 passengers in 1983, 2,770 in 1984, was almost fully booked for 1985, and probably will be full for 1986. It runs on Saturdays in the former Pullman Car Co's best tradition of exclusiveness, discipline, pride in the job, elegant menus exquisitely served, and the notion that railway passengers are more important than the staff.

Fortunately BR has many dedicated officers who think Pullman comfort is splendid and have enthusiastically welcomed the return of private enterprise Pullmans to its tracks, in particular the charter activities of Pullman Rail on tours of all Britain, especially Scotland. In the southeast of England, that most comfortable of all Pullman trains runs twice weekly to a regular timetable between London and Folkestone Harbour, named the Venice Simplon-Orient-Express, though only one of the cars in it has ever been to Venice (in 1926). Its passengers proceed to 'Serenissima' in an equally sumptuous continental train.

Though many people may wish only to read about the Pullman Cars, it is the rest of the Pullman Heritage that I wish to mention here. George Mortimer Pullman's 1874 former Midland Railway operation had ended, but those between London and Brighton of 1875 were bought in 1908 from his Executors by Lord Dalziel, who founded the (British) Pullman Car Co. Though operations were almost totally suspended during World War 2, the company's flourishing contracts with the Southern and London & North Eastern Railways (including the just-revived 'Yorkshire Pullman') lasted until 1962. The company's survival was wholly accidental, as the Government forgot to nationalise Pullman in 1947 along with the railways. So for 15 years this private enterprise (paying 15% dividend in its last years) remained on nationalised BR, who in 1960 extended Pullman operations to the London Midland and Western BR Regions, where they had never been tolerated by railway officers since the early 1930s.

On 1 January 1963 the Pullman Car Co Ltd was absorbed by British Transport Hotels Ltd (itself a subsidiary of the newly established British Railways Board) which, through the British Rail Catering Services wing, was then also responsible for ordinary BR restaurant car services. Eventually, management of the Pullman services too, passed to BRCS (known as Travellers-Fare from 1973) in which, however, a number of ex-Pullman employees were to hold key positions over the years. But the Pullman mystique was adulterated, and the number of Pullman trains declined, until just one survived: the London Midland Region's 'Manchester Pullman'. This lingered on until 1983, when there was talk of removing the Pullman name and the supplementary fare, and calling it the 'Manchester Executive'. The regular passengers were loyal to their Pullman train, however, and did not want it changed. So in October of that year BR gave the cars new livery, improved interiors and some good publicity, and in no time at all passenger receipts went up by some 20%. Thus the future of BR's own Pullmans was restored at the eleventh hour, and now new BR Pullmans are being built at Derby, the self same city that had assembled George M. Pullman's cars from parts sent from the USA in 1874!

'Pullman means Super, the best' wrote Sir John Elliot, the company's last Chairman in the Foreword to my book *Pullman in Europe*. This 1960s book made first class passengers aware that they were missing something under railway Nationalisation. At last BR is redressing the balance, providing Pullman lounges at King's Cross, Leeds, Euston, Manchester and elsewhere into the bargain, so that Pullman passengers no longer have to queue with the second class ticket holders.

Brian Haresnape brings the trained eye of a successful artist to the story of the Pullman Car in Britain, and as author of *BR 1948-1978 A Journey by Design* is well qualified to describe the Pullman Cars' artistic craftsmanship and style, which he does with aplomb.

Fliquet,
Jersey, CI
1985

7

7
Car No 279 *Hazel* entered service in the 1933 'Southern Belle' set, renamed the 'Brighton Belle' in 1934, this vehicle is now preserved at the Black Bull Inn, Moulton, North Yorkshire. *IAL*

Introduction, The Pullman Heritage

On the first day of June 1874, the Midland Railway Co introduced a new Pullman Car train, between Bradford and London St Pancras. Without knowing it, the Midland also introduced a new word to the British vocabulary, a word which was in fact a name, and a name that in North America was already becoming a household word! For whether we speak of the man, George Mortimer Pullman, or the railway carriage for which he was to become world-famous, we speak of Pullman as representative of a whole attitude to life whilst travelling by rail. Pullman's contribution started in America and was born of necessity, at a time when overland travel was fast developing, but full of rigour and discomfort. This book is an examination of his contribution, and that of his successors. It is not a technical treatise about how Pullman Cars were built and operated; that has already been written by others. It does not survey in detail the good quality of the food and wines, of the dedicated service and of the consideration for the passengers, which are hallmarks of the name Pullman; that also has been most ably written by others. In these pages I am primarily concerned with the development of the art and style of the Pullman Car at first in America and then in Britain in particular, and this is viewed against a backdrop of changing social times and tastes.

The initial idea of this book came to me whilst dining in a Pullman car, in 1978. The car concerned was *Malaga*, and it had not turned a wheel since 1963! The fact that we commenced our lunch in Shepperton station, and finished it there, was no reflection upon British Rail, because *Malaga* is alongside their tracks and does not belong to them any more. Car *Malaga* was once included in the postwar 'Golden Arrow' service, and was refurbished in 1949 for use by HM King George VI for his journeys on the Southern Region. It now belongs to the publisher of this book; Ian Allan. It is one of a great many vintage Pullman cars (having, in this case been constructed in 1922) which survive today in various ownerships; a story which is recounted in Chapter 19. Fellow guests of Ian Allan at this particular luncheon included Mr David McKenna, then a member of the British Railways Board, and Chairman of the BR Design Panel, and Mr James Cousins the BRB's Director of Industrial Design. We were discussing the launching of my book *BR 1948-1978 A Journey by Design*, and when the conversation inevitably got around to Pullman cars there was a slightly embarrassing moment when David McKenna recounted that it was he, as General Manager of the Southern Region at the time, who had authorised the withdrawal of the last Pullman cars still in SR operation, in 1972, when the world famous 'Brighton Belle' and 'Golden Arrow' ceased to run. I will refer to this again in Chapter 18; armed with the first hand account of what took place, given to me by Mr McKenna.

There was no question in anyone's mind at that excellent 1978 luncheon party that there could be a revival of Pullman cars on BR, although some modern cars were still earning enough revenue to remain in service on the London Midland Region, of which more anon. As we sampled the delightful environment of *Malaga* — truly fit

8

8
George Mortimer Pullman (1831-1897), creator of the Pullman Car Co, the leader in luxury train travel in the mid-19th century. *Pullman Car Co*

9
The heading of an 1877 share certificate, which depicts St Pancras station, London, on the left and the Detroit workshops on the right, whilst in the centre is an incredibly elaborate monogram. The British train shown (apparently running down the Euston Road!) has obviously been drawn by an American artist, without due regard to the accurate details of the Pullman cars actually built for the Midland Railway — the nearest one being identified as a 'Pullman Parlor Car' named the *President*, and running on six-wheel bogies, such a car never ran in Britain. *Doug Lindsay collection*

9

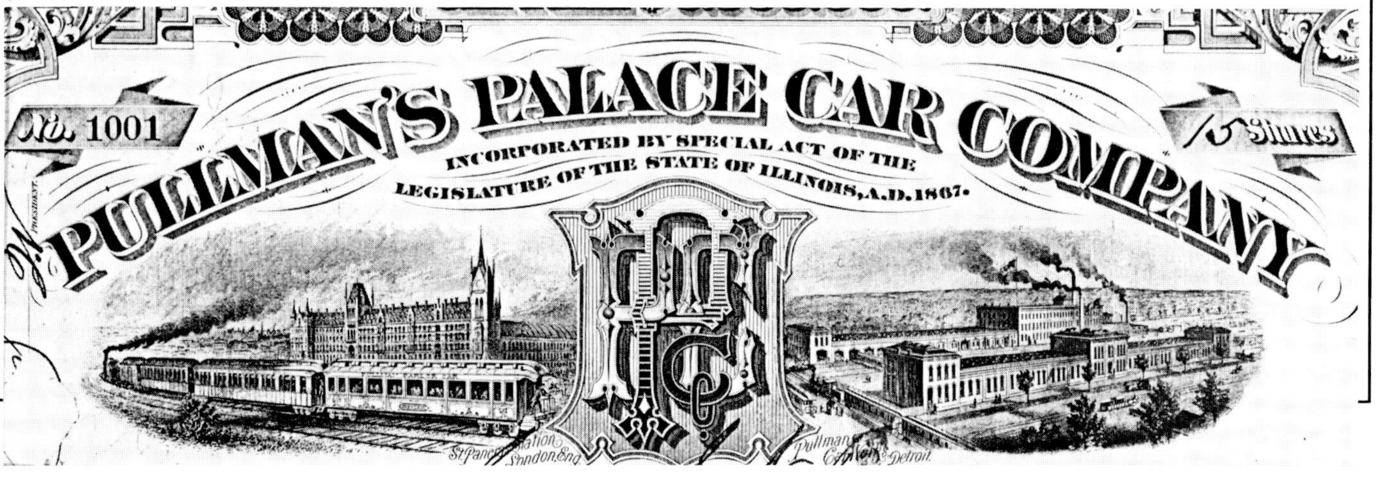

10

11

for a king! — it was with an obvious air of nostalgia, and perhaps more than we would care to admit with a feeling of regret at their passing.

Later that day I made a search in the Ian Allan Library, to find what had already been written on the subject.

There were quite a number of small booklets devoted to particular Pullman trains, and a pleasing little tribute entitled *Pullman and Perfection* published by Ian Allan in 1948, and compiled by F. Burtt and W. Beckerlegge. Pride of place went to George Behrend's splendid 1962 *Pullman in Europe*, also published by Ian Allan Ltd. In his book, Behrend traces the complex story of Pullman car operations after George Mortimer Pullman had brought them to Europe, and he gives a vivid picture of the problems and competition that arose there. In 360 pages, Behrend gives a fascinating and complete review of virtually every aspect of Pullman car operation up to the year 1962, and there is a table at the end of the book listing every Pullman car known to have operated in Europe, with names, numbers, building dates, dimensions and other relevent details. This 'bible' of the European Pullmans cannot be bettered as a standard reference work, and I am grateful indeed to have been able to use it as such, and to have had the benefit of George Behrend's advice during the preparation of this present work; for which he has also very kindly supplied the Foreword.

Several books have included brief details of the life and career of George Mortimer Pullman, none to my knowledge have fully examined the factors leading up to the introduction of Pullman cars in America (except in the briefest of ways, and nearly always without realising that Pullman was neither the first to introduce sleeping cars, nor indeed the only one to do so!). From time to time Pullman has been credited with a lot of things he did not invent, one of them being the gangway between carriages for example, and in my text I am careful to establish the true origins of features used by Pullman for his railroad cars, where known. G. M. Pullman was, to quote C. Hamilton Ellis: 'one per cent inventor, nine per cent improver and 90 per cent businessman'. He was also by all accounts a very religious and law abiding man, and one with a sincere

10
American railroad travel in the 1860s, showing the sort of conditions upon which G. M. Pullman sought to improve. The conductor and brakesman (with lantern) try to calm a drunkard, whilst passengers endeavour in vain to relax in the low-backed hard seating. The idea of producing proper sleeping berths was already around when Pullman commenced an interest, but it was Pullman whose name was destined to go down in history as initiator of luxury mid-19th century rail travel. *Author's collection*

11
In Britain it was the daytime Pullman layout which eventually became popular, and his sleeping cars were gradually eclipsed by the compartment-style layouts of other designers. On the London Brighton & South Coast Railway a fine tradition of Pullman day service was established, and continued by Mr Davison (later Lord) Dalziel, who purchased the British part of the Pullman Co in 1907. Pictured here is the interior of the 'Southern Belle' car *Princess Helen*, introduced by Dalziel in 1908. This was the first British-built Pullman train, running between London (Victoria) and Brighton. This delightful scene is one of the artist Fortunino Matania's illustrations for the original commemorative booklet, the original being in oils on canvas. Travelling in style indeed! *Author's collection*

love of beautiful craftsmanship. His contribution to the American railroads (and in due course to railways in far-off parts of the world) was that of an enterprising hotelier, intent on giving his customers the highest quality accommodation, with the finest possible service. The fact that he chose to put his hotels upon wheels has been to our lasting benefit, provided that we accept the idea of paying a supplement for the privilege of Pullman travel. The supplementary fare was the basis of his business enterprise, and has remained the hallmark of Pullman service to the present day.

In passing, it should perhaps be observed that it there was one aspect of Pullman services that was not always greeted with enthusiasm by the travelling public, it was undoubtedly the payment of a supplementary fare for their use. As Charles Long, a former Pullman employee, observed in a letter to the author: 'Pullman clientele did not always pay their supplementary fares as a matter of choice. From this point of view, the "Brighton Belle" in particular, was a highly unpopular train with many of the day trippers who used it. Having bought cheap excursion tickets they generally did not take kindly to being required to pay two shillings a head (1960) for the doubtful privilege of a cramped second class seat in a rough riding motor coach. I can certainly recall the Pullman Car Co being abused for perpetrating a "typical nationalised industry rip-off" at a time when the NUR was campaigning against the extension of private enterprise'. Although I take Mr Long's point, I also believe there were many people who considered the supplement to be excellent value for money; myself included.

The most remarkable fact is that, had I chosen to complete this book somewhat sooner, it would then have read like an obituary! Since the Midland Railway had taken the initiative back in 1874, at times the tides of fortune had ebbed and flowed quite dramatically for Britain's Pullmans, and there had been times when all seemed lost, and when the surviving Pullmans were regarded rather like old and faithful but tired workhorses, put out to pasture to end their days, because no one quite had the heart to kill them off completely. At the time when I conceived the idea of this book, this was the feeling once again. What has happened since then is one of the most encouraging stories of enterprise and enthusiasm I have ever personally encountered. The story of the Pullman revival — for that is in effect what has taken place — forms the latter part of this book, and it is still an unfinished story. The first part of the book examines the factors that led to George Mortimer Pullman's enterprise, and how the Pullman car came to Britain. The story then revolves around the subsequent history of the cars up to and including the takeover of the British Pullmans by British Railways in 1962, and the entry into service of the 'Manchester Pullman' in April 1966 — the time that 'new blood' was injected into the Pullman image by BR. What transpired after that date was a reflection upon both national, and individual attitudes, and it very nearly sounded the ultimate death sentence for Pullmans — at least on British Rail's part. I choose the word 'nearly' with care, because so many factors weighed

12

12
Perhaps the best-loved, and most-missed, of all the British Pullman trains was the 'Brighton Belle' (originally the 'Southern Belle') and it is seen here when running a 24hr non-stop proving trial, ready for the commencement of the Southern Railway's electrified service to the South Coast in 1933. Both train and staff survived the 24hr test in flying colours; having journeyed from London to Brighton and back 12 times non-stop! *BR*

against the retention of Pullman cars at one stage, that it was little short of a miracle that the one service, the 'Manchester Pullman' survived to keep the name in the public eye. More than once the rumour got around that its days were numbered whilst I was writing, and as I said earlier, I began to think I was writing an obituary for the BR Pullmans!

However, at the same time that BR's Pullmans were threatened with extinction, the remarkable phenomenon of private preservation had begun to make its presence really known. It had earlier been spurred on by the dramatic surgery of Britain's railway routes and services that had been overseen by Dr (later Lord) Beeching. If there was one factor that the good Doctor had overlooked when performing his operations it was the sheer strength and dedication of railway enthusiasts, who refused to stand by like helpless amputees, awaiting each fresh cruel turn of the blade! Beeching was himself a sort of preservationist — with a brief to keep open the bits of BR that *paid*, in the eyes of the accountants and with hindsight he did a reasonable, if unpleasant, job. But there was no room for sentiment in his operations and he alienated the

13

14

13
The all-Pullman 'Golden Arrow' in 1954, with SR Bulleid 'Merchant Navy' Pacific No 35028 *Clan Line* in charge; seen at Chislehurst. Competition from the airlines, the private car and cross-Channel car-ferries steadily eroded the profitability of this famous train in postwar years. *Brian Morrison*

14
And then there was one! British Rail's sole surviving Pullman train, the 'Manchester Pullman', kept the famous name alive in the early 1980s, although there were persistent rumours that it was to be withdrawn and replaced by an 'Executive' service. This was mainly due to the age of the Pullman cars themselves, which could no longer be thought of as the most luxurious first class carriages available. However, the standards of personal care and service which the staff provided kept the train's popularity going. In BR grey and blue livery, the nameless Pullman cars (only the word Pullman was on the sides) are seen here passing Willesden Junction, hauled by electric locomotive No 86205 *City of Lancaster*, on a Manchester-London turn on 12 May 1980. *Colin J. Marsden*

enthusiasts — the very people most dedicated to the railway cause. It was a reaction on their part against the Beeching administration that led to the creation of the now numerous privately preserved railways, and the preservation of the necessary rolling stock to operate them, including Pullman cars. From this preservation movement there has now grown a new element: namely the restoration of Pullmans, privately operated over BR rails, as a luxury leisure activity. I refer of course in particular to the now well-established 'Venice Simplon-Orient-Express', which has done so much to restore the image of the luxury Pullman car in the public's eye, also to the private 'Pullman Rail' organisation.

I have already mentioned the 'Manchester Pullman' as being for some years the sole surviving BR-operated service, and this is fully described in the chapters that follow. What I would like to emphasise here is that this particular train kept going against the odds for some years, because of the regular patronage and support it received. Regular users were forcefully opposed to attempts to change the 'Manchester Pullman' into yet another anonymous InterCity train, and they voted with their first class tickets *and* supplementary Pullman fares! Now, an entirely new generation of Pullman cars has been constructed for the BR InterCity services, appropriately commencing with the 'Manchester Pullman' in May 1985. The pendulum has swung back in full favour of George Mortimer Pullman's ideals, and a new generation of travellers will now have the opportunity to travel in style — Pullman style! It is an ideal moment to record a birth in the family, instead of the death which seemed imminent!

I use the word 'family' deliberately, for that is the feeling which has grown up over the years, first of all in everyday railway Pullman operations, and more recently in the fields of preservation, and the active reinstatement of Pullman cars: to a feeling of belonging to a family with a very proud heritage indeed. I became aware of this in the course of my researches, whilst contacting people in all walks of life and occupations who, for one reason or another, had a part of the story to tell. Almost without exception, my requests for information, pictures, specific details or permission to use material in their possession have been met with unfailing enthusiasm for the subject and a willingness to co-operate. Not only has this made my task very pleasant, it has thrown much new light on to the underlying appeal of the Pullman car and its traditions. Pullman represents a visible manifestation of a belief in the qualities of life, together with the associated values of good design and workmanship, of attention to detail and of respect for the customer — who after all foots the bill! People were proud to work on Pullman cars in the past, and they still are today that in itself speaks volumes for the achievements of George Mortimer Pullman. The heading of an American Pullman dining car menu of the 1880s carried the following:

'As you journey through life, live by the Way.'

I always like to think that Pullman himself had that put there! Apart from *Pullman in Europe* by George Behrend, two more recent books must be mentioned, and in both cases they are highly recommended as further reading upon the subject. The first to appear, in 1983 was entitled simply *Pullman* by Julian Morel; published by David & Charles. This gives a remarkable account of the British Pullman Car Co, its services, cars and traditions. Julian Morel served with the company for 21 years, as catering superintendent. As a result he has been able to bring first-hand experience and knowledge to his very readable book, and there is a great deal of useful background information about the way the Company was run. Inevitably there is an emphasis upon the catering aspect, and very fascinating that story is, as seen through his eyes. I am most grateful to Mr Morel for his support for my own project, and I have had some very helpful letters from him. In this book I will not attempt to retread the same ground, and the reader in search of typical menus, of car working arrangements, staffing, servicing and supplying, special events, and all the other day-to-day features of Pullman operating will find all their needs satisfied in Mr Morel's book. There is also a list of British Pullman cars of the past, similar to that provided by Behrend, and particularly useful is an Appendix listing the identity and whereabouts of all the preserved Pullmans which the present writer found invaluable. It came as a surprise indeed to discover that approximately 100 vehicles fall into this latter category, a truly remarkable

15
The decline, and virtual fall of British Pullman services during the 1970s suddenly left the railway enthusiast with a gap to fill. Scores of old Pullman cars were rescued for preservation, and the search continues today. Here is the lovely old 12-wheeler *Cambria* built in 1920, which was rescued by the Kent & East Sussex Railway in 1980, after 40 years of neglect, standing in all weathers as a stationary departmental vehicle in railway yards near Sheffield. One day it will probably be running again — thanks to the enthusiasm and devotion of those who wish to retain the Pullman heritage. *Brian Stephenson*

15

proportion of the total fleet of postwar years, and certainly yet another testimony to the popularity of all things Pullman.

The second book to appear recently is *The American Pullman Cars of the Midland Railway*, by J. B. Radford, published by Ian Allan Ltd. This is an in-depth account of the first Pullmans to reach England, together with their subsequent Midland Railway brethren, and the story comes right up to date to reveal that three original car bodies have survived quite remarkably intact, and have been secured for preservation to be undertaken as a long term project. In effect, Mr Radford devotes a whole book, complete with excellent drawings and nine very detailed Appendices, to that portion of Pullman car history which only occupies a small part of this present work. Need I say more than this to the reader with a thirst for all available details of this interesting period? I have been unable to avoid some duplication of material and illustrations in this instance, because of the extreme scarcity of possible alternatives to what are basic and important pictures, or reliable contemporary factual details, without which my own book would appear incomplete. I offer here my apologies for this duplication of illustrations, and I would appeal to any reader who possesses, or knows the whereabouts of, unpublished or little known pictures of the Pullman cars of the Midland Railway to contact either myself, or Mr Radford, via Ian Allan Ltd.

Other books referred to in the course of compiling my script, and which I think might be of interest to the reader, are listed in a Bibliography, and there is also a full list of Credits and Acknowledgements. Here, I would just like to say a very sincere word of thanks to everyone who has, in some way or another helped to make this book possible, and to make the writing of it so enjoyable.

Now, dear reader, I invite you to join me on a journey which commences as one of endurance and rigour, but which later becomes, thanks to George Mortimer Pullman, an altogether more acceptable one, which is at times tinged with an aura of luxury and romance, of enterprise and dignity, of quality and nostalgia. Pullman has become legendary — indeed the man himself already was in his own lifetime — and the legend is alive and well. The name Pullman ranks almost unique, alongside that other famous quality marque of the transport world — Rolls-Royce. Indeed even the makers of motor cars and luxury road coaches, the builders of blocks of prestigious flats, the owners of cinemas and restaurants, and other necessary facets of our time, have frequently borrowed the name Pullman, in order to denote quality. However, for me, Pullman rests firmly on the rails, and that is how he will be remembered, I am sure, in the history books of future generations. Here, then, the story of the plush pleasures of Pullman and how they came to Britain.

Brian Haresnape, FRSA NDD, Gassin, Saint Tropez, France. November 1985.

Part One:

Genesis in America and Britain

CHAPTER 1

Journeys of Endurance

George Mortimer Pullman was born at Brocton, New York on 3 March 1831, some 15 months after the official opening of the Baltimore & Ohio Railroad's first passenger train service, hauled by horses, the first railroad to provide a train service for fare-paying passengers in America. In fact it was opened before the British Liverpool & Manchester Railway, which is however generally claimed to be the first specifically designed to carry fare paying passengers; this opened to the public on 15 September 1830, nine months after the B&O. Prior to these two events some passengers had been conveyed by the freight railroads, but not as their priority as a commercial enterprise, and very much at the passengers' own risk! The Stockton & Darlington Railway had improvised a passenger carriage for the opening ceremony on 27 September 1825 (as well as using open wagons to carry passengers), which was a road stage coach mounted upon flanged wheels. Although the steam locomotive *Locomotion* hauled a large number of passengers on this auspicious occasion, in everyday workings fare-paying passengers were considered of secondary importance and were conveyed in horse-drawn open wagons. By 1829 a covered carriage, looking very much like a wooden garden shed on wheels, had been provided to give more protection against the elements and thus the purpose-built railway carriage was born.

The Baltimore & Ohio Railroad seems to have copied the Stockton & Darlington's methods, and the first passenger trains were formed of four open wagons drawn by a horse, with no shelter of any description. The first service was between Mount Clare and Carrollton Viaduct (close to Baltimore) and passengers were charged nine cents to make the journey 'there and back' a total of some three miles. The first months of service were basically sold as excursion trips for pleasure, to sample the novelty of travelling upon a railroad, and to view the civil engineering works involved. As the track was extended and the journey offered became longer, the B&O wisely decided to build some covered or 'closed' cars to provide protection in wet weather. These closed cars* bore a striking resemblance to the primitive Stockton & Darlington affair already mentioned, being virtually sheds on wheels, with no seats, but a central table to place packages or baggage on. Passengers had to cling to each other for stability as the horse hauled them along the tracks, because there was no form of springing used. Seats arranged around the sides were soon considered necessary!

*Readers will perhaps be confused by the fact that so far I have used the terms wagon, carriage and car. In general American usage the the 'car' is the carriage, and as this has been traditionally applied to Pullmans, even in Britain, in this book I will use this term henceforth for Pullmans, and American carriages. Wagons, generally speaking, are for the conveyance of merchandise and minerals, although to begin which they were considered satisfactory for poor passengers of the lower orders in life!

The idea of travelling by train caught the public's imagination very quickly indeed, and by the summer of 1830 the Baltimore *Gazette* was reporting as follows:

'Notwithstanding the great heat of the weather for the last three weeks, the amount of weekly travel on the railroad has not diminished, the average receipts being above one thousand dollars per week. In the hottest time of the hottest days the quick motion of the cars causes a current of air which renders the ride at all times agreeable. In many instances strangers passing through Baltimore, or visiting it, postpone their departure for a day and sometimes longer, to enjoy the pleasure of an additional ride on the railroad. We only repeat the general sentiment when we say, it is the most delightful of all kinds of travelling.'

It must have been a pleasant enough experience, with none of the smoke and steam of the railroad of the future, and taken at a speed where very little dust was likely to be raised from the passage of the cars. Yet to come was the real adventure of steam railroad travel, which was to open up the continent of America, just as it did for Britain and Europe. These first B&O trains were 'joyrides'. Nevertheless it was not long before the B&O decided that some form of springing was going to be necessary, as journeys lengthened, and steam locomotives were introduced.

The development of the steam locomotive as an alternative to the horse proceeded at a furious pace during the first

16

16
The mutation from stagecoach to railway carriage in Britain took place over a span of some 10 years, and chiefly involved the provision of slightly longer, and more robust four-wheeled underframes, upon which three or perhaps four compartments were grouped together. The roof had provision for luggage, and a dickey seat for the train guard. Sprung buffers and link couplings made the connection between each carriage, with additional safety chains in case of coupling breakage. The style of bodywork was still basically of road stagecoach design, as seen here in this composite first and second class 1846 Stockton & Darlington Railway carriage.
National Railway Museum

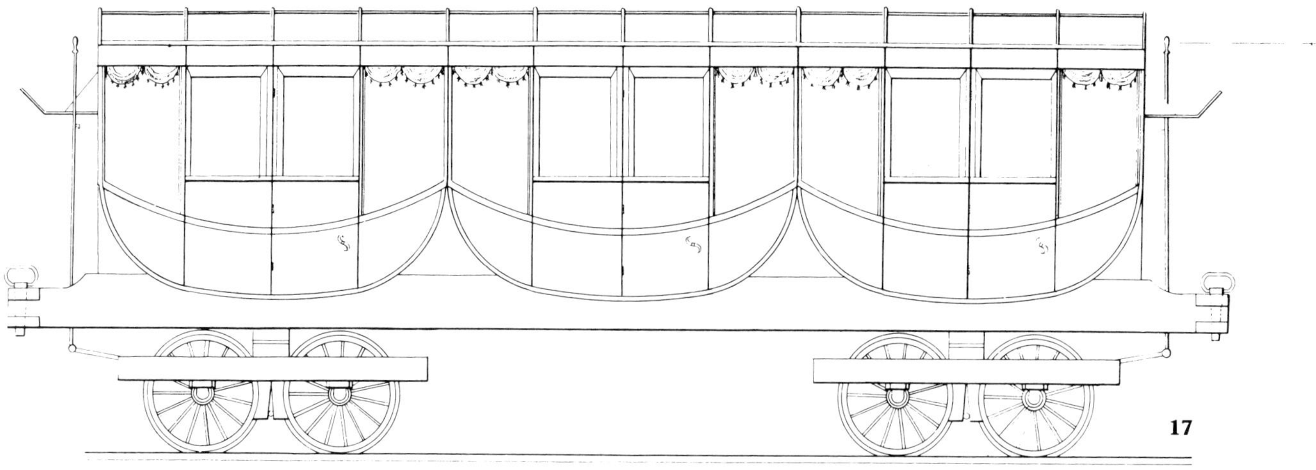

17

10 or so years of G. M. Pullman's life, and the spread of new railroads was the wonder of the age. By the time the young George Pullman was able to sit up and take notice of the world around him, the era of the steam railroad had arrived in America, and the faithful horse was being displaced as the premier mode of haulage available to man on land. As the railroads progressed, the stage coaches declined, and the expertise of road coachbuilding was to become more and more available to the demands of the railroad engineers. It was not long before some really beautiful examples of coachwork took to the rails, although the wheels and suspension upon which they were mounted were still somewhat lacking in refinement, to say the least. The Americans certainly knew how to build stagecoaches, and to adapt this skill to the railroads was a natural evolution, in order to satisfy a changing mode of transport. However, it was soon to become apparent to all concerned that operating passenger trains on steam railroads was full of new perils and rigours, and as journeys became longer, and speeds higher, the railroad passenger car was found to be inadequate in a number of important respects; because the builders lacked railroad experience. It was for example soon discovered that lengthening the stagecoach-style body, in order to carry more passengers, led to very poor riding, with a sort of to-and-fro 'seesaw' motion setting in as speed increased. The four wheels became inadequate to cope, and axles broke, causing disastrous derailments as well as giving much wear and tear on the tracks. Nevertheless, long four-wheel cars remained much in evidence until the late 1830s, when a double-truck (bogie), eight-wheel car design began to displace them; (most four-wheelers having gone by 1850 or thereabouts).

Here we should draw a comparison with contemporary British developments, because it was at this stage that the American railroad car began to assume its own particular layout and character, leaving the British builders to follow a much less ambitious course for a great many years to come. In Britain, the four-wheel and six-wheel non-bogie layouts were steadfastly developed, with a compartment style interior, and with no corridors or gangways between each car (carriage). The British railway traveller (except the poorest who rode in what were basically still open wagons) was conveyed in stolid privacy and with a door adjacent to his seat, exactly as the road stagecoaches had provided. The earliest four-wheel American cars followed a similar pattern, but seem to have permitted a considerable number of additional passengers to travel on the roof! This is a habit still to be found in some Asiatic railway operations. There also appears to have been little resistance to the idea of passengers clinging to the outsides of the car body, using available footsteps and grab-handles, and accidents were relatively numerous and of a rather gruesome nature. One man who was obliged to risk his life in this fashion was the conductor (guard) who patrolled the outside length of the train by means of narrow footboards. This arrangement was necessary because, in the compartment-(stagecoach-) style layout the passengers were effectively marooned once the train was in motion, with no means of communication in emergency. Hence the reassuring visits by the conductor, en route.

An alternative layout was tried out in 1835 by the Philadelphia, Germantown & Norristown Railroad, who joined two of these four-wheel cars together, and stripped out the compartment arrangement, providing instead cushioned longitudinal seats, with a centre gangway. Entry to these long cars was by the ends, and in some the saloon-style layout apparently included a bar at one end and a ladies' saloon at the other; with very elegant and comfortable furnishings. This two car arrangement (making in effect one large car) with saloon

17
The American railroads had almost immediately lengthened the underframes of their passenger cars; at first still using only four wheels. These rode so badly that 'double trucks' (primitive four-wheeled bogies) were soon introduced in an effort to achieve stability. This Imlay-bodied stagecoach style car still featured separate compartments, with side access doors similar to British and Continental designs of the period.

18
The idea of making the car body one long saloon quickly followed and was the start of the breakaway between American and British practice. This early car, with sprung four-wheeled trucks, had side catwalks and end platforms (for the train conductor), complete with handbrakes at each end. *Baltimore Municipal Museum*

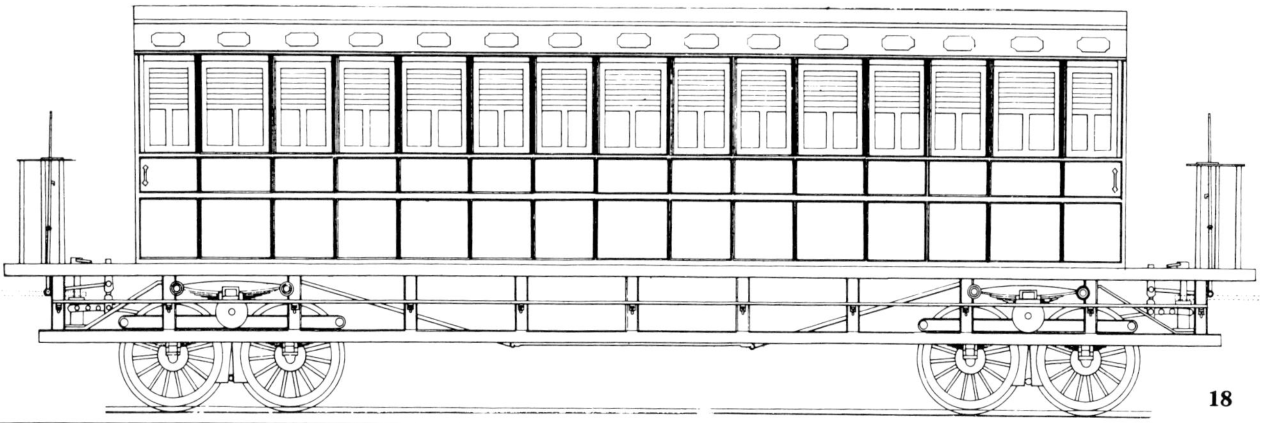

18

19

interior, heralded the shape of things to come in America. Names were carried by these cars (a relic of stagecoach practice) and the two just mentioned, with bar and ladies' saloons, had the names *Victoria* and *President*. The practice of naming cars had in fact been used by the Baltimore & Ohio right from the start, and it was in due course to become a feature of the sumptuous Pullman cars; but I anticipate my story.

It was without question the development of the eight-wheel car, with double trucks (bogies) that gave the American designers the scope to produce a more purposeful layout than hitherto possible. One designer in particular, Ross Winans of the Baltimore & Ohio, must be credited with this advance in design. His car had the name *Columbus* and appeared on the B&O rails on 4 July 1831. It was built in the company's own workshops under Winans' personal supervision, from his own drawings. This car, the first eight-wheel passenger car to run anywhere in the world, had a body twice the length of existing four-wheelers, and was mounted upon two four-wheel trucks with Winans' patented friction wheels. There were no springs, and the body was mounted on each truck (bogie) by means of a bolster and centre bolt, upon which the truck could swivel.

The greater length of the eight-wheel car was to allow a rearrangement of the accomodation within (with passengers' luggage still carried on the roof to begin with). A saloon type layout with end doors was envisaged by Winans, and these end doors were to be served by an open platform, which allowed the conductor, and the passengers, to pass from one car to the next. This was, however, to prove an extremely hazardous feat, because the link and pin coupling between cars could not control the considerable sway and motion between each car, in particular on curved sections of track. Following a number of accidents in which passengers either fell off completely, or got severely crushed between two car ends, it became necessary to restrict this movement of passengers whilst the train was actually in motion! An iron verandah with waist height railings was then added to give more protection to people standing on these open ends, as was their habit, despite the warnings.

The centre-aisle layout which was to become known as the 'day car' in America, with two-by-two seating arranged each side, and with windows adjacent was not immediately favoured, however.

19
This 1834 double truck car already displays the general features which came to be recognised as the typical American 'day car' with a saloon interior, verandah ends with doors opening into the saloon, handbrake (with conductor shown holding the wheel), sprung bogies, large roof-top assemblies for the candle illumination, and central stovepipe for the heater.
Baltimore Municipal Museum

Experience with *Columbus* led the B&O chief engineer Jonathan Knight to report to the company that he strongly recommended the use of springs for the eight-wheeled cars. This the company approved and quickly sought a suitable design from inventors of the day. Another result of experience with the first eight-wheeler was revealed in 1832, one which casts an interesting sidelight upon one of the American social customs of the day. This was the habit of tobacco-chewing and subsequent spitting ('expectorating') into spittoons, or on the floor. Thus, when the B&O directors were considering the design of new eight-wheelers, they opposed the centre aisle layout with seats for two persons each side, on the grounds that passengers would be able to expectorate at will upon the extent of floor provided by the aisle, and would soon turn it into one long spittoon. As a result the car design was rearranged with doors on the sides, with half height divisions backing the seats across the full width of the car (a semi-open/compartment layout) and the conductor could only reach the passengers by means of a narrow catwalk (footboard) down each side of the car. It is not recorded in what fashion the passengers expectorated in this design of car! Spittoons were doubtless provided.

Fortunately, common sense soon prevailed, and the advantages of the centre-aisle layout were recognised to outweigh these objections, and the 'day car' was to become the standard American layout; whereas the British steadfastly continued to favour the compartment arrangement. Winans developed his four-wheeled truck design so that it could carry much longer bodies, and it was fitted with adequate springing and braking to bear the weight. The 'day car' was to have many variations upon the basic theme in the course of American railroad development, but the principle has remained true to Winan's concept, and it has had an enormous influence upon the travelling habits of Americans, from its inception. One quick reaction was that the 'open' layout exposed ladies and children to the noxious habits of the male traveller, in particular tobacco-chewing and its aftermath. As a result a number of railroads introduced either complete 'ladies' cars' (in which men could only travel if accompanying the lady) or 'ladies' retiring rooms' in one section of a day car, offering a discreet retreat if required. There was objection to this from some operators who saw the loss of revenue-earning space, or the provision of half-filled railroad cars specially for women, as poor business.

A very good description of a new train of cars, built in 1842 for the Auburn & Rochester Railroad, at a price of around $1,700 each, by Davenport & Bridges of Cambridge, Mass was given in the Rochester *Evening Post* at the time of their introduction:

'Yankee ingenuity is rarely more pleasantly exemplified than in the luxurious arrangement for railroad travelling of which we now have in Rochester a magnificent specimen in the splendid train of passenger cars just launched on the Auburn & Rochester Railroad. There are six cars designed to form two trains. The cars are each 28 feet long and 8 feet wide. The seats are well stuffed and admirably arranged with arms for each chair and changeable backs that will allow the passengers to change front to rear. The size of the car forms a pleasant room, handsomely painted, with floor matting; with windows secure from jarring and with curtains to shield from the blazing sun. We should have said rooms, for in four out of the six cars (the others being designed for way passengers) there is a ladies' apartment with luxurious sofas for seats and in recesses may be found a washstand and other conveniences. The arrangement for ladies we consider the greatest improvement and it will remedy some serious objections that have hitherto existed against railroad travelling on the part of families, especially where any of the members are in delicate health. The ladies may have their choice either of a sofa in their own apartment or a seat in the main saloon of the car as their health or inclination may require. The cars are hung on springs and are of such large size that they are freed from most of the jar and especially from the swinging motion so disagreeable to most railroads. The lamp of each car is so placed as to light inside and out and last but not least, the breaks are so arranged so to be applied readily and with great power.

'On the whole it would be difficult to imagine any improvement that could be desired, though we dare say, these down-easters will rig up some new notion ere long that will furnish board and lodging as well as a mere passage on the railroad.'

The last observation made was indeed prophetic! At that time George Mortimer Pullman was 11 years old.

The Rochester *Evening Post* report, above, mentioned 'a washstand and other conveniences'. As journeys lengthened, the needs of passengers became more respected by the railroads, and toilet facilities had to be incorporated into the Winans' day car layout. There was also a need for interior illumination, other than by the windows, and of some system of heating and ventilation. In fact, at a time when inventors seem to have been in great abundance in America, a good many 'patent devices' were produced to satisfy these needs — with varying degrees of success. It also became apparent that passengers wanted to relax themselves by 'putting their feet up' whilst travelling. This could cause great annoyance to the traveller in front of the offending pair of feet which were perched upon his seat-back! As yet journeys were not so long that all-night travel was really necessary, and most passengers were quite satisfied with day travel only. Indeed, an all-day journey upon the hard seats of a 'day car' with constant jolting and jarring as the cars rode the indifferent quality tracks beneath them, sent many in search of a comfortable bed for the night in the nearest hotel along the line!

I mention the indifferent quality tracks here, because they played a major role in the many accidents that befell the early American railroad passenger. Some of the routes were built in great haste, to capitalise upon the 'bonanza' of the railroad business, and the tracks were too lightly laid and soon became distorted by the passage of trains and by the cold weather. The heavy unsprung masses of the early steam locomotives and four-wheeled cars, and even the early-type Winans' trucks must have played havoc upon the iron rails. Locomotives were frequently derailed, in a good many instances by cattle which strayed upon the unfenced railroads, and a journey was one not only of discomfort but also one of potential danger to life and limb! Nevertheless the American public accepted this, in a pioneering spirit of adventure, and the railroads flourished. When Winans' 'day cars' were the prime form of land travel available, the American traveller willingly placed his life in his hands to travel at hitherto unknown speed, over previously unattainable daily distances. The fact that derailments, collisions, snow bound tracks, stray cattle, and even hostile Indians, were likely accompaniments to his journey seems to have been regarded as an acceptable price to pay for mobility and progress.

To heat the car interior a stove was sometimes, but by no means always, carried (sometimes it was portable) and this, plus keeping all the windows closed seems to have produced an almost overpowering fug when a car was filled to capacity. A traveller in the Eastern States in 1839 however recorded the following:

'The car was seated for fifty-six passengers and contained at least thirty. There was no aperture for ventilation and when night came on the company insisted on shutting every window to keep out the cold. A few who, like us, preferred cool air to suffocation congregated at one end where we opened two windows for our relief.' (From *Notes on the United States of North America during a Phrenological Visit in 1838-40*. George Combe, Philadelphia, 1841)

The presence of a stove did not always mean an improvement in travelling conditions it would seem, to judge from this 1846 British traveller's observations, which give a very vivid picture of the risks encountered travelling between Boston and New York:

'The carriage in which I found myself seated consisted of one great compartment constructed to accommodate sixty people. It was like a small church on wheels. At either end was a door leading to a railed platform in the open air, and from door to door stretched a narrow aisle on either side of which was a row of seats, wanting only bookboards to make them look exactly like pews, each being capable of seating two reasonably sized persons. The car was so lofty that the tallest man present could promenade up and down the aisle with his hat on.

'In Winter two or three seats are removed from one side to make way for a small stove and as I was rather late in taking my place the only vacant seat I could find was one adjoining this portable fireplace. My immediate companion was a gentlemanly looking man under forty and directly opposite him sat a lady of about two hundred and fifty pounds weight, whose travelling stock consisted of a carpet bag almost as plump as herself, which, as she was bringing herself to a comfortable bearing, she consigned to the safekeeping of the gentleman who sat beside me. He could not put it on the floor which was moist with expectoration nor could he put it on the stove which was already getting red hot, so he had no alternative but to carry it the whole night upon his knee, as the ladies are used to such attentions in America.

'Finding the heat of the stove uncomfortable, I repaired to one of the platforms attached to the car, where I enjoyed myself in the open air, smoking a cigar and observing the country through which we passed I had not been long engaged in such reflections when from the next car emerged the conductor, who is in America a peripatetic functionary.

'Having received my ticket, he was about to enter the car I had just quitted when he stopped short and, having permitted his quid to change sides in his mouth, he observed that it was "tarnation cold" and inquired if I preferred standing where I was to being seated inside. "It's a poor choice between being frozen and being roasted," I observed. He looked at me again and then intimated that he was merely acting as a friend in telling me I would be safer inside.

' "Is there any danger?" I inquired.

' "We do sometimes run off the rail, that's all," said he without the slightest emotion and then passed into the car. I preferred a slow broil by the stove to being crushed to death and repaired to my place and submitted to it until the train reach Worcester.' (From *The Western World; or, Travels in the United States in 1846/7*, Alex. Mackay, London, 1850.)

These early cars had only one or two candles, in special burners, to cast illumination when travelling in the dark. Although travellers could of course carry with them supplies of food and drink (the latter often of a strong variety!) the general practice was to eat at stations along the route. A vivid picture of what this entailed was given by an 1838 traveller:

'At every fifteen miles of the railroads there are refreshment-rooms. The cars stop, all the doors are thrown open, and out rush all, the passengers like boys out of school, and crowd round the tables to solace themselves with pies, patties, cakes, hard-boiled eggs, hams, custards and a variety of railroad luxuries too numerous to mention. The bell rings for departure, in they all hurry with their hands and mouths full, and off they go again until the next stopping-place induces them to relieve the monotony of the journey by masticating without being hungry.' (From *A Diary in America*, Captain Frederick Marryat.)

Delays en route (and these were plentiful!) would result in all the passengers raiding the nearest bar room or general store, to pass the time, as this 1839 traveller recorded:

'About ten miles from Springfield we came to a dead stop and the whole train stood motionless for three hours, enlivened only by occasional walks in the sunshine and visits to a cakestore, the whole stock of eatables in which was in time consumed, the price of them having risen from hour to hour in proportion to the demand.' (From *Notes on the United States of North America during a Phrenological Visit in 1838-40*.)

One can only reflect that the 'good humour' of these early railroad travellers must have soured a good deal as time progressed!

The greatest hazard to these pioneer railroads, and their passengers, seems without doubt to have been due to obstacles upon the tracks, and accidents and derailments were so commonplace that they scarcely warranted a special report. Take these examples, and imagine the public outcry that would result if they happened again now! All date from the 1838-46 period:

'The railroads are not fenced on the sides so as to keep the cattle off them and it appears as if the cattle who range the woods are very partial to take their naps on the roads, probably from their being drier than the other portions of the soil. It is impossible to say how many cows have been cut into atoms by trains.' (1838, from *A Diary in America*.)

'On October 29 we left Springfield and started for Worcester by the railroad, which has been opened since we travelled to Springfield a month ago. Yesterday a stray horse had its legs and head cut off on this railroad by the engine and the night before a carter had left a cart with stones standing on the track against which a train loaded with merchandise had run in the dark and been smashed to pieces. We hoped to be more fortunate and were so; but, although we encountered no danger, our patience was sufficiently tried.' (1839,from *Notes on the United States of North America during a Phrenological Visit in 1838-40*.)

Unhappily, it was not only the animal population of America that strayed on to the tracks, as this horrific story relates:

'On quitting the steamboat at Newcastle we took the railroad for twenty-two miles to Frenchtown which thus connects the Delaware and Chesapeake Bays. We had made about two-thirds of our journey when at one of the road crossings a violent jolt accompanied by a loud crash made all the passengers start and considerably alarmed some of them. The continued rapidity of our movement however, satisfied all that no accident had occurred to the cars, and in a quarter of an hour the train stopped close to the water side at Frenchtown. As we stepped out I went up to the conductor and engineer, who stood together on the platform of the locomotive, and inquired the cause of the sudden shock we had experienced.

' "Well, it was in going over a chaise and horse," replied one of them coolly.

20

20
Once established in style the 'day car' became the norm for the American railroad passenger, irrespective of class or creed, race or heritage. Within the large open saloon layout everyone had to occupy their place and their time as best they could. The stove (right-hand foreground) was either to be desired or avoided, according to the time of the day, and month of the year. This Baltimore & Ohio Railroad day car shows quite a genteel collection of passengers, almost the same as might be found in a contemporary first class British carriage. It was by no means typical! (From the *Illustrated London News*, 6 April 1861.)
Baltimore & Ohio Railroad

' "There was no one in the chaise?" asked I, anxiously.
' "Oh, yes, there were two ladies."
' "Were they thrown out?"
' "I guess they were and pretty well smashed up, too."
' "Good God! and why didn't you stop the train? Can't you send back to know what state they're in?"
' "Well, mister, I recon' they're in the State of Delaware; but you'd better jump into the steamer there or you are like to lose your passage."
'With these words the conductor turned to some other inquirer.
'Many of the passengers agreed with us that it was inhuman of the conductor not to have stopped the train and look after the injured persons. Others remarked that it wouldn't have done any good and that the train was obliged to be on time or have delayed the steamer for ten minutes or more. This was unanswerable; the subject dropped. But a few days afterward I saw in a Baltimore paper a paragraph stating that one of the ladies had been killed, the other badly wounded, the horse smashed and the chaise broken to pieces. The miracle was that the train was not shaken off the track.' (1840, from *Civilized America*, Thomas Colley Grattan, London 1859.)

One has to wonder at the fortitude of the locomotive crews who performed such involuntary feats of slaughter. Some light is thrown upon one man's reaction in this report:

'I left Baltimore by the late night train for Washington. For two-thirds of the way we went on smoothly enough, but when within ten miles of Washington a violent jerk to the whole train appraised us that we had run against something, not however, sufficiently formidable to bring us to a sudden halt or to dislodge us from the line. The engineer gradually slackened speed and on stopping the train we discovered that we had run against a cow which had been lying on the line.
' "I can stand a hog but them cows are the devil to pay," said the stoker as he proceeded with the help of some others to drag the carcass off the engine and deposit it by the side of the line.
'It was two in the morning when we reached Washington. A more miserable station than we were ushered into can scarcely be conceived.' (1846, from *The Western World: or, Travels in the United States in 1846/7*, Alex. Mackay, London, 1850.)

If the actual railroad journey and the car in which one travelled were so fraught with problems and perils, the depot (railroad station) at the beginning, at interchanges, and at the end of the journey, could also be a nightmare. There was a particular problem with passengers' baggage, it seems. Here are some more eye-witness accounts (all by the same person):

'As there was then only one train a day for the West and as for the first forty miles two railways were blended into one, the bustle and confusion which occurred at the station before starting was perfectly indescribable. Everybody was getting to the wrong car and everybody's luggage into the wrong van. At length, after a hubbub which would have been more amusing had it been less intense, the long heavy train started at four P.M. for Worcester.' (1846 from *The Western World; or, Travels in the United States in 1846/7*, Alex. Mackay, London, 1850.)

Later he wrote:

'For some minutes after we arrived it appeared to me as if Bedlam had been let loose upon the station, or dépôt, as it is universally called in America. To give a true picture of the confusion — the rushing to and fro — and the noise with which all this was accompanied is impossible. Some pounced upon the refreshment-room as if they fancied it the up-train and in danger of an immediate start; others flew about frantically giving orders which there was no one to obey; while by far the greater number were assuring themselves of the safety of their baggage. This was very necessary inasmuch as the line here branched off into two; the one proceeding to Albany and the other to Norwich en route to New York.
'It is by no means an uncommon thing for a passenger to find that his baggage has from this point taken an independent course for itself. This sometimes arises from the baggage being put into the wrong car and at others from the cars themselves being put on the wrong line. Sometimes the separations are most heart-rending, husbands and wives, parents and children being sent off in different directions. This was the case with a lady in the carriage immediately behind that in which I sat. She had been torn from both her husband and her handbox and consoled herself by abusing the conductor, until at last he was goaded into telling her that she had no business to get into the wrong train; from which he derived but little satisfaction as she insisted the whole way that it was the train that was going wrong.' (*Ibid.*)

In this instance there is also another reference to the over-heating caused by the use of a stove in the car:

'My destination on leaving New York was Philadelphia. We were conveyed across the Hudson to Jersey City, on arriving at which the passengers jumped in crowds upon the floating ship, where we landed and fled. I followed the breathless and panting crowd into a large unfinished building which I found to be the railroad station. Once within the station, the hurry-scurry if possible increased; men jostling each other and rushing in at every available aperture into the cars like so many maniacs. On venturing inside one of the cars I discovered the cause of the tumult. It appears that in Winter there is a choice of seats, the preferable ones being such as are not too near or too far from the stove and the race was for these seats.' (*Ibid.*)

And here we glimpse something of the added annoyance caused by 'touts':

'It had never been my lot to encounter such a hubbub as saluted us on entering the station in Baltimore. There was not a hotel in town that was not represented by one or two Negroes who did the touting for it, each having the name of his boniface displayed on a band around his hat.' (*Ibid.*)

One final picture of these journeys of endurance, to which the American traveller became accustomed in the first 15 or so years of railroading, must suffice. It also gives a good description of an 1853 'day car'; and there is a particularly amusing end to the report:

'Now come we to the climacteric of travelling — the railroad cars. The cars are of an extraordinary length and width, capable of accommodating about fifty-six people, having on either side fourteen seats holding two people each. There is no room to deposit any small article under one's feet and even if there was they would be spit on to a certainty; indeed, even in the event of dropping money on the floor no decent person could venture to pick it up unless he put on an old glove.
'The moment the cars start a string of filthy lads stream in offering for sale sweetmeats, apples, books and other important wares, and they are succeeded by travellers who, if they find no other accommodations, stand up in the middle of the cars and spit away. It is utterly impossible to mistake an American for anyone else. He has his feet on the seat next him, which he turns over for the purpose, or, if it is occupied, he sits with his knees "let in" to the back of it. He either sucks a piece of sweetmeat, bites a piece of wood or chews a bit of tobacco and invariably reads a newspaper.
'In Summer they have blinds to the numerous windows on each side and in many trains they hand round tumblers of cold water, while in Winter there is a stove in each car plentifully supplied with fuel.

21
Flamboyant in style, but functional in layout, the appearance of the American railroad train soon differed greatly from its British counterpart, due to the very different conditions prevailing. Traversing vast unfenced territories, running on indifferently-laid tracks, and burning wood or poor quality coals, the steam locomotive, its tender, and the passenger cars, all used bogies in order to keep the train upon the tracks. There was an enormous headlight, a warning bell, a melodious but penetrating whistle and a huge cowcatcher — all designed to protect both humans and animals, too often in vain. This Currier & Ives print portrays well the typical train; complete with the huge spark-arresting chimney on the 4-4-0 locomotive. In the background is a paddle-driven river steamboat; the ornate interior, typical of these, was to inspire George Mortimer Pullman in due course. *Author's collection*

21

'On reaching the dépôt you deposit your luggage with the baggage master who fastens a check to each separate article, giving you a corresponding check, and all your effects will be safely delivered or in case of accident made good by the company. In America a small quantity of travelling equipage goes a long way. In Georgia, especially, the luggage considered essential for a journey of any distance consists of a shirt collar and a pair of spurs, and when I asked a native of that State how he managed about a shirt he said, "I can't spare the time to wait while they're cleaning it and if I could it's almost as cheap to buy a new one."

'The occurrence of accidents is alarming in the highest degree and there can be no doubt of there being hundreds, the casualties of which have never been made known, it being deemed advisable not to agitate people's nerves to fever pitch. A friend of ours near whose country residence the Harlem Railroad runs invited a few friends to dine with him, bringing down a fine piece of roasting beef. When they quitted the cars, which had again got into motion, it was discovered that the beef had been forgotten and was almost out of sight.

' "We shall lose our dinner after all," said a guest.

' "No we shan't," said the host, "for the train generally meets with an accident about a mile further on and we shall get it again."

'They ran on and sure enough an accident did happen. On coming up with the cars and ascertaining that the beef was in safety, without troubling themselves about the passengers, they recovered their meat and walked off with it.' (From *Old England and New England*, Alfred Bunn. London, 1853.)

When that amusing event took place, George Mortimer Pullman, the son of a general mechanic, and the third of a family of 10 children, was 22 years of age. He had left school at the age of 14, when he started his first job as a clerk in a small general store in Westfield, New York, a job which lasted for three years. He then went to work for his older brother, Albert, at Albion, New York, who was an established cabinet maker. This gave him an insight into the true skills of the woodworker — that of an artist in his craft — and it evidently inspired him to see the beautiful designs and decorations that his brother produced. However, the business was not sufficient to pay for both brothers, and George Mortimer was forced to seek another job. This was at the time that the Erie Canal was being widened through the town of Albion, and somehow he gained himself a job as a building contractor, whose first big task was to move back complete buildings — houses and other premises — to make room for the new wider canal. This experience — in itself a remarkable feat — gained him further similar contracts in Chicago over the next few years. By now he was earning considerable money, and establishing a reputation for enterprise, skill and sheer industry. He had a chance to turn his attentions to new endeavours, and to indulge in his flair for creative invention.

It was perhaps inevitable that the railroad — still very much the wonder of the age — should attract Pullman as a potential source of development for his own career. It was, as I have attempted to show in this chapter, still very much a question of a journey of endurance to go by train and this was the challenge that George Mortimer Pullman recognised.

CHAPTER 2

Building on Dreams

By the time that George Mortimer Pullman was established in Chicago (where he was to meet and marry his wife, Hattie Sanger), he was already well acquainted with railroad travel, having acted as a travelling salesman during the time he worked for his elder brother, the cabinet maker. Nonetheless legend persists that it was one railroad journey in particular that was to foster his interest in improving railroad car design. The story goes that George Mortimer and Hattie Pullman were going on honeymoon from Buffalo to Westfield in 1853, a distance of some 58 miles which they chose to cover by night, in a sleeping car. It was apparently an extremely cold night, and the newly-weds could not sleep for the discomfort. We are told that Mr Pullman spent most of the journey dreaming or thinking of ways to improve the accommodation provided.* It was certainly not the ideal start of a honeymoon!

(I would also add that according to another source, it was whilst alone on a bad night journey from Chicago to New York in 1855 that G. M. Pullman first conceived his ideas for a sleeping car.)

In order to understand the discomforts endured by Mr and Mrs Pullman on that memorable night, we have to retrace our steps a little, because until now I have concentrated upon the development of the Winans' 'day car', and yet here we have a reference to a 'sleeping car'.

I have shown in the first chapter how the majority of Americans preferred to use the railroad by day, and then find rather more comfortable land-based alternatives for the night unless delayed en route. However some night travel *did* exist, and as early as 1836 the Cumberland Valley Railroad had decided to make proper arrangements for male night travellers. They introduced a 'sleeping car' in the winter of that year; which is the first one on record. This had four sections of the car body fitted up to form beds, or more accurately 'couchettes', which were simply shelves to stretch out upon, without bedclothes and with only a pillow and a thin mattress. In 1843 the New York & Erie Railroad produced similar facilities, but it is of interest to note that at this time most trains were not supposed to run after dark and the Erie cars, called 'Diamond' cars had seats which converted into couches for passengers who wanted to doze during the daytime journey. In the same State of New York, the trains from Albany, westward, stopped overnight at Syracuse, whilst those from Buffalo for Albany stopped over from supper time to breakfast time at Auburn. However, by this period, there were railroad cars on the Pennsylvania and Maryland lines which had seats which could be converted into a reasonably comfortable bed, by turning them face to face, and placing slats across as a support for a mattress.

*Sadly for romantic legend, when this book was already at proof stage an American correspondent wrote to suggest that this story was a latterday public relations invention. George and Hattie Pullman were not in fact married until 1867!

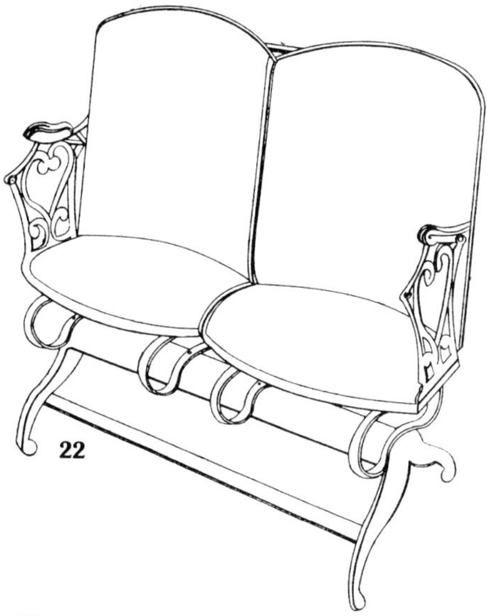

22
Iron-framed individual day car seat for two, with curved backs and very little padding. A whole day of travelling, with constant jolting and jarring, was a formidable experience in hard seats such as these. This seat was designed by T. E. Warren and patented in July 1850. *Author's collection*

23
C. P. Bailey's reclineable car seat, patented in 1854, and giving a measure of comfort to the feet of the passengers, but no support for the back of the neck or the head. Improved, buttoned, upholstery is indicated in the drawing. *Author's collection*

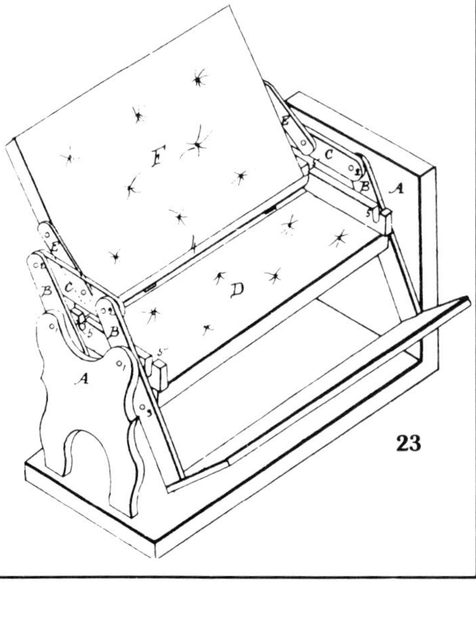

The seats of a typical Winans' day car of the mid-1850s were usually iron framed, with plush upholstery, giving a solid rather than particularly comfortable support. The evolution of the day car seat had passed through a number of stages, starting as simple wooden boards; then covered with oil-cloth, leather or carpet; then stuffed with horsehair, or straw padded; finally gaining interior springs with hair or plush covering. The idea of making the backs movable, to face the direction of travel, had been introduced, but the back was low (just the height of an ordinary chair), and did not support the passenger's head. As a result various portable headrests were invented and sold by pedlars at the stations. The idea of a reclining seat had been introduced in 1852 and a number of inventors produced variations upon the theme, the best being by Mr C. P. Bailey in 1854.

Night travel became more accepted as the railroads extended their tracks and journey times were lengthened. To travel at night in what was really a 'day car', with these hard seats — reclining or otherwise — was an ordeal best avoided if possible! The early alternative of a convertible bunk car, already mentioned, was soon adopted by most railroads who ran trains at night, and by 1860 more than 20 types of 'sleeping car' had been patented and some of these were in regular service. It was in one such car (exact version not known) that Mr and Mrs Pullman made their historic journey. It was almost certainly only lit by candles, and apparently it had no stove, although these were quite commonplace by that date.

The general layout for lighting was two candles, one each end of the car, not bright enough for night reading, and the railroads continued to use candles rather than oil lamps (long after oil was the usual form of home lighting) because there was less risk of fire, particularly in an accident. As it was the habit to keep all the windows closed at night, to retain the heat of the stove, the interior soon became unbearably stuffy and the only fresh air was when the conductor passed from car to car, opening the end doors, which of course blasted cold air into the interior in fine style. There were other reasons for keeping the windows and doors closed, in particular to prevent the smoke and sparks of the locomotive's chimney from entering the car and adding to the atmospheric pollution; and there was also the problem of dust.

The dust was raised from the trackbed by the passage of the train, and in hot weather, travelling across the plains, this dust could become a veritable cloud surrounded the passenger cars for miles on end. If the weather required open windows to avoid being boiled, the dust (mixed with smoke and perhaps live sparks from the locomotive chimney) entered the cars, causing much annoyance. It is recorded that by the mid-1840s knowledgeable passengers wore a special long coat of linen, or similar material, known as a 'duster', as protection of their day or night clothes beneath. A great many schemes to reduced the dust nuisance were considered by the railroads, and some weird and wonderful inventions appeared, only to be discarded again as ineffective, or troublesome. Many of these ideas included the use of blocks of ice, or fountains of water to clean or cool the air entering the car interior, with stove heated air replacing it when required at night or in wintertime.

The change from wood-burning to coal for the locomotives, and better ballasting of the railroad tracks eventually reduced this problem, but not before it had reached such serious proportions that passenger business began to decline because of the discomfort caused by dust. In summer, a man, or a boy (often a Negro) was employed by many of the railroads to walk through the train dispensing iced water from a tin can into tumblers for passengers at the company's expense.

The introduction of the reclining seat certainly gave *some* added comfort, and it was not long before inventors came up with ingenious schemes to convert these seats into beds, by using the seat backs folded down into a horizontal position, with the cushions placed like a mattress. As this was in a day car, it lacked privacy, and could only be described as a compromise solution to the demand for a bed whilst travelling by night. In fact the reclining seat probably offered more comfort than these crude designs of convertible bed *cum* seat. By raising the height of the reclining seat back sufficiently to support the passenger's neck and head, whilst reclining, some positive comfort was at last offered, as this 1856 traveller reported:

24

'The train stopped twenty minutes at Springfield for an early supper. Bodily refreshment is never lost sight of in the arrangements of American travelling. Almost immediately after leaving Springfield night fell and before long it was quite dark. The seats in the cars of this line are the least disagreeable I met with in America; the backs are sufficiently high to form an easy rest for the head and the footboard adjusts itself to give your legs repose. The result of all this, especially after supper and in the dark, which is only partially dispelled by the car lamps, is that most people fell asleep, so the last part of the journey was quiet enough.' (From *America by River and Rail*, William Ferguson, London, 1856.)

That of course was a journey which ended in the hours of darkness, not an all-night vigil. Less fortunate was this all-night passenger in 1857, who held very positive views about the inability to sleep on the train!:

'After travelling all night from Cumberland, Md, we arrived at the little dingy, dull city of Wheeling in Western Virginia before daylight, and hastened to our beds to snatch the sleep which is next to impossible to win or even to woo in the hot, frowsy, uncomfortable railway car, containing from fifty to sixty people and a demoniacal furnace burning anthracite coal. Without a proper place to stow away one's hat, with no convenience even to repose the head or back except to the ordinary height of a chair, with a current of cold air continually streaming in and rendered necessary by the sulphurous heat of the furnace, and with the constant slamming of the doors at either end of the car as the conductor goes in or out, or some weary passenger steps on to the platform to have a smoke, the passenger must indeed be dead beat who can sleep or even doze in a railroad car.' (From *Life and Liberty in America*, Charles Mackay, New York, 1859.)

24
As the railroad tracks were extended, and journey times lengthened, the idea of travelling by night became more or less accepted. The layout of the day car was wholly impractical for night use, and provoked such scenes as this, from *Harper's Illustrated Weekly* of 2 October 1858, depicting night travel on the New York Central.
Author's collection

Because it was generally recognised to be virtually impossible to sleep in an American railroad car at the time of the Pullmans' honeymoon trip, this problem was to become the focus of attention by a number of clever inventors, of whom Pullman himself was soon to be one. Most schemes involved converting day seats into night beds, or couches (*couchettes*), and the use of curtains to provide privacy. In December 1856 Theodore Tuttle Woodruff patented

25
The interior of an 1859 Wagner sleeping car on the New York Central, with the seats and bunks arranged for night use, complete with curtains. The artist has exaggerated the width of the centre aisle and ceiling in order to better portray the layout. In reality the two sides of the car, each with its sleeping arrangements were much closer together. The stove in the foreground was a subject of much complaint at night, creating an almost unbearable atmosphere. (From *Frank Leslie's Illustrated Weekly*, April, 1859.) *Author's collection*

25

a layout which gave more consideration to the privacy and needs of night travellers. However, before looking at the Woodruff design, a few general observations about the railroad cars of the period are necessary. The American cars were generally much larger and longer than their British contemporaries, although for the most part running on the same gauge tracks, of 4ft 8½in. A typical feature of the American day car was the high roof, sometimes with pothole ventilators and a stovepipe. As time progressed, the cars became more spacious in internal dimensions, and could carry between 40 and 60 passengers, all of one class. (Some larger cars had appeared with capacities up to 75, for example on the broad gauge Erie Railroad.) By the year 1850, the size of these cars, and their weight was such that six-wheeled trucks (bogies), were being introduced for the latest cars, with wooden framed construction. It was the sheer size of the cars that gave inventors room to play with when designing sleeping arrangements.

Theodore Tuttle Woodruff's design, just mentioned, divided the interior of the car into sections, with the seats permanently fixed in facing pairs. Two 'berths' (a now familiar term used to describe an area for sleeping in), a lower and a middle one, could be made in rather complicated fashion from the day seats, with pivoted seat cushions. Most significant was the top berth, which was made up of hinged frames that folded away into the ceiling by day. Only the spaciousness of the day cars' high roof layout had made this last feature possible. The bottom berth was wide enough for two people, but the middle and the top ones could only cope with one person each. Woodruff's cars first appeared on the New York Central Railroad in 1858, and very quickly spread to eight of the western railroads. A feature was the charge of a supplementary fare for the use of a sleeping berth. The New York Central also introduced some sleeping cars designed by Mr Webster Wagner in the same year, 1858, but these seem to have been of a less substantial design, whereby the bottom double berth was created by placing the day seat backs in a horizontal position, and then above this were two single berths made of light cane-bottomed frames, which were fastened by night to the side of the car, but were stored under the roof during the daytime. Proper mattresses and bedding were however supplied, and the berths were shielded by curtains. It was not long before inventors looked to ways to provide solid partitions instead of curtains, to give more privacy. And there was certainly no shortage of inventors at this time, for this was when George Mortimer Pullman first revealed his interest in the design of sleeping cars.

In Chicago Pullman had become an established contractor, and apparently a very wealthy man as a result. The story of the Tremont House Hotel is legendary, and shows much of the skill and enterprise of the man whose name was to become world famous. Pullman specialised in the removal of badly sited buildings at a time when expansion was the keyword in Chicago. He developed a technique of jacking-up the entire building and then easing it bodily to its new site. The Tremont House Hotel had to be moved from an insanitary low-lying swamp area to a drier location where proper plumbing could be installed. Apparently Pullman jacked up the whole four-storey building and moved it without breaking a single pane of glass, and the hotel was not even closed during the operation! The legend has it that the chambermaids did not even smash the crockery they were handling during the move, and the clients spilled no beer!

By 1859 Pullman had conceived his idea for a converted day coach, based upon his observations of the many shortcomings of the existing designs of sleeping car. Apparently he tried to sell his idea to the local Chicago, Alton & St Louis Railroad, who turned it down as uneconomical. Not

26
George Mortimer Pullman's first venture into the railroad sleeping car business was by means of converting the interior layout of an existing Chicago & Alton Railroad day car, at his own expense. This illustration depicts a latterday replica of this historic car, which was Chicago & Alton No 9. *Pullman Car Co*

26

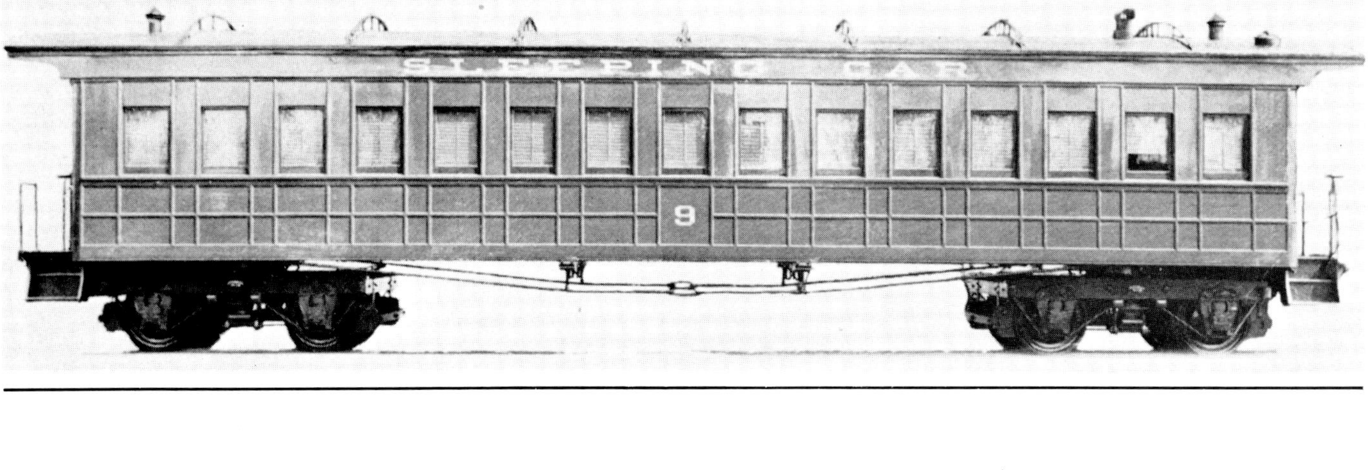

to be outdone he purchased a day car from the company and then at his own expense he converted it to a sleeping car. This was Car No 9, and apparently it convinced the Chicago & Alton that his ideas were sound. Pullman seems to have done a great deal of the actual conversion work himself; aided by a skilled carpenter called Leonard Seibert. A second car No 19 was similarly converted, and both had 20 berths, 10 upper and 10 lower. The lower berth was created by the use of pivoted day seat backs, whereas the platform-style upper berth was lowered from the roof when required, by pulleys. Pullman paid much for the conversion work, some $2,000 for the two cars, despite doing most of the work himself along with Leonard Seibert.

It is said that no real plan or blueprint existed at the start of the remodelling of car No 9. This car was a 44ft long day car with the usual open ends, and two four wheel trucks. The roof had a slight arc, in the manner of its day, but provided considerably more headroom than was normal. Pullman fitted it out with reversible seats whose backs could be let down to form beds, arranged longitudinally with the car; these seats were upholstered in plush. Oil lamps were fitted, but for some reason not used, and candles had to be provided as well; there was an adequate stove for heating. The upper berths were raised up into a position just under the ceiling when not in use. No drawings of No 9 have come to light — probably because none were ever completed! — and the original layout can only be surmised from a much later drawing of 1864, which is probably a refined version of the original design, produced for patent office purposes.

This first Pullman sleeping car was accepted, by the Chicago & Alton, to run on its tracks between Bloomington and Chicago, and entered service on the night of 1 September 1859. A supplementary charge of 50cents was made for its use, and passengers were supplied with a pillow and blankets, but no sheets. A Pullman conductor was in charge of the car, the first of the now famous clan. In fact, he was a white man, by the name of Jonathan L. Barnes, but later many Pullman conductors (or porters, as they were called in America) were to be recruited from Negroes and it was a job that gave them much status. Mr Barnes lived to be 85 years of age, and in his old age he recounted his early experiences with the first Pullman sleeping car. Apparently, the first trip only carried three fare-paying passengers, who all went to bed wearing their boots, until Mr Barnes insisted that they were taken off! Mr Barnes was himself taken off sometime later, when the railroad decided it was cheaper to use their own staff, but the principle of the Pullman Porter, or Conductor, had been established in George Pullman's mind. A total of three cars was converted for the Chicago & Alton line, and it is recorded that the passengers preferred them to alternatives which existed, which were no doubt older, despite having to pay a supplement.

On one occasion Conductor Barnes spoke with George Mortimer Pullman about Car No 9, remarking that it was a 'fine car'. The brief response came: 'It ought to be. It cost enough'. There can be no doubt that Pullman's experiences when working for his brother Albert had shown him the value of good workmanship and fine woodwork. To this he added, at his expense, the best tastes of the period, in the belief that when offered the best, the public would be prepared to pay for it. It was a move in the right direction, but Pullman felt frustrated by not being in full control of the Chicago & Alton cars. He perceived that in order to fulfil his ideals, and to make money, he would have to be more involved in the actual operating of the cars built to his design.

At the time that George Mortimer Pullman was busy reconstructing Car No 9 to his ideas at the Chicago & Alton's Bloomington workshops, two other sleeping car builders were very active. I have already mentioned them both — Theodore Tuttle Woodruff and Webster T. Wagner. Both could lay claim to have 'invented' the sleeping car, but so far as I can discern neither they, nor indeed Pullman, ever made such an extravagant claim upon their own behalf. By some curious twist of fate, history was later to grant G. M. Pullman much fame at the expense of the other two, although at the time both the others already had cars in operation on other railroads, and in Wagner's case, they were already highly developed in stylish design when Pullman was just beginning. Webster Wagner was associated with the Vanderbilt family, and they granted him an exclusive right to build and operate cars to his design on the New York Central and its associated lines. Wagner seems to have obtained the licences of a good many new patent devices, and his cars were equipped with all the latest gadgetry, as well as being very ornately decorated. Fortunately, good contemporary accounts of travel in cars of both Woodruff and Wagner design have survived. Here is the Woodruff car as seen in 1859, and apparently not considered to be much of an improvement upon an earlier layout, described first:

'Coming up the Erie I tried a sleeping-car for the third time and not very successfully. We all retired at ten o'clock with a fair allowance of open windows and virtuous resolutions, but the rain poured, the night was chill and damp, and soon every orifice for the admission of external air, save the two or three humbug ventilators overhead, was shut and a mephitic atmosphere produced. After gasping awhile like a netted fish on a hot sandbank I rose to enter my solemn protest against all sleeping-cars not provided with abundant and indefeasible means of ventilation. I tried one two nights later on the Michigan Southern Road (Woodruff car) which served much better but still far from perfect. It is very true that no arrangement can secure a healthy circulation of air by night in any passenger-car while the popular ignorance is so dense that the great majority imagine any atmosphere healthy that is neither too cold nor too hot, but the builders of cars have no right to be ignorant of the laws of life with which they tamper and two or three presentments by grand juries of the makers of unventilated cars, especially sleeping-cars, as guilty of manslaughter would exert a most salutary influence. I commend this public duty to the immediate consideration of jurors and prosecutors.' (From *An Overland Journey from New York to San Francisco*, Horace Greeley, New York, 1860.)

By the year 1864 Woodruff had lost interest in sleeping cars, and sold out and retired; assigning all his interests to O. W. Childs of the Central Transportation Company.

This description of a Wagner sleeping car dates from 1860:

'I was going from Albany to Buffalo. The conductor seeing me walking about the platform and eyeing the several cars said to me with sagacious forethought, "Sleeping-car, mister? Going through, stranger?" I replied that I was and followed the man into the last car, which was lettered in large letters on a sunflower ground, Albany and Buffalo Sleeping-Car. I entered and found that it was like the ordinary railway-car with the usual filter and stove and with the seats arranged in the usual manner. I asked how much extra I must pay for a bed. "Single high, twenty-five cents, double low, half a dollar," said the officer on duty. I ordered a single high and as I paid my twenty-five cents the bell on the engine began to get restless. Ten other persons entered and ordered beds and paid for them with more or less complaint and wrangling.

'When we were between Little Falls and Herkeimer the officer of the sleeping-car entered and called out, "Now then, misters, if you please, get up from your seats and allow me to make up the beds." Two by two we arose and with quick hand the nimble Yankee turned over every other seat so as to reverse the back and make two seats facing each other. He shut the windows and pulled up the shutters, leaving for ventilation the slip of perforated zinc open at the top of each. Then he stripped up the cushions and unfastened from beneath each seat a light cane-bottomed frame which he fastened to the side of the car at suitable heights and covered each with cushions and blankets.

'I went out on the balcony to be out of the way and when I came back the whole place was transformed. There was no longer an aisle of double seats but the cabin of a small steamer with curtained berths and closed portholes. The bottom berths were wide and comfortable with room to roll and turn. There were two high berths to choose from; both wicker trays, ledged in, cushioned and rugged. The one was about a foot and a half higher than the other and I chose the top one as being nearer the zinc ventilator. Some of the passengers had turned in and were snoring. Others, like young crows balancing on spring boughs, were swinging their legs from the wicker trays and peeling off their stockings or struggling to get rid of their boots.

'I clambered to my perch and found it was like lying on one's back on a narrow plank. If I turned my back to the car wall the motion of the train bumped me off my bed altogether and if I turned my face to the wall I felt a horrible sensation of being likely to roll backwards into the aisle, so I lay on my back and settled the question. It was like trying to sleep on the back of a runaway horse. At each place the train stopped there was the clashing of the bell and if I peered through the zinc ventilator into the outer darkness a flying scud of sparks from the engine did not serve to divest my mind of all chances of being burnt alive. Then there were blazes of pine torches as we neared a station, more bell clamour and jumbling sounds of baggage, slamming doors and itinerant conductors.

'At last I fell into a precarious and fragile sleep that lasted until daylight returned. One by one the passengers woke up, yawned, and stretched themselves. There was something suspicious in the haste in which we all flopped out of bed, for no really comfortable bed was ever left with such coarse ingratitude. Presently the attendant entered and proceeded to readjust the seats, and through the opened windows there came a draft of pure air that freshened our frowsy and dishevelled crew. We repaired to the washing-room which contained one dirty brush fastened to the wall by a chain and a basin. There was very little water and the basin had to be tilted to one side to get enough for even a scanty rinsing.

'By this time we were near Buffalo and breakfast. The train slackened and stopped and out poured the hungry swarms. Five gongs of five opposition breakfast places banged and thundered for our custom. In a minute I was seated with some thirty other hungry souls stowing away white piles of hominy, pink shavings of corned beef and bowls of stewed oysters, while a Negro boy waved a plume brush of wild turkey feathers over my head to keep off the greedy flies.' (From *Criss-Cross Journeys*, Walter Thornbury, London, 1873.)

We shall return to Webster Wagner again later in this chapter, but first it is necessary to trace the career of George M. Pullman over a few more of these extremely fertile years of American railroad history.

If Pullman was somewhat disappointed with his first sleeping car venture, he probably began to devise new schemes quite quickly, for such was the nature of the man, who could never be called 'a loser'! However, the outbreak of the American Civil War played another dud card into his hand in 1861 because the Chicago & Alton rolling stock was requisitioned by the Unionists, including his sleeping cars, leaving Pullman without business or gain from all his enterprise and hard work. As a result he left Chicago in 1862 and went to the mining regions of Colorado, in seach of money. Characteristically, he did not join the gold-rush itself, but instead set up a general store at a place with the splendid name of 'Pike's Peak or Bust'. By selling provisions of all kinds to the gold prospectors (who were not for the most part lucky in their searches), he amassed a considerable sum of money, and when the war ended he returned to Chicago with around $20,000, and a renewed desire to enter the railroad business as an entrepeneur. During his time as a store-keeper, dealing with the everyday needs and complaints of his customers, Pullman seems to have recognised hidden strengths within himself, enabling him to deal with the public in exactly the manner of a hotel keeper or restaurateur — able to please, and not get himself shot, in the vitriolic climate of the 'gold rush'! During idle moments at his Colorado store, Pullman apparently sketched out his revised ideas for a sleeping car, and when back in Chicago in 1863 he enlisted the help of a friend, by the name of Ben Field, to perfect the plans for building — at Pullman's own cost — his 'dream' sleeping car. Pullman and Field patented a sleeping car arrangement in April 1864, at a time when a number of other patent ideas were produced by various inventors, some of whom quickly sold the patents. Some of their ideas seem to have been purchased, or otherwise acquired by Pullman for use in his own plans. The 1864 drawing did not have a clerestory roof arrangement, and would therefore have perpetuated the stuffy atmosphere during the night; especially in winter. However, when Pullman produced a second drawing, patented in 1865, the clerestory roof, of Wagner's design, was an important feature.

The adoption (though not the invention) of the clerestory roof must be credited to Webster Wagner, who was apparently very worried by the many complaints he received from his customers about the poor ventilation in his sleeping cars. With the assistance of a Mr Alba Smith, and after studying some of the many weird and wonderful 1850s ideas for air-conditioned or better-ventilated railroad cars, Wagner devised a car-ventilator system, which was patented in September 1862. This had a central portion of the roof raised to form a clerestory above the two side portions of the roof. Along the vertical sides supporting the raised central portion were a series of openings, each with a pivoted shutter that could be adjusted to direct the air currents created by the movement of the train, thus exhausting the stale hot air that gathered under the car roof, and allowing some fresh air to displace it. This raised deck, or clerestory roof, was henceforth to become a characteristic feature of American railroad cars for many decades, finally being replaced by a flush-topped roof once again when full air-conditioning became normal practice. Once established as an adequate ventilation system it was but a simple matter to install small windows — or 'deck lights' — into the scheme, thereby giving extra natural illumination to the car interior. Deck lights had first apeared on an Imlay-designed car called *Victory* in 1836, but the idea had not been followed up in America, although in Britain shortly afterwards the clerestory roof had appeared on some early broad gauge Great Western Railway saloon four-wheeled 'posting carriages', possibly the first use of this feature anywhere else, and just as in America, it was not immediately accepted for further use.

Another feature made its appearance on the Chicago Burlington & Quincy Railroad, from 1859 onwards was the 16-wheel passenger car. Presumably in search of greater stability of ride upon the still indifferent quality tracks, and to cope with steadily increasing tonnage, the 16-wheel layout had two four-wheel trucks each end, which supported a swivelling sub-frame at both extremities of the car body. This 16-wheel layout must have caught Pullman's eye, and together with the clerestory roof, he adopted it when finishing the plans for his prototype car — a car that could certainly be described as the 'car of his dreams', far in excess of all prior experiments in luxury and practicability.

27

28

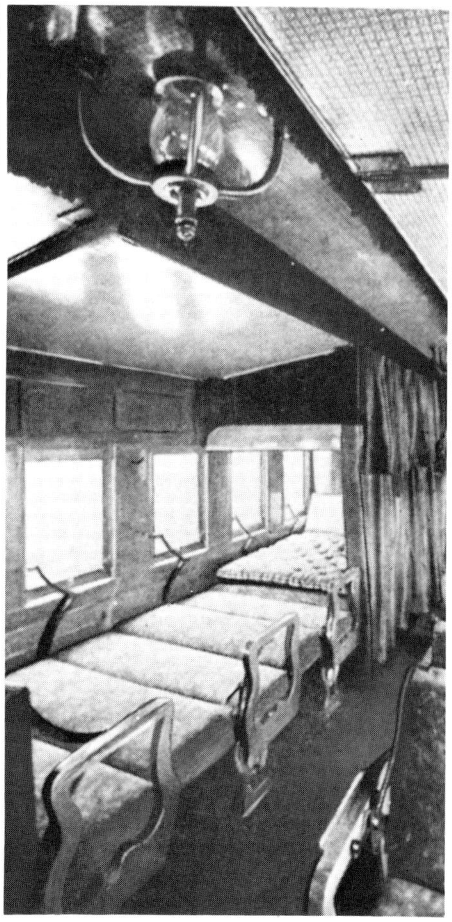

27-28
These two views are supposed to be of the interior of Chicago & Alton sleeping car No 9, and purport to show how it appeared as arranged for day (left) and night (right) use. In fact, the photographs date from much later, but are probably a fairly accurate representation of this original Pullman layout. It was a reconstruction for display at Chicago World's Fair.
Both: Pullman Car Co

CHAPTER 3

The Dream becomes Reality

The year 1863 must have been a busy time for George Mortimer Pullman and Ben Field, as they worked together to perfect their sleeping car layout, and they evidently made a careful study of contemporary developments, at a time when inventors were pouring out new ideas and seeking patents for them. Pullman and Field were quick to realise the advantages offered by the clerestory roof, and as we have seen they changed their drawings to include it. They also acquired an interest in an 1858 patent sleeping car design by a Mr Eli Wheeler (exactly when is not certain) because a feature of the Wheeler design, a sliding partition at the back of each seat, stowable by day, was used in their design. Here it should be observed that so many ideas were being produced, and sometimes of such basic similarity, that it is sometimes virtually impossible to establish their exact origins; some inventors not even gaining patents, but nonetheless publicising their ideas in one way or another.

The sleeping car design of Pullman and Field was completed in time for construction to commence by the middle of 1863, with Pullman ready to invest his $20,000 in the project (at a time when $4,000 was an average price for a railroad sleeping car to other people's designs). Before commencing the story of this prototype car, which Pullman named the *Pioneer* — for fairly obvious reasons — a few observations are necessary about railroad car construction of the period, because neither Pullman nor Field were experienced car builders, and this was to have some repercussions on *Pioneer*, as we shall see in a moment.

A typical railroad car of the early 1860s was basically of wooden integral construction, although various experiments and patents had been produced for metal-bodied or metal-framed cars since 1846, when a design for a wrought-iron car by H. L. Lewis had appeared in the journal *Scientific American*. During the 1850s a metal-framed car design by Dr B. J. LaMothe was constructed, and in due course several railroads experimented with it. The advantages, particularly in case of accident, were considerable; but so were the disadvantages! The poor state of the tracks on most railroads would soon play havoc with a steel body, twisting it permanently out of shape, whilst the noise level would have been greatly increased by the drum effect. Heat insulation to prevent excessive condensation forming on interior surfaces, was also a problem. Metal-bodied cars were certainly a real possibility when Pullman was working on the design of *Pioneer*, but wooden construction was the norm, despite its fire risk and tendency to disintegrate and splinter in an accident. Another factor was the abundant supplies of good quality timber that were available to American car builders at that time.

The use of timber was considerably cheaper than metal (although experiments with metal did continue and the first all-metal Pullman sleeping car eventually appeared in 1907), and it was perhaps only logical that Pullman and Field should choose to use timber construction for *Pioneer*; but a factor which was already becoming apparent to many car builders had to reckon with — namely an increase of weight without a corresponding increase in strength. Pullman's design was an ambitious one, involving considerable quantities of timber and other materials for the interior layout, as well as for the actual body of the car. To carry the weight, they chose the recently-introduced 16-wheel arrangement with improved springing and solid rubber shock absorbers. To get rigidity of the body

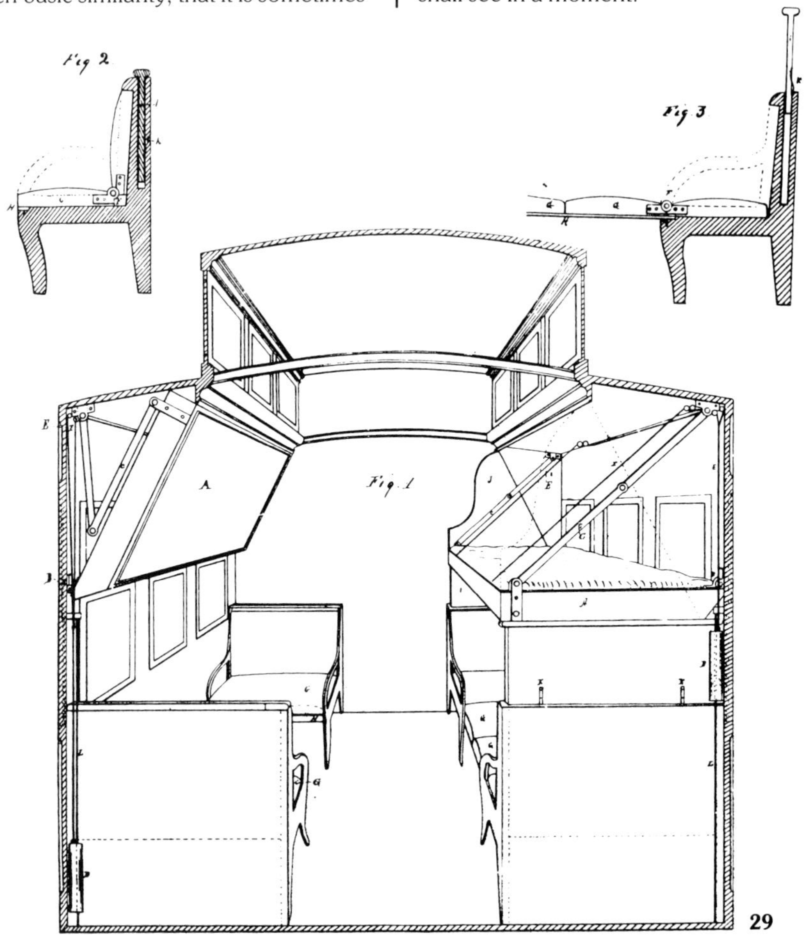

29

September 1865 patent drawing of the Pullman and Field sleeping car arrangement, complete with clerestory roof, as produced for the first real Pullman sleeping car *Pioneer* in that year. *Author's collection*

30

they used the current practice with massive balks of solid timber, which were incorporated into the framework of the sides and floor which was then stressed with adjustable iron tie bars, between the bogies, under the middle of the car. *Pioneer* was a truly massive car for its time, and in this respect Pullman proved to be too ambitious! In his search for the 'best' he made the car as large as he thought possible, in order to accommodate all the fittings he desired, with the end result that *Pioneer* was built longer, wider and taller than any standard gauge railroad car produced to that date.

The mists of time have discreetly shrouded the exact story of how Pullman and Field came to build a railroad car with the apparent approval of the Chicago & Alton Railroad, which when completed proved to be out of gauge for the very same railroad's route! One may surmise that somewhere a misunderstanding arose as to the exact dimensions which were permissible, or perhaps the truth was that no one had bothered to check out this loading gauge factor properly before design commenced. Loading gauge is the railroad's term for the clearances through which its trains can be moved with safety. The main factors are: the width between the train and buildings alongside the tracks; the internal clearances of cuttings and tunnels; the dimensions of bridges over the route, or bridges over which trains must travel — plus, of course, the clearance between one train and another when passing in opposite directions on parallel tracks, and on curves.

Whatever the reasons, a first trial run of the *Pioneer* in the summer of 1864 on the Chicago & Alton Railroad, immediately showed that it was foul of the loading gauge, exactly how much is not known, but it was probably only a matter of mere inches, rather than feet. The result was a blunt refusal of the new car by the Alton line, and *Pioneer* was put away to rest in a Chicago shed, where it soon gained the unenviable nickname of 'Pullman's Folly'.

It would be interesting to know more about the reactions of George Pullman and Ben Field (a former Senator of New York) to this seemingly disastrous setback to their project. I have read several books which claim that the platforms were out of gauge for the car, but I find this difficult to believe, in view of the low height of American railroad platforms — which are virtually on ground level! If bridges were involved, it is quite probable that at that time these were of timber construction, and therefore reasonably easy to modify; and the same would certainly apply to most buildings of the period likely to be found alongside the railroad tracks. It may have been a sheer case of bloody-mindedness on someone's part, because — as we will shortly see — the line was subsequently altered to accommodate *Pioneer*, and in a very short space of time!

30
General layout of the sleeping car *Pioneer*, as built by Pullman & Field in 1865, and regarded as the first modern design of sleeping car. It was longer, higher and wider than predecessors (initially causing loading gauge clearance problems along the Chicago & Alton routes), and it was mounted upon no fewer than *sixteen* wheels at a time when four and six-wheelers without bogies were commonplace in Britain. This relatively modern artist's rendition is very accurate as to original style and arrangement, showing the raised clerestory roof, with ventilators and candle lamps, and deck lights, and the folding upper berths. The heating was from a hot air furnace under the floor. There were two compartments at each end, eight 'sections' and a roomy washroom; black walnut woodwork with much inlay was used and many mirrors. The vehicle was fully carpeted, and featured French plush upholstery, good beds and ample bedding.

However, before recounting how *Pioneer* once again rolled on Chicago & Alton tracks, it is a pertinent moment to describe the new car itself, which was by any standards a magnificent creation of its time. *Pioneer* weighed more than 27 tons, a lot for one car, and it was 58ft in length. The body of the car was divided into eight sections, plus two compartments and two washrooms, one of each at each end of the car. The very ornate interior was finished in black walnut, embellished with inlaid marquetry, and some gilt-edged mirrors were placed in the panelling. Rich red carpets covered the floor and the seats were upholstered in plush. The use of the clerestory arrangement, with candle light in silver lamps to each section, gave adequate light, ventilation and spaciousness to everyone. In sheer style and elegance *Pioneer* apparently matched the best examples of the ornate and elaborate decor to be found on the river steamboats of the period, which had themselves much taken the public's fancy.

During the autumn and winter of 1864 'Pullman's Folly' rested in its Chicago shed, but we can rest assured that Pullman himself was very active! With some $20,000 invested in the project, he was far too much of a businessman to accept such an easy defeat. A man who had found ways to move entire buildings without harming them was surely not going to be persuaded that a railroad could not adjust its loading gauge by a few inches in order to accommodate his new sleeping car. There must have been some negotiations between Pullman and the C&A Railroad during this period,and no doubt in the course of time a suitable arrangement would have been made to get *Pioneer* back on the rails. However, fate now played an ace card into Pullman's hand for a change, and in a most dramatic manner.

In April 1865 President Abraham Lincoln was assassinated, much to the horror of the nation. His burial was to take place at his birthplace of Springfield, with a special funeral train to take the body there, which had to make frequent stops en route. The former President's widow, Mrs Lincoln, collapsed under the strain of the event, near Chicago, and it became necessary to transport her safely (along with the cortege) from Chicago to Springfield, in privacy. The man given responsibility for making the travel arrangements was Colonel James Bowen, the President of the Third National Bank of Chicago, of which Pullman was also a director. Colonel Bowen was, of course, only too well aware of 'Pullman's Folly' lying idle in its shed — the most pristine and magnificent of railroad cars, and surely fit for the purpose of transporting the ailing Mrs Lincoln to Springfield. No doubt at Colonel Bowen's suggestion, Pullman offered the use of his car *Pioneer* to Mrs Lincoln who apparently knew about the car already, and she accepted. Any objections to the Pullman car that still existed on the Chicago & Alton, were overnight forgotten! In the greatest haste, a team of workmen was sent along the line, checking loading gauge clearances and, where necessary, making the alterations which would allow *Pioneer*, with Mrs Lincoln on board, to pass safely. On 2 May 1865 the funeral special left Chicago for Springfield, and included in the train was the Pullman car *Pioneer*. Crowds gathered all along the route, and inevitably the superb car made an immediate impression; providing much publicity for George Mortimer Pullman. Ironically, as a result of this tragic event the Chicago & Alton Railroad loading gauge was henceforth able to accept the car as Pullman had originally planned, and the contractural agreement between them to operate it was honoured.

President Grant, who succeeded Lincoln, was then offered the use of *Pioneer* by George Pullman for a Presidential journey from his home town of Detroit to Galena, which Grant accepted. This was a masterpiece of strategy on Pullman's part, because the Chicago & North Western Railway also had to alter their loading gauge to accept the Presidential train! With such important passengers the future of *Pioneer*, and of all the cars that followed, and of George Mortimer Pullman himself, was assured, for no railroad was going to deny the prestige associated with the Pullman sleeping car. There were, however, some powerful rivals, including the already well-established operations of Webster Wagner, backed by the wealthy Vanderbilt family, which became known as the Wagner Palace Car Co. In the course of his lifetime, George Mortimer Pullman was to buy out virtually all the competition and merge them with his own operations, except for those of Wagner. Three years after Pullman's death the two giant companies of necessity were finally merged.

Encouraged by the success of *Pioneer*, Pullman evidently lent the car to several railroads to carry various notables around the country, and each railroad ensured that its own loading gauge was acceptable to the big car. Pullman sought exclusive contracts to operate his cars on the various railroads which had no existing sleeper services, and if he could not get exclusivity he did profit-sharing deals, basically in order to keep his rivals at bay. The charge for a sleeping berth on a railroad car had by then risen to one dollar and a half. When *Pioneer* was ready for service, Pullman characteristically raised it to two dollars for his car. When more Pullman cars were built and in service some railroads were surprised to see the passengers paying out $2 for a Pullman, when they could travel in one of the railroad's own sleepers for $1.50, attached to the same train!

With ample financial support now forthcoming, Pullman built new cars, of greater length and additional comfort, to start with for use on the Chicago & Alton route and the Michigan Central Railroad, with whom he had also quickly obtained a contract. Six cars followed *Pioneer* for these early operations. These were named: *Atlantic*, *Aurora*, *City of Chicago*, *Maritana*, *Omaha* and *Pacific*. Each car was as opulent and luxurious as the others and each was staffed by Pullman's carefully chosen men. The word soon got around that a Pullman sleeper was the finest thing on wheels, and the service offered matched the quality of the cars. Pullman's arrangement was a simple one, whereby he provided the car and staffed it; the railroad for its part had to haul it and look after the running gear, the wheels and brakes, etc. The normal fare for the journey went to the railroad as usual, but the Pullman supplementary charge went to Pullman exclusively.

Early in 1867 George Mortimer Pullman, by then firmly established in the railroad sleeping car business with some 35-40 cars running, created the Pullman Co, with various subsidiaries including the Pullman Palace Car Co Inc to operate the cars, and the Pullman Standard Car Co to build them. The use of the word 'Palace' seems to have been in vogue at the time and it is not certain that Pullman was the first to apply it to railroad cars; he was certainly not the only one to use it. 'Palace' was an expression of grandeur, of the finest luxury, ie 'fit for a King'. It was certainly a prophetic use of the word by Pullman, because in due course his cars were indeed considered suitable for the highest royal patronage; as we shall see in later chapters.

31

31
Contemporary woodcut of the interior of *Pioneer* (almost certainly, because of the candle-lamps) showing it arranged for day use with the upper berths folded up against the deckrail of the clerestory. *Ian Allan Library*

Having established a flourishing company, with a reputation for the best, Pullman began to spread his wings and reveal the true scope of his talents as a businessman, inventor, visionary and true entrepeneur. His slogan — 'Improvement is our Watchword' — was carefully chosen. In the same year, 1867 Pullman first introduced his cars to Canada, a sign of things to come, and at the same time he also entered into the railroad catering business, with what he chose to call 'Hotel Cars', of which more anon. Just as in the case of the sleeping car, Pullman did not actually *invent* the restaurant, or dining, car, but he was destined to be the first to give it a proper status, with a service of quality, and value for money. Dining cars had been proposed in America as far back as 1838, and the idea had been looked at again several times since, but it was not until 1863 that two 'restaurant cars' were placed in service, on the Philadelphia, Wilmington & Baltimore Railroad. These were basically ordinary day coaches of the Winans type but with a central cross partition, which divided one half, which was used as a smoking saloon, from the other half which formed an eating bar. To create this the ordinary seats were removed, and a bar counter installed, together with a steam box which carried hot food, pre-cooked at station restaurants. It is thought that this eating bar was patronised only by male passengers. It must have been reasonably successful, because the two cars were in use for several years, predating Pullman's interest in the idea.

What Pullman himself recognised was that a potential source of additional revenue was being tapped by the wayside station restaurants, where by tradition the trains were halted to allow the passengers to eat, and which I have already described in Chapter 2. The situation by the mid-1860s was that express long distance trains were delayed hours in a day solely for the purpose of eating, at restaurants which — with a truly captive clientele — ranged from the justly famous and the acceptably good, to the indescribable and third rate, where the food was inedible. The railroads themselves owned a lot of these refreshment saloons, but some were let out to unscrupulous concessionaires, over whom they had but little control on quality and service. Stories of the mad scramble to gain a place and eat a meal in the space of 20min or less only to find it deliberately served so hot that most of it was left (to be served again to the next train load!) are legion, and a part of American folklore. One story often retold was of the 'speciality' of the day, which passengers on the train could order in advance by telling the conductor on board, who had the message sent by telegraph to the station ahead, stating how many portions were required. If the 'dish of the day' was chicken, that also gave the order to start killing the required numbers to cook. If travellers' tales are to be believed, the chicken was sometimes served to the tables 'still kicking'!

32

The origins of the now accepted American 'fast food' method of satisfying the human stomach can obviously be traced to these railroad restaurants, and those of the competitors, adjacent to the tracks. If the weather was bad it must have been difficult to leave the warmth of the railroad car in search of sustenance, and of course the elderly and disabled were virtually ignored because they took too long. The mid-1850s seem to have witnessed the lowest ebb of the railroad restaurant service (after which by all accounts they made a determined effort to improve, in order to keep their clientele) and this report in the *New York Times* of 10 June 1857 was truly damning:

'If there is any word in the English language more shamefully misused than another it is the word refreshment, as applied to the hurry scurry of eating and drinking at railroad stations. The dreary places in which the painful and unhealthy performances take place are called Refreshment Saloons, but there could not be a more inappropriate designation for such abominations of desolation. Directors of railroads appear to have an idea that travellers are destitute of stomach; that eating and drinking are not at all necessary to human beings bound on long journeys, and that nothing more is required than to put them through their misery in as brief a time as possible. It is expected that three or four hundred men, women and children, some of whom must, of necessity, be feeble folk and unaccustomed to roughing it, and all of whom have been used to the

32
The same interior, arranged for night use, with the seats forming the lower berths and (on the left) the upper berths lowered to a horizontal position. Thick curtains gave some privacy to the occupants of the berths; although for a woman dressed in the fashion depicted, the task of undressing alone behind the curtain and on the bed was almost impossible!
Illustrated London News

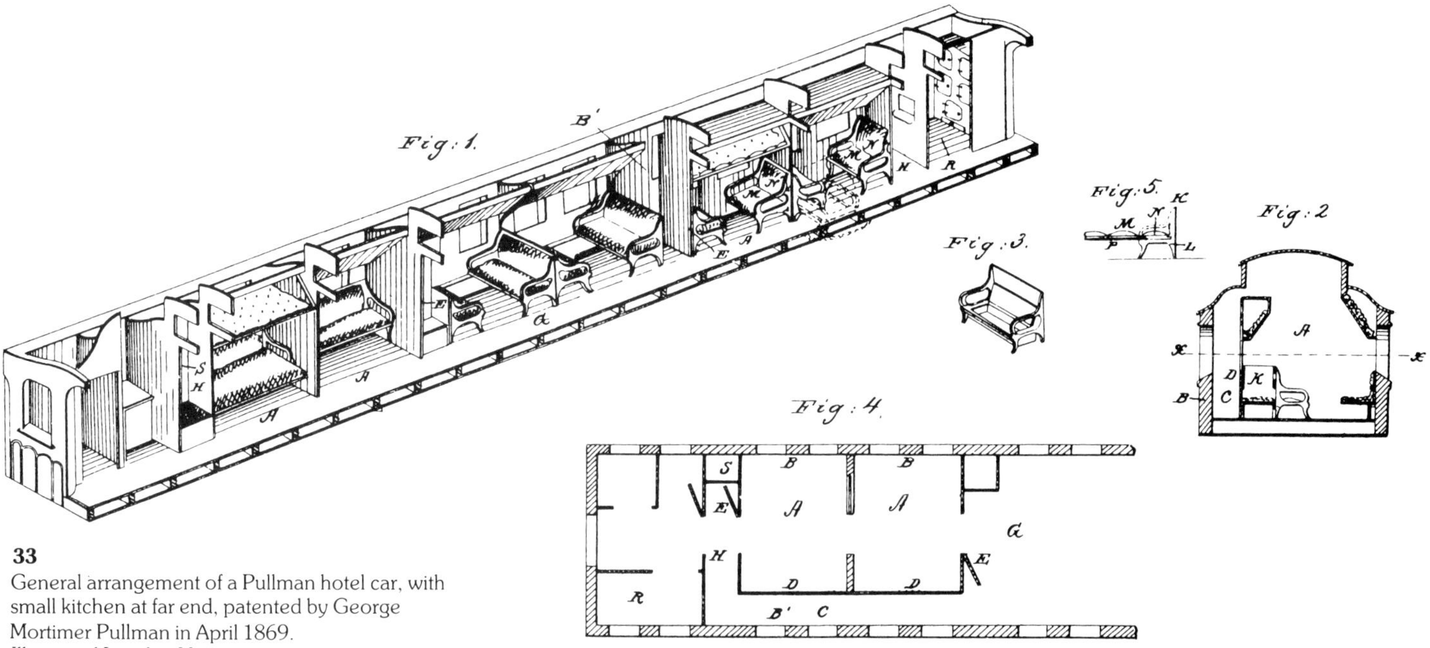

33
General arrangement of a Pullman hotel car, with small kitchen at far end, patented by George Mortimer Pullman in April 1869.
Illustrated London News

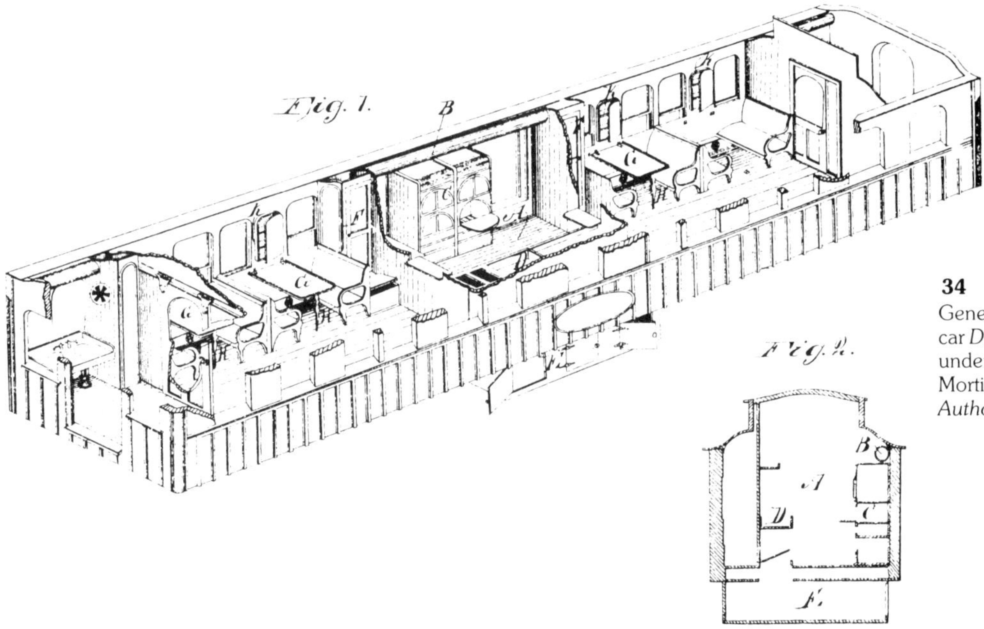

34
General arrangement of the first Pullman dining car *Delmonico*, with central kitchen and underfloor storage, patented by George Mortimer Pullman in April 1869.
Author's collection

decencies and comforts of orderly homes, can be whirled half a day over a dusty road, with hot cinders flying in their faces; and then, when they approach a station dying with weariness, hunger and thirst, longing for an opportunity to bathe their faces at least before partaking of their much needed refreshments, that they shall rush out helter-skelter into a dismal, long room and dispatch a supper, breakfast or dinner in fifteen minutes. The consequences of such savage and unnatural feeding are not reported by telegraph as railroad disasters; but if a faithful account were taken of them we are afraid they would be found much more serious than any that are caused by the smashing of cars, or the breaking of bridges. The traveller who has been riding all night in a dusty and crowded car, unable to sleep, and half suffocated with smoke and foul air, will be suddenly roused from his half lethargic condition by hearing the scream of the steam whistle, which tells of the near approach to a station; but before the train stops, the door of the car opens, and the conductor shouts at the top of his voice, "Pogramville — fifteen minutes for breakfast!"

'Here is a prospect for a weary and hungry traveller to whom fifteen minutes would be brief time enough for ablutions. But washing is out of the question, even if all the conveniences were at hand, and he rushes into the "saloon" where he is offered a choice of fried ham and eggs, or tough beefsteak soaked in bad butter, tea and coffee, stale bread; the inevitable custard pie and pound cake, are also at his service; but half the fifteen minutes allowed for breakfast having been lost while waiting for a turn at one of the two washbasins, the bewildered traveller makes a hasty grab at whatever comes within his reach, and hurries back to his seat, to discover before he reaches the end of his journey, that he has laid the foundation for a fit of dyspepsia, which may lead to a disease of the lungs or a fever.

'A company that expends thirty millions of dollars in building and equipping a road, ought not to begrudge a few hundred dollars in furnishing a suitable place for travellers to refresh themselves at, nor deny them the necessary time to do it in. Refreshment saloons ought to be under the charge of the railroad company, for passengers ought not to be put at the mercy of grasping and ignorant men, who are regardless of everything but their own profit. There are very few roads in the whole country on which there is a station where a comfortable meal can be obtained, or where the necessary accommodations are to be found which passengers require when they stop. The Erie road has the most comfortable and spacious cars of any railroad in the country, and we have no doubt that it would be a great accession of travel if there were decent station-houses on the line for the refreshment of passengers, and the arrangement of the time-tables were such that they could be enjoyed. As our great lines of railroad are laid through wild and half cultivated regions, where the common refinements of city life are not to be looked for, the railroad managers should themselves make the necessary provision for the refreshment of passengers which they have a right to expect. As affairs are now arranged, a few days of railroad travelling are sure to end in a fit of sickness to all excepting those who have hearty constitutions, and are accustomed to the very roughest and toughest manner of living'.

Of course some people preferred to take their own food with them, but as journeys lengthened this became more of a problem, and the idea of 'stretching one's legs' at a stop was more appealing. On board train there was no shortage of peddlers, who also plagued passengers at every wayside station. Pullman realised that his supplementary fare-paying passengers could be protected from this chaotic situation if he provided meals on board the comfortable cars. In 1867 as already mentioned, he produced his first 'hotel car' named *President* for the Great Western Railway of Canada. It was a sleeping car of his latest pattern, with a small kitchen added, located at one end, and large enough only to serve the actual occupants of the car, with preparation by a professional chef. To serve the meals tables were set up between the seats, but only when required, and Pullman added all the finery of linen tablecloths and elegant crockery, that the best hotels of the day would consider necessary.

It was a logical step to provide dining facilities for the occupants of other cars, and in the following year, 1868, Pullman introduced his first dining car, open to all the train's passengers, for the Chicago & Alton line. This was given the name *Delmonico*, after a famous New York restaurant, and further dining cars quickly followed it into service, because from the public's viewpoint they were a great success. To reach the dining car it was necessary for the conductor to guide passengers over the perilous gap between the two adjacent open car ends; a practice that was disliked by both the passengers and the more safety-conscious railroad officials. Although the adoption of the Miller 'knuckle' coupler (later the buckeye coupling) was to make the join between cars more stable than the previous link and pin arrangement, it was still a chilling experience to cross the gap above the coupler, with the track clearly visible beneath and rushing past, with all the noise and dust of a train at speed thrust upon one, and without any more protection than an iron verandah rail to cling to! In these early dining cars, the kitchen was located in the centre of the car, because Pullman himself believed 'that in whatever direction the car may be travelling at least half of it will be in advance of the kitchen, the odours of which are borne by the draught toward the rear'.

The success of the dining cars was eventually to lead to the closure of the majority of station dining rooms, known as 'eating stations' but it is of interest to record here that Pullman himself found the cars to be unprofitable and after a few years the American Pullman Co opted out of the dining car business, leaving it to others, chiefly the railroads themselves who mostly operated them (and still do) at a loss. Notable in this account is the fact that the large meal involved was *breakfast*:

'Thirty miles beyond Julesburg on the Union Pacific Railroad we came to Sidney, an eating-station and therefore a place of importance. We stopped half an hour to hurry out and get breakfast. Tearing into the station house past a man standing at the doorway who was vigorously proclaiming on a loud sounding gong that the feast was ready prepared and spread within, we found ourselves in a room furnished with many neatly arrayed tables. We had plenty of the gentler sex to wait on us. There was plenty of time to eat and plenty of room to spare. There were given us eight little dishes apiece containing hot beefsteak, two slices of cold roast antelope, a bit of cold chicken, ham and poached eggs, a couple of boiled potatoes, two sticks of sweet corn, stewed tomatoes and four thin buckwheat hot cakes laid one on top of the other, to be eaten with golden syrup the last thing of all.

'We were all served alike; everyone was given the same as his neighbour. Knives and forks were lamentably scarce, as usual. One knife and fork to each to last through the whole meal is the order of the day out here. Cold tea in tumblers with a quantity of sugar added seemed to constitute the popular beverage, if it was not cold milk. There was hot tea and coffee for those who preferred. All ate as if for their very lives and the result was that we were all through together a quarter of an hour before it was time to start. For this and for every meal, except two, along the route of the Pacific Railroad from Omaha to San Francisco the charge is one dollar.' (1878) (From *Through America*, W. G. Marshall, London, 1882)

With sleeping cars, hotel cars and dining cars all in service the Pullman Co was truly flourishing by 1869, and when on 10 May that year the famous 'golden spike' was driven, linking the Union Pacific and the Central Pacific Railroads and making journeys from the Atlantic coast to the Pacific coast possible, the demands for Pullman cars increased dramatically! Added to the list were 'private cars', built to Pullman's highest standards of design and comfort, and purchased from him, to be used by their wealthy owners over the various railroads, affording complete privacy to important business people and other notables of the time in quite splendid luxury on wheels.

A very good description of an early encounter with the sumptuous Pullman

35
Pullman hotel car *City of New York*; mounted on 16 wheels, as reproduced in a wood engraving for the British *Illustrated London News* of October 1869. This gives an authentic record of the ornamental style of the wooden bodywork and clerestory, and the elaborate window treatment. Note the lettering 'Pullman's Sleeping Car Line', and the oval nameboard on the centre of the car waist. The eight-wheel bogie was designed by Mr C. F. Allen, a master mechanic on the Chicago, Burlington & Quincy Railroad. He also patented the compressed paper wheel centres that G. M. Pullman used for his early cars, including this one. *Illustrated London News*

35

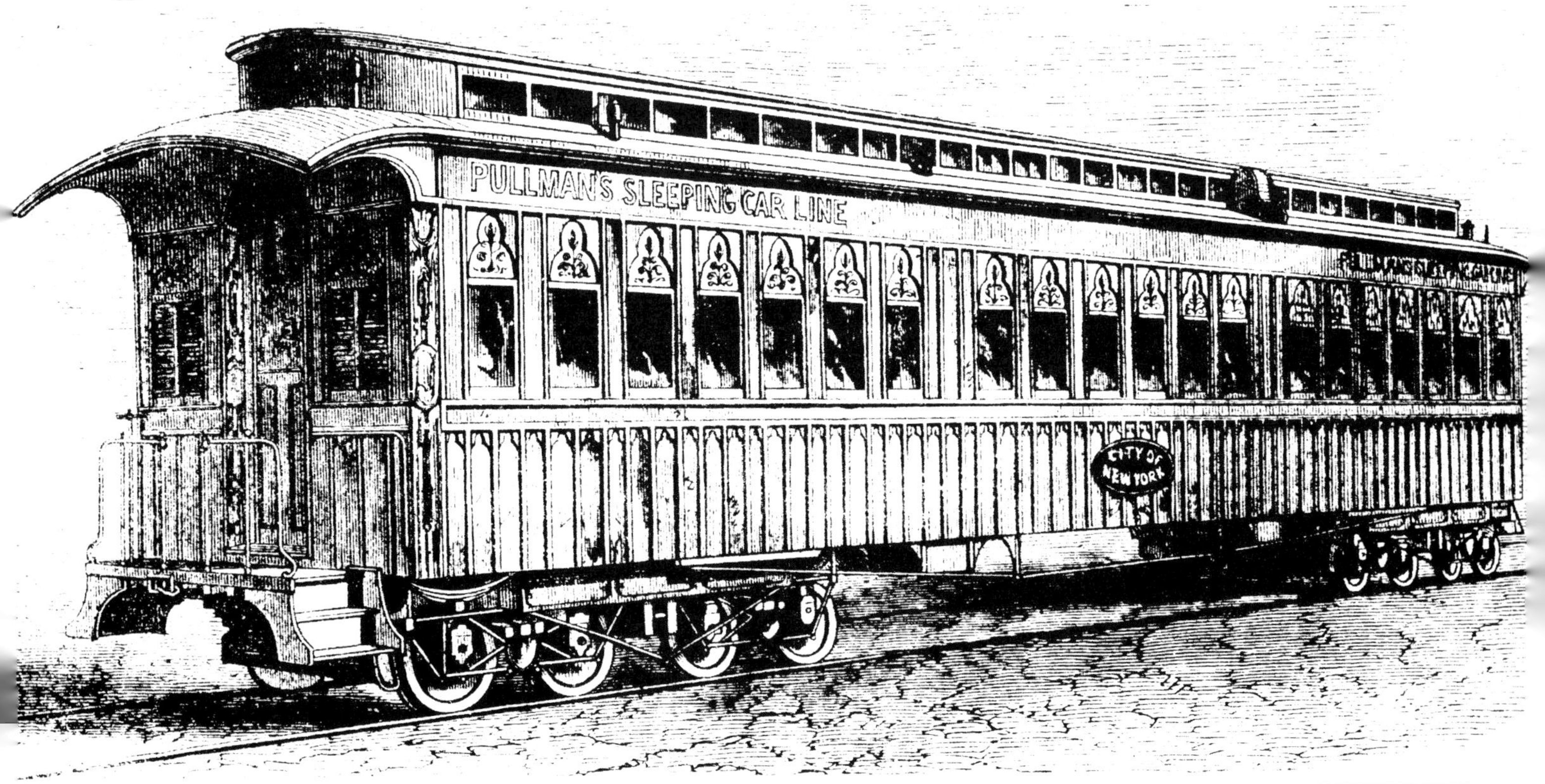

palace cars was published in London in 1870, concerning a journey made the previous year:

'Before beginning my journey by rail in September, 1869, from the Atlantic coast to the Pacific slope, I had to ascertain various particulars as to the route. There was no difficulty in purchasing a through ticket. In most of the hotels and in numerous shops the tickets of any railroad in the United States can be bought. The traveller sees innumerable advertisements in which the Union Pacific Railroad is conspicuous but in which the names of various lines are enumerated as being in connection with it. He reads in one that the Allentown Line is the shortest and best, in another that the Great Central Route is without a rival, or he may see the advantages of the Erie extolled to the skies. As the fare in all cases is the same the puzzle consists in ascertaining the respective merits of the competing lines. He learns that in any event he must first reach Chicago. If he is curious to become acquainted with the far famed Pullman cars he will decide to travel by the Great Central Route and that is the one I chose.

'The traveller bound for the Far West starts from New York in the evening by the Pacific Express and on the morning of the following day arrives at Rochester where Pullman palace-cars are attached to the train. I had heard much said in praise of the Pullman car but I was unprepared for the reality. The first trip in one of these cars forms an epoch in a traveller's life. No royal personage can be more comfortably housed than the occupant of a Pullman car, provided the car be a hotel one. In the train by which I travelled one out of three sleeping-cars was of the latter description.

'The hotel-car is divided into sections forming staterooms wherein parties of four can be accommodated. Between these rooms are seats arranged in the usual way. At the rear is a kitchen, which, though small, contains every appliance necessary for cooking purposes. There are water-tanks in which is stored a supply of water for washing and drinking sufficient to last the journey. A wine cellar contains the liquors which are likely to be in demand and an icehouse preserves ice for the gratification of those who prefer cold beverages. At stated intervals the conductor walks around taking passengers' orders, who make their selections from the bill-of-fare.

'The choice is by no means small. Five different kinds of bread, four sorts of cold meat, six hot dishes, to say nothing of eggs cooked in seven different ways and all the seasonable vegetables and fruits, form a variety from which the most dainty eater might easily find something to tickle his palate and the ravenous to satisfy his appetite. The meal is served on a table temporarily fixed to the side of the car and removed when no longer required. To breakfast, dine and sup in this style while the train is speeding at the rate of nearly thirty miles an hour is a sensation of which the novelty is not greater than the comfort. An additional zest is given to the good things by the thought that the passengers in the other cars must rush out when the refreshment station is reached and hastily swallow an ill-cooked meal.' (From *Westward by Rail*, W. F. Rae, London, 1870)

In addition to the hotel car *President*, which went immediately to Canada, other early hotel cars were *City of Boston*, *City of New York*, *Kalamazoo* and *Western World*. It was the *Illustrated London News* of October 1869 that published a detailed account of the American Pullmans then in service in the USA, and this was illustrated with both exteriors and interiors, in exquisitely detailed wood engravings. The exterior of the *City of New York* hotel car showed the quite striking elegance of the coachbuilding, and the 16-wheel layout was very well depicted. The wooden body followed the general style adopted for new cars of the period, with matchboarded sides below waist level, but the clerestory was exceptionally elegant and had rounded ends. An oval plaque carried the car name in the centre of the matchboarding, and there was the slogan 'Pullman's Sleeping Car Line' in gilt letters at each end of the car, above the windows. Exact livery details are not known, but the colour was probably deep umber brown (slightly greenish) with elaborate gilt decoration on every portion of the panelling. Railroad cars of the period were often in green livery, or Tuscan red, but it is recorded that some of Pullman's rivals adopted brighter colours to draw attention to themselves, and the Central Pacific Railroad's 'Silver Palace Cars', owned and operated by them west of Ogden, once the transcontinental route was opened, were painted in bright canary yellow, with all the interior fitting finished in white metal to imitate silver.

When George Mortimer Pullman created the Pullman Co in 1867 he established his first main workshop at Elmira, New York. At that time he was 38 years old and already a millionaire, this latter achievement being due to his new business association with a former rival, the Scottish-born financier Andrew Carnegie. The demand for new Pullman cars forced him for some time to go to outside car builders, and to avoid this he then established further shops at Detroit (closing down the New York shop) and at Wilmington, Delaware and San Francisco. By the early 1870s the name Pullman was known across the continent of America as a symbol of quality and service to the railroad customer, but it was still an unknown quantity in Britain and Europe, apart from the published reports filtering back from visitors to America.

36
The grand lifestyle of a Pullman palace hotel car on the Union Pacific Railroad in 1869, with a negro attendant serving refreshments. (From *Harper's Illustrated Weekly*, May 1869.) *Author's collection*

36

CHAPTER 4

Britain and America Compared

Having shown how George Mortimer established his business in America, it is necessary to retrace our steps, and draw a comparison between Britain and America since the early days of the steam railway. If an American had arrived in Britain at any time between the mid-1840s and the early 1870s and decided to travel by railroad, or railway as the British preferred to call it*, he would have found some surprising contrasts to the travelling conditions he was accustomed to on his home railroads. Since the earliest days the carriage of passengers by rail in Britain had been segregated into various classes, according to one's social status, with the price charged for the journey varying accordingly.

In Chapter 1 I have shown how the American railroad car developed away from its original similarity to the British counterpart, in order to meet the somewhat different requirements of American travellers and the railroad operators. By the mid-1840s the American traveller was already housed within a quite lengthy wooden-bodied carriage, with a roof over his head and a seat of at least basic comfort. There was possibly a stove in winter and candles to light the interior after dark. The tracks were bad and derailments pretty common, but efforts were being made to improve the riding of the cars by the use of bogies in place of the rigid four-wheel layout. All classes of traveller shared the same accommodation, and they were supervised by the train's conductor who moved from car to car en route.

In Britain at this time only the wealthier, first class traveller could expect to be offered anything approaching the standard of accommodation available to *all* in America. So poor was the standard offered to the common masses that in August 1844, Gladstone's Railway Regulation Act received the Royal Assent. This obliged the railway companies to provide weatherproof travelling accommodation for third class passengers, once a day. Many companies literally obeyed the word of the law and provided one such train a day, but at night or else in the early hours of the morning — thereby doing their best to discourage third class travel. These unearthly timed trains became known as the 'Parliamentary trains' and their slow and weary progress through the countryside was matched only by the cramped, unlit, and poorly ventilated carriages used. This well intentioned Act of Parliament did not abolish the old open carriages; on the contrary, they were utilised for passengers foolish enough to choose third class travel on trains other than the one 'Parliamentary' a day! They were also used for excursion traffic; presumably a trip to the seaside could be a reasonably gay affair in one of these open carriages — weather permitting.

However, the Act did force the companies to improve their second class accommodation, and low partitions divided these into semi-compartments, with covered bodies, but poor ventilation and illumination remained their lot.

To begin with, first class passengers who wanted night time lighting facilities brought these with them, in the form of an individual candle in a holder which could be hooked to a suitable spot on the compartment wall close to their heads. But by the early 1840s lamps burning rape oil were used in some carriages. Placed in position through a hole in the roof, the lamps could be removed as necessary for refilling. The waterproofing of these potholes was not very effective, and sorry was the plight of passengers should the lamp not have been replaced properly on a wet and stormy night! Some companies decided that one lamp might well be shared by two compartments if it

37

37
The interior of a British first class compartment of the late 1840s/early 1850s, still showing road vehicle influence. The first class was quite comfortable and well furnished, with buttoned upholstery, headrests between seats, and padded armrests. Note the silk top hat placed in the silk cords provided above the seats and the fabric straps by the windows, which are curtained. (Painting by Abraham Solomons entitled *The Return*, exhibited at the Royal Academy in 1854.) *National Railway Museum*

*Henceforth in this story, whenever it concerns Britain I shall refer to the railway rather than the railroad: bogies rather than trucks, and carriages rather than cars, *except* for Pullmans which have always been called cars in Britain as well as in the United States.

38
The interior of a third class carriage of the mid-19th century on the Stockton & Darlington Railway, still with bare wooden walls and seats, and depicting a cross section of the poorer classes of the day. Someone plays a penny whistle, whilst the baby howls, and no doubt the fat woman, who takes up a lot of seat, has a particularly smelly cheese in her basket! (From a contemporary painting.) *National Railway Museum.*

38

were placed in a position cut out of the wall between the two. The housings for these rape oil lamps were a prominent feature on the roofs of carriages so equipped. On some of the enclosed third class carriages introduced as a result of Gladstone's Act, one lamp sufficed to illuminate the whole carriage, which was of open layout crammed with bench seats. Another innovation was provision of heating in cold weather, by means of hot water footwarmers.

The British were not addicted to the American habit of tobacco chewing, but there was the problem of men who wanted to smoke. The railway companies discouraged smoking on trains and in their stations to begin with, but in due course they had to give way to public pressure and some special compartments were included in some carriages; whilst the Eastern Counties Railway had already introduced a special smoking saloon in 1846. Saloons were also favoured by the wealthier people with large families, as they could hire the complete carriage and thereby ensure privacy whilst travelling. These family saloons were a popular development of the late 1860s (although the Great Western Railway had some in its early days known as 'posting carriages', as mentioned in Chapter 2). Victorian families tended to be large, and the saloon layout catered well for these when travelling complete with servants. There was also the very desirable provision of toilet facilities — no doubt a large family blessed this! But at the time such sanitary provisions were the exception rather than the rule.

During the period 1840-1860 the design and the method of construction of British carriage bodies advanced slowly, with some increase in length (adding extra compartments) but this was limited by the rigid wheelbase used. The lighting was still rape oil, which offered dismal illumination at best. Some improvements in the welfare of second class passengers was achieved with such luxuries as stiff cushions to sit upon and quarterlight windows flanking the door. The first class remained stodgily comfortable, with panels and mouldings picked out rather crudely in gold, probably a mirror, ponderous buttoned-in upholstery, curtains, and a carpet. The inside of the door would usually be padded with a thick buttoned leather squab. The exterior was beginning to lose its road coach ancestry and panelling was more rectangular with possibly a curved treatment remaining at each end of the carriage side. Slotted 'venetian' ventilators were inserted above the windows.

Some increase in the width and height of carriages began to be noticeable in the 1860s, but there can have been little incentive to make any great improvement in this respect whilst lighting and heating remained so poor. Nevertheless six-wheeled carriages with larger bodies began to replace the four wheelers, only the GWR used six wheelers from the outset). Although a Mr Joseph Wright had patented bogie carriage designs as early as 1844, British companies were not to take up the idea for many years.

Sanitary facilities were unknown (except in the saloons), to ordinary passengers of any class, and at this time sleeping on a train was by means of an improvised arrangement of seat cushions across the

39

40

39
By the late 1860s, the typical British four-wheel compartment carriage was considerably longer, but still with a low roof and pot-oil lamps. This Midland Railway example (photographed in 1874) was probably originally built as a composite first and second class (the second class being for the servants of the wealthy in the first) and was later altered to first and 'improved third', as described in Chapter 7. Provision for luggage was in a small end compartment. Continuous brakes were fitted by 1874.
National Railway Museum

40
Pullman Palace drawing room car *Dayton*, built by Barney & Smith Manufacturing Co of Dayton, Ohio. A classic example of the style adopted by G. M. Pullman for his cars and (apart from the use of six-wheel bogies) very much the exotic and elaborate vision that was presented to British eyes by the introduction of Pullman cars on the Midland Railway. What a striking contrast to the preceding picture! *Author's collection*

compartment, with bed linen hired from the guard. Not many people availed themselves of this amenity, however, and for the majority of stalwarts who had perforce to journey overnight, rugs about the person were advisable.

Our American traveller would have found the British trains perhaps 'quaint' and remarkably small both outside and inside and, in the earliest days, sometimes with no room for his bulky luggage except on the roof. Most noticeable would be the different behaviour of the passengers, according to class. If he (wisely) chose to travel first class, then he would have found himself in a small, low-roofed compartment with five, or perhaps less, other occupants. As there was no corridor linking each compartment in a carriage, and merely side doors for getting on and off at stations, he would be 'trapped' for the duration of his journey. Attempts at conversation would quite possibly be met with obvious disapproval or stony silence, and the voluble American would perforce have to read his newspapers or look out of the window! The passage of the train would seem to him to be quite fast and rather bumpy, in the four- or six-wheeler, but he would not experience the violent swaying so characteristic of American trains of the period. If he was observant, he would notice that there were fences alongside the tracks, and that the locomotives did not have cow catchers fitted. No food was available on the train, unless he brought his own with him, but there was possibly a stop scheduled to be made at one station en route, long enough to eat in the dining room, as in America.

If our American was of only modest means, and chose to travel among the poor in the third class, he would then be thrown in with a more talkative crowd, with endless screaming children a very distinct possibility, and (one suspects) an overpowering smell of stale urine. Draughts would abound within the half-partitioned open carriage, with plenty of soot and sparks in the air, and the passage through tunnels would be particularly unpleasant. If he chose the compromise of second class there would be some cloth covering and padding to his seat (wood was still commonplace for the third class), and the partition would reach to the ceiling. At least 11 other people might be crammed into his compartment, which was no more spacious than that provided for the fewer first class passengers — who of course paid a higher fare. The railway companies so favoured the higher fares that they went to the extremes of putting third class trains into sidings en route to allow first and second class passenger trains to overtake them. It was James Joseph Allport, General Manager of the Midland Railway, who observed: 'I have felt saddened to see third class passengers shunted on to a siding in cold, bitter weather — a train containing amongst others lightly-clad women and children — for the convenience of allowing the more comfortable and warmly clad passengers (travelling first and second class) to pass them. I have even known third class trains to be shunted into a siding to allow express goods to pass'. (We shall hear more about Mr Allport a little later.)

A British visitor to America in 1871 made this report, which gives a good contrast to the British conditions just described:

'The average speed on the American lines is about twenty miles an hour. The express trains rarely exceed thirty miles. Nominally, there is but one class of passengers and one scale of fare. Every traveller takes his place where he has a fancy except that there is a car reserved for ladies and for gentlemen accompanying them. But though in theory all are equal there are practically various classes of passengers. On the main lines there are cheap trains for immigrants. There are attached to most trains drawing-room cars, reclining-chair cars and sleeping-cars, for which additional charge is made. Some of these select cars belong to speculating builders or companies who purchase the privilege of attaching them to the trains and make their profit by the additional charge.

'The most notable of these speculations are the Pullman hotel-cars. In one of these I travelled from Niagara to Chicago, leaving in the forenoon and arriving on the morning of the following day. Two of us chartered a compartment like the cabin of a ship with a comfortable sofa above which a board was fixed at night, so as to form a second sleeping-berth. The beds were regularly made, boots put outside the door for cleaning, and hot water brought in the morning by an active black boy. Meals were served on a table carried into the cabin. The bill-of-fare contained many items and at moderate charges. For lamb chop or mutton chop and tomato sauce the price was seventy-five cents; fresh mackeral, fifty cents; omelet with ham, forty cents and a spring chicken cost a dollar. There was ample choice of vegetables, fruit and relishes with five or six kinds of wine in the *carte*. A cup of French coffee, tea or chocolate was fifteen cents. The kitchen, clean and commodious, had every appliance for cooking, and the dressing compartment was equally convenient. Some of the drawing-room cars are as luxurious as those of royal or imperial carriages on European lines, with mirrors, lounges, chandeliers, piano and bookcases. A novelty in some of the cars on the Pacific Line is an outside balcony from which the scenery can be surveyed.

'In America the ingenuity of the directors seems to be exerted for the comfort of the poorest classes. The carriages are large and well ventilated, with stoves for wintry weather and Venetian blinds and sun shades at every window in summer. The seats are roomy and well cushioned. For parcels or small packages there are plenty of pegs and wire racks, and there is a barrel of filtered water, iced in summer, at one end of the car. Notwithstanding all this, there is room for improvement. The jumping and jolting on many of the lines is terrible. The rails are laid upon big wooden sleepers which seem often of needless irregularity in level. Great is the dust and glare in hot weather and the draft in cold weather. Except in the larger towns and at the hotel stations the dépôts are generally very plain wooden structures, with few fixings and at night cheerless and

41

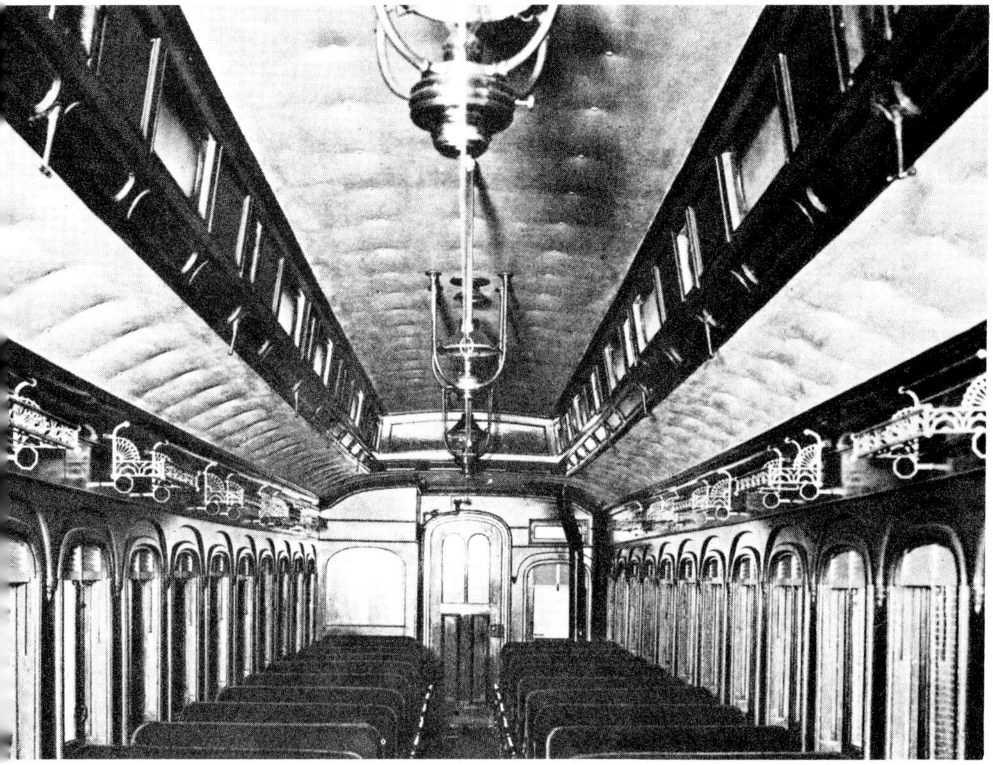

42

ill lighted.' (From *Across the Ferry*, James Macaulay, London, 1887.)

Although by no means perfect, it seems that the Americans were considerably more advanced in the art of railway passenger carrying by this time, when the train was the fastest available means of mass transport in man's history and was virtually without any competition. The eyes of the British began to turn to the work of the American railroads, and to the achievements of George Mortimer Pullman in particular, with some envy, as more and more first-hand accounts reached them. The *Illustrated London News* feature of 1869, already mentioned, must have been read by a great many British railway company chairmen, shareholders, directors and managers alike, as well as by a great many of the more educated passengers of the period. There may have been sceptics, of course, because some of the accounts published seemed far too romantic to be true, and it is in a way quite refreshing to come across this highly critical 1873 account; but note that this actually refers to the rival 'Silver Palace' cars, and not as erroneously stated by the writer, to Pullmans. That error apart, there are some interesting observations, in particular about the way passengers' luggage, or baggage, was treated at that time.

'We took the Erie Railroad from New York to Niagara as it seemed by the map to pass through a more picturesque country than the other routes, and on a dull damp morning ensconced ourselves in a Pullman silver palace drawing-room and hotel-car. Such a high sounding title was enough to take one's breath away. It reminded me of a palace scene in the Arabian nights. The reality, however, was simply a long railway carriage with chairs like music stools with arms; some nets of a size quite incapable of holding anything and a brilliant array of spittoons. Nor must I forget the filter of water, with one glass or tin cup chained to it, out of which everybody from the peanut boy and porters were supposed to drink. For the privilege of entering these luxurious structures an extra three dollars per day had to be paid.

'At Port Jervis a most voluble old lady entered the car bringing with her an extraordinary assortment of luggage. She had two bandboxes, a dog, a flower pot, an umbrella, a jug of milk, a lunch basket, a parcel of figs and a boy of fifteen whose age she gave at the ticket-office as nine. I heard her tell her nearest neighbour of the fact and she chuckled over it as if she had performed a most virtuous action.

41
The interior of a typical late 19th century American 'coach' (third class) day car. The seats had swing-over backs and would normally be arranged so that passengers faced the direction of travel. Note the flue pipe from the heating boiler in the right hand corner of the saloon. *MR*

42
A characteristic US sleeping car interior of the 1870s. The right hand side is in the day position, but the 'sections' on the left have been made up for the night — with the seats forming the basis of the lower berths and the upper berths swung down from the ceiling. The only privacy is provided by the heavy curtains.

'As our sleeping-car afforded no signs of sleeping accommodations I waited with curiosity for bedtime. It came at last and the chambermaid in the shape of a black man entered and asked me to move as he was going to make up the beds. He then made up the beds, all the arrangements of which were very good in themselves but there was no possible way of making use of them with the smallest degree of comfort. If there had only been a few hooks for hanging hats, coats and other garments upon and a place to put boots in it would have been something to be thankful for. For usually boots were to be found in the morning at different ends of the car or else on the wrong person's feet.

'The space between the upper and lower berth does not allow of the occupant of the bed sitting up; the double windows are kept shut that he may not be smothered in dust and ashes and the night is passed in the most luxurious misery. The man who slept in the berth next to mine snored frightfully; in fact, night was made hideous by the unmusical sounds issuing from all parts of the car. The horrors of that first night in a Pullman car are indelibly impressed on my mind. The atmosphere ran a close heat with that of the Black Hole of Calcutta. On my asking the porter why he kept a fire burning all night he said he had to sit up and it would never do for him to catch cold.

'If going to bed is misery, getting up in the morning is simply agony. If you are late you have to wait some hours before you can get a turn at the one washbasin. If you are early you have to stand outside on the platform in the dust and smoke until the beds are once more metamorphosed into seats, there being no room or other place to retire to until that operation is performed.

'We chose the Burlington and Quincy Railroad for the trip from Chicago to Omaha and again had the pleasure of looking forward to another night in a sleeping-car. On one section our car bumped about in a most extraordinary manner. On looking back at the long lines of rails over which we had just passed they seemed so unevenly laid, so rugged with ups and downs that it was a wonder how we kept on the rails at all. One of the passengers remarked to the conductor that the road seemed rather rough. "Well, yes" he answered, "trains do jump about some, but it's enough for us if they keep between the fences." However, we managed not only to keep between the fences but actually on the track and gradually approached the town of Council Bluffs. We very soon had enough of Council Bluffs and crossed the muddy river on a fine iron bridge to Omaha.

'I was glad to get away from Omaha without having my luggage entirely knocked to pieces, as Omaha is celebrated for its baggage-smashers. There were several boxes lying about whose owners could only recognise them by their contents. Many a handsome box must have left Chicago as a Saratoga and arrived at San Fransisco mere lumber.'

His final observation could equally well have applied to a long distance 1873 British railway journey in third class accommodation.

'I have often wondered how people who cannot afford the high fares of the Pullman cars get through the long journey from New York to San Fransisco. The ordinary cars are without carpets, the seats are often without cushions and not a place for a sick man or child to lie down day or night and no room for change of position. Crowded indiscriminately with whites, blacks and Chinese, people of cultivated minds and habits have to herd with the vulgar and low-mannered and the journey altogether must be as near an approach to a seven days' purgatory as is possible.' (From *Western Wanderings*, J. W. Boddam-Whetham, London, 1874.)

A most significant event had taken place in Britain a year prior to this particular record of American railroad travel. As from 1 April 1872 the Midland Railway Co Board of Directors had decreed that henceforth third class passengers should be admitted to *all* Midland trains. This was at the instigation of Mr James (later Sir) James Joseph Allport, the Midland Railway General Manager, whose personal observations about the plight of the third class traveller have already been mentioned.

This move by Allport was only part of an ambitious plan he had conceived to improve the overall standards of passenger travel on the Midland Railway, including those available to the first class traveller. Here it must be remembered that in those days the private railways were in fierce competition with one another, often running to similar destinations over virtually parallel routes; hence any improvement was welcomed which could give a possible advantage over the rivals, and attract more passengers and revenue. Allport had obviously read many accounts of what had taken place in America in recent years, and in the autumn of 1872, with leave of absence granted by the Midland Board of Directors, he visited America to see for himself what was being achieved.

In America, James Allport travelled some 6,000 miles on the various railroads, and obviously spent a good deal of time travelling in Pullman cars, and no doubt also those of Pullman's competitors of the period. We can judge the sort of conditions he found from the various eye-witness accounts I have already included, and there can be no doubt that the Pullmans were particularly impressive, although there were many other aspects of American railroad operation that would also have made him draw comparisons with his own Midland Railway. To begin with there was the sheer mileage involved, with journeys that lasted days rather than hours, and with all life's requirements — eating, sleeping, washing, etc — having to be catered for en route. There was the (in theory at least) one class system, which could only be avoided by paying the supplement to travel in a Pullman, or similar, privately-owned car. There was the ornate interior decor and elaborate styling of the cars. The locomotives of the period must have looked strange to his eyes, with their cowcatchers on the front and huge spark-arrestors mounted on the chimney. Stations did not have raised platforms, and everyone seemed to be placed at constant risk of life, by crossing the lines on the level. Locomotives carried huge headlights, and constantly rang warning bells, to give some notice to the public and animals alike.

It was in so many ways a very different prospect of rail travel to that then offered to the British traveller, and in one respect it was, Allport realised, unquestionably superior because the Pullman car had brought a sense of dignity and civilised life-style to those prepared to pay for it. He believed that his Midland Railway customers included many who would equally welcome such an advance in comfort and amenity whilst travelling, and who were prepared to pay for it. Allport arranged to meet George Mortimer Pullman in Chicago, and one suspects that he had planned that even even before departing British shores!

43

43
A typical American railroad scene of the early 1870s, such as the visiting Midland Railway General Manager, James Allport, must have witnessed. This carefully detailed drawing depicts the 'Lightning' expresses of the Erie Railroad, at an important station, with crowds on the platform (left-hand background). The left-hand locomotive is an imposing Mogul 2-6-0 with a baggage car behind the tender and a lengthy string of cars attached. Centre of the picture is a 4-4-0 collecting its train from the yards, and on the right is a 'Pullman Palace Drawing Room & Sleeping Car' with the conductor working the handbrake on the car end platform. The Erie Railroad was broad gauge at this period. *Author's collection*

CHAPTER 5

Mr Pullman comes to Britain

George Mortimer Pullman was invited to Britain by the Midland Railway (MR) General Manager, James Allport, and the latter carefully paved the way for him in advance. When Allport returned from America to his Midland Railway post, he proposed the idea of using Pullman cars to work special trains. In particular he must have been looking ahead to the completion of the Settle & Carlisle line, which would give the MR a direct route to Scotland. Until this line was opened the rival London & North Western Railway held the advantage over that traffic. With the opening of the Settle & Carlisle, Allport envisaged a through service of sleeping cars from London. The Pullman layout must have seemed ideal, and in any case there were other existing MR services on which the use of Pullman cars could score an edge over their rivals. James Allport convinced the Directors of the Midland Railway that they should meet Pullman, and afterwards present him, complete with plans and models of his cars, to the shareholders at their next half-yearly meeting. This was due to take place on 15 February 1873.

At the meeting, the Midland Railway shareholders were first of all given an account of George Mortimer Pullman's American achievements, and how he had created high standards of comfort for the passengers. The shareholders were then told that Mr Pullman was himself present at the meeting, and would state his proposals personally. This announcement was greeted with some cheering and applause.

Pullman's proposal was exactly along the lines he normally adopted when approaching an American railroad, in that he would be prepared to build the cars (and ship them to England in this case) at his own financial risk, and to cover all the costs of operating them, including the selection and supply of the attendants for the cars, as many as were required by the service operated. In return for this he would require permission to charge his 'Pullman supplement' on top of the normal fare for the journey. He also wanted an exclusive 15-year contract with the MR to provide these services, which would include sleeping cars, dining cars and parlour cars.

Although the Midland Railway shareholders accepted Pullman's offer, they made a proviso of considerable significance, requiring him also to design and build some carriages to provide *non*-supplementary fare accommodation for all three classes of passenger. Outwardly these were to resemble the Pullman cars, and they were always to be included as part of a train which was otherwise made up of Pullman's own supplementary fare special cars. The MR was to purchase these non-supplementary 'mock-Pullmans' outright from him when they were built. The reasoning behind this proviso was the desire by Allport, and the MR directors, to improve the general standards of comfort offered to travellers, as already mentioned in the previous chapter. Thus, whilst offering the de luxe Pullmans to those prepared to pay a supplement to use them, the MR evidently did not wish to widen still further the gap between classes and their accommodation. For his own part, George Mortimer Pullman made it clear that he wished to remain free to secure similar contracts with other railway companies, and this was apparently agreed. He saw the MR contract as opening a door for further contracts, not only in Britain but in Europe as well.

Because the British loading gauge was somewhat more restricted in height and width than the dimensions that American railroads considered to be normal, some adjustment of dimensions was necessary. One can rest assured that Pullman had learned this lesson from the troubles with *Pioneer*! Nevertheless, the cars designed for the Midland were essentially to the American style of the period, both in terms of method of construction and in appearance, being essentially scaled-down versions of what the Detroit workshops were building for home use. The method of construction favoured by Pullman was still the wooden integral type with heavy timber side trusses and solebars and with metal tie bars. This wooden body was mounted upon two four-wheel bogies of typical American design, with frames and most structural parts made of wood and with the ironwork bolted on. Only the wheels themselves differed from American practice, because Derby chose to use Mansell-pattern 3ft 6in diameter wheels (with wooden centres), a type very popular in Britain, and larger than the typical car wheels of American railroads.* It is interesting to note that Pullman had by that time ceased to use the 16-wheel layout, in favour of four- or six-wheel wooden framed bogies. In the case of the British Pullmans it is possible that the 16-wheel double bogies would in any case have fouled the loading gauge clearances below platform height, but it is more probable that Pullman had by then realised that 16 wheels, eight axles and eight axleboxes created in-built maintenance problems of their own!

The form of construction, with the heavy timber side trusses built into the bodyside below the windows, meant that the windows could not be lowered into the bodyside, as was normal practice. Instead, they slid *upwards* into the fascia panel; an arrangement that restricted the amount of aperture and which must have been very draughty. It is not clear how the windows were held in the open position, but a system of balance weights seems probable.

The agreement with Pullman was signed on 18 February 1873 and it involved the Pullman Co workshops at Detroit as builders of the complete cars, which were then to be partially dismantled and packed into crates for shipment to England, where they would be reassembled in a special shed to be built for the purpose, at Derby. The Detroit works built the cars with amazing speed and the first were ready for shipment by June, and by December they were already being assembled in their special shed, close to Derby Midland Railway station.

The first Pullman sleeping car for the Midland Railway was appropriately named *Midland* and this was ready for trial running by 21 January 1874; less than one year after the agreement had been signed. This was followed by two more sleeping cars, *Excelsior* in February, and then *Enterprise*, together with two drawing room, or parlour cars, named *Victoria* and *Britannia*. By March, demonstration running was in progress with examples of both sleeping and parlour cars, of which more anon.

The sleeping car *Midland* must have created a minor sensation when it first

*Pullman used patent pressed-paper wheel centres — instead of wooden segments or iron — for some of his early cars.

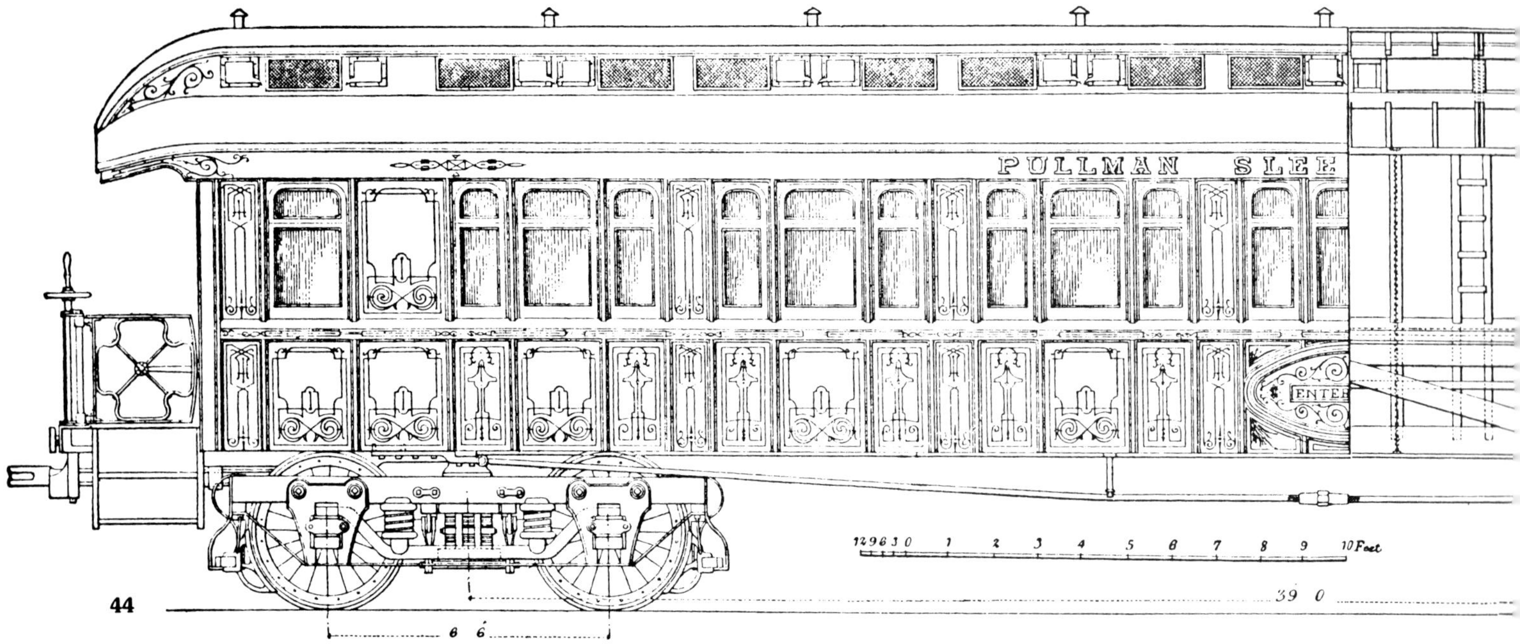

44

emerged into public view! Nearly everything about it was different to the existing MR carriages, even the colour scheme, of deep mahogany or umber brown with amazingly ornate gold leaf decoration on every panel. It was taller than normal, and the clerestory roof emphasised this feature, and there were no doors along the sides, just in the centre at each end, and reached by an open verandah with a domed roof canopy and with iron railings; moreover, it was mounted upon bogies! *Midland* is often claimed to be the first standard gauge bogie carriage to run in Great Britain (the narrow gauge Festiniog Railway had an earlier experimental bogie carriage), although in the same year, 1874, the Great Northern Railway also ran trials with a bogie carriage. The MR could however lay claim to having introduced the first complete train of bogie carriages to run in Britain, and it was also remarkable as a train in being fitted with American-style Miller centre couplers and continuous brakes, for which Derby locomotive works had to adapt the tender of selected locomotives, to enable them to be coupled to, and haul, the new Pullman trains.

In his excellent book *American Pullman Cars of the Midland Railway*, J. B. Radford includes the following contemporary description from the *Derby Mercury* of 28 January 1874.

'The Pullman cars on the Midland Railway

'One of the results of the recent visit of Mr Allport, General Manager of the Midland Railway Company, to Canada and the United States, was an arrangement by which the celebrated Pullman Palace Cars are to be placed within the reach of travellers on the Midland line. This luxurious carriage owes its origins to the gentleman whose name it bears, and who was led to contrive the means for making the long journeys across the American continent less wearying to travellers. Ten years ago Mr Longstreet, who is now erecting similar cars at the Midland Railway station in this town, built the first and now there are between six and seven hundred travelling on the railways in the United States. The ingenious contrivances by which a carriage full of seats is turned into a sleeping car are the subject of patents, and these are now the property of the Pullman Car Company, which owns all the existing carriages of this nature. This

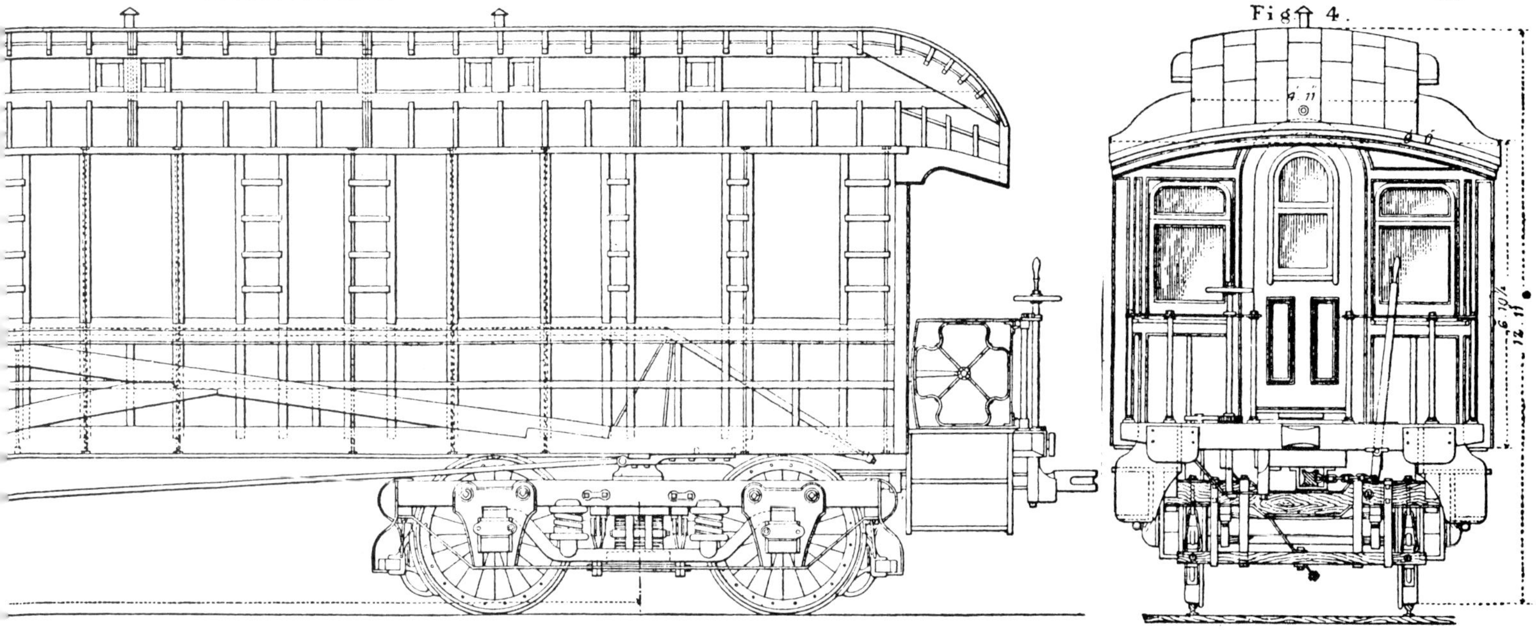

45

44

This detailed engraving depicts every structural and decorative detail of the Pullman wooden-bodied sleeping car *Enterprise*, built at Detriot USA, and reassembled at Derby, England in 1874 for the Midland Railway.

45

Interior of a Midland Railway Pullman sleeping car, arranged for day use, with upper berths stowed into the ceiling either side of the clerestory. *National Railway Museum*

Company has entered into an arrangement with the directors of the Midland Railway to run 16 carriages on their line charging passengers a fixed tariff for their use, in addition to the ordinary fare. Three of these carriages are now near completion at the Midland Railway station, Derby. The whole of the work has been done in the United States, and brought over here to be put together.

'The cars are 50ft in length, with the entrance at each end, and a passage direct through them, down the centre. They run on two trucks, which are square frames with double pairs of wheels, and are fixed at each end of the car, upon a pivot on which they turn, so that the double pairs of wheels adjust themselves readily to a curve, just as the fore wheels of an ordinary carriage, and there is less danger of their leaving the rails. They rest on a double set of springs, elliptical springs at right angles to the rails, and four circular coiled springs, one at each angle. The cars are strengthened with iron trusses. The exterior appearance is handsome and striking. It is however inside that the special features of the Pullman car are presented. They are of two kinds, Day or Parlour Cars, and Sleeping Cars. The former are fitted with two rows of handsome easy chairs, covered with crimson velvet, between which the passage extends. These are fixed on silver-plated pivots, so that the traveller can turn round easily, and face his fellow passengers. The Sleeping Cars are adapted both for day and night. By day they present handsome double seats, facing each other at right angles to and on each side of the passage. To turn these into beds there is an ingenious contrivance by which the opposite seats are brought forward so as to cover the space between them, and the backs are lowered, and thus the whole forms an enclosure in which two persons may lie. Just above a graceful curve covers during the day a place for stowing the bedclothes, and when the front is lowered an upper tier of beds is formed, each of which is ample for a single person. In each class of car there is at the end a warming apparatus. A coil of pipes passes around a fire, and these being filled with water it is heated and constantly circulates under the seats occupied by the

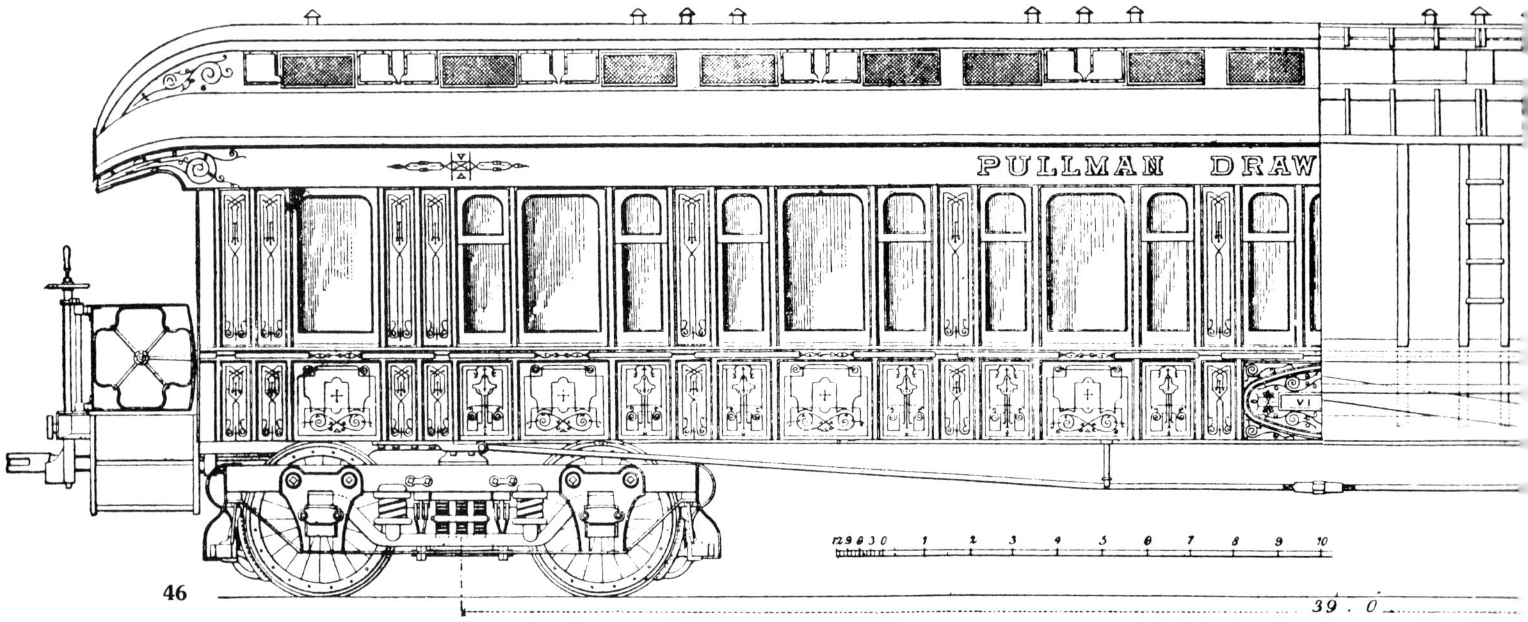

46

passengers. In each of the cars there are two private apartments, one of which will accommodate five persons, or which may be thrown together for a larger family or party. In the day cars each of these contains two arm chairs and a couch; in the sleeping cars the seats can easily be turned into beds. There are ladies' and gentlemen's retiring rooms, with every convenience, and a supply of water for drinking, and hot water for washing. Some of the American cars for the longest journeys are supplied with cooking kitchens. The interior fittings of the carriages are exceedingly elegant. They are of American black walnut, polished, with panels and other parts veneered with French knotted walnut or burl, and figured decorations in gold are employed with good effect. Almost the whole of the metal fittings, except the lamp frames, are silver-plated, the floors are covered with Wilton carpets and everything is graceful and handsome, whilst considerable ingenuity is displayed in every part of the construction. The provision for ventilation is very complete, and under perfect control. The ventilators are double, one opens towards the front, and by the motion of the train secures the entrance of a current of fresh air into the car, the other opens to the rear, and serves as a means of exhausting the air from it. They are opened or closed independently, and can be adjusted to any degree desired. The cars being the property of the Pullman Car Company, the charge for their use will be an addition to the ordinary fare. In the United States, reducing the amounts to English money, the extra charge for the Parlour Cars is for 50 miles two shillings; for 100 miles three shillings; and for 400 miles four shillings. The sleeping cars are charged eight shillings a night for a double berth, which would hold two persons, though one would be more comfortable. For the journey from New York to Chicago, a distance of 1,000 miles, and which occupies two nights and a day, the charge for the use of the Sleeping Car, in addition to the ordinary fare, if five dollars, or £1. As the English railways are narrower than those in the United States, the seats are three inches narrower on each side in these English cars, being three feet instead of three feet three inches. Besides the increased comfort, these carriages must be very favourable to conversation. People can walk erect along the passage and can look out and breathe the air at the ends as well as by the windows.

'Mr Longstreet is also erecting for the Midland Company several cars on the model of those ordinarily used in the United States, for first, second, and third class passengers. These are of the same length as the Pullman cars, 50ft, with a passage down the centre and seats on each side.* The advantage of these cars has often been noticed. There is a complete communication between all parts of the train. The carriages are connected by a self-locking apparatus, and they can be separated with ease without descending from the carriage, so that there is no coupling nor uncoupling. The small platforms at the end by which the carriages are entered come within a few inches of each other, and it is easy to step from one carriage to the next, and in this way the conductor traverses the whole train. As there are only two entrances for a large car, no doubt egress will be rather slower, but there will be less danger of passengers suffering from accidents by getting out before the train stops. The third class trains will have divisions at each end, appropriated to passengers' luggage, or, as the Americans call it, baggage. We believe the Midland Company is disposed to try on these trains the American plan of checking luggage, by which passengers entrust their luggage before starting to an officer of the company, receiving for each article a check corresponding in number with one placed upon it.'

*There were slight variations in vehicle body length and internal layout for the sleeping cars, the longer ones such as *Enterprise* having four more beds than the shorter ones, such as *Excelsior*.

The method of heating described was the Baker solid fuel stove used in American cars of the period, whereas in Britain passengers was still supplied with 'footwarmers' at stations, a portable flat pot filled with hot water, which of course soon went cold. A surprising omission from that otherwise excellent contemporary account is the lack of any description of the method of lighting the cars. This was by means of bronze 'Argand' kerosene lamps, suspended from the centre of the clerestory roof, in elaborate holders. Although these gave much better illumination in the hours of darkness than the existing British pot oil lamps, burning rape oil, the 'Argand' kerosene lamp (used widely in America at that time) was noted for its tendency to flicker and smoke when subjected to draughts; whilst according to American accounts their fumes quite often filled the car interior.

Inside *Midland*, the main saloon had 10 sofa sections which comprised pairs of longitudinal seats facing each other, which were capable of comfortably seating two passengers during daytime use. Beyond this were two private state-rooms each having a sofa section below the window and a pair of single seats which could be turned to face each other and the seat squab pulled out allowing part of the seat back to fall, thus providing a further berth for night-time use. The sofa seat squabs could be similarly pulled out towards the centre of the car, allowing part of the back again to fall to provide a lower sleeping berth. The upper berths, normally raised out of sight up against the roof corners, could be lowered on pulleys for night-time use, and were hinged at the body side. At the state room end of *Midland* a ladies' toilet and water closet was provided at one side and the Baker solid fuel heating stove in a recess at the other. The gentlemen's toilet and water closet were installed each side of the gangway at the other end of the car beyond the saloon. Water for the toilets was raised from low level tanks below the floor. The windows along the bodyside comprised one large and two small ones for each sofa section length and could only be opened upwards, due to the trussing of the vehicle below the window level, and were each fitted with roller blinds.

I have already mentioned the 'drawing room' or 'parlour cars' of which the first to be completed was *Victoria*, in March 1874. This layout was a new one, unknown to American travellers, and was presumably an idea of George Mortimer Pullman to cater for clients on short daytime journeys, where the distance travelled was not long enough to warrant the provision of a sleeping berth; nonetheless offering his superior accommodation, and Pullman service with supplementary fare. *Victoria's* spacious open saloon was fitted with 17 individual armchairs mounted on a pivot so that each could be revolved to face in any direction and also tilted backwards if required. Upholstery was moquette with a floral pattern, some of the chair backs being buttoned and others plain. Beyond the open saloon, and served from a side corridor, were two private state-rooms each containing two chairs and a cross sofa seating five passengers in all. At this end of the car was the usual Baker solid fuel-fired heating stove in its own compartment whilst at the other end at the sides of the short

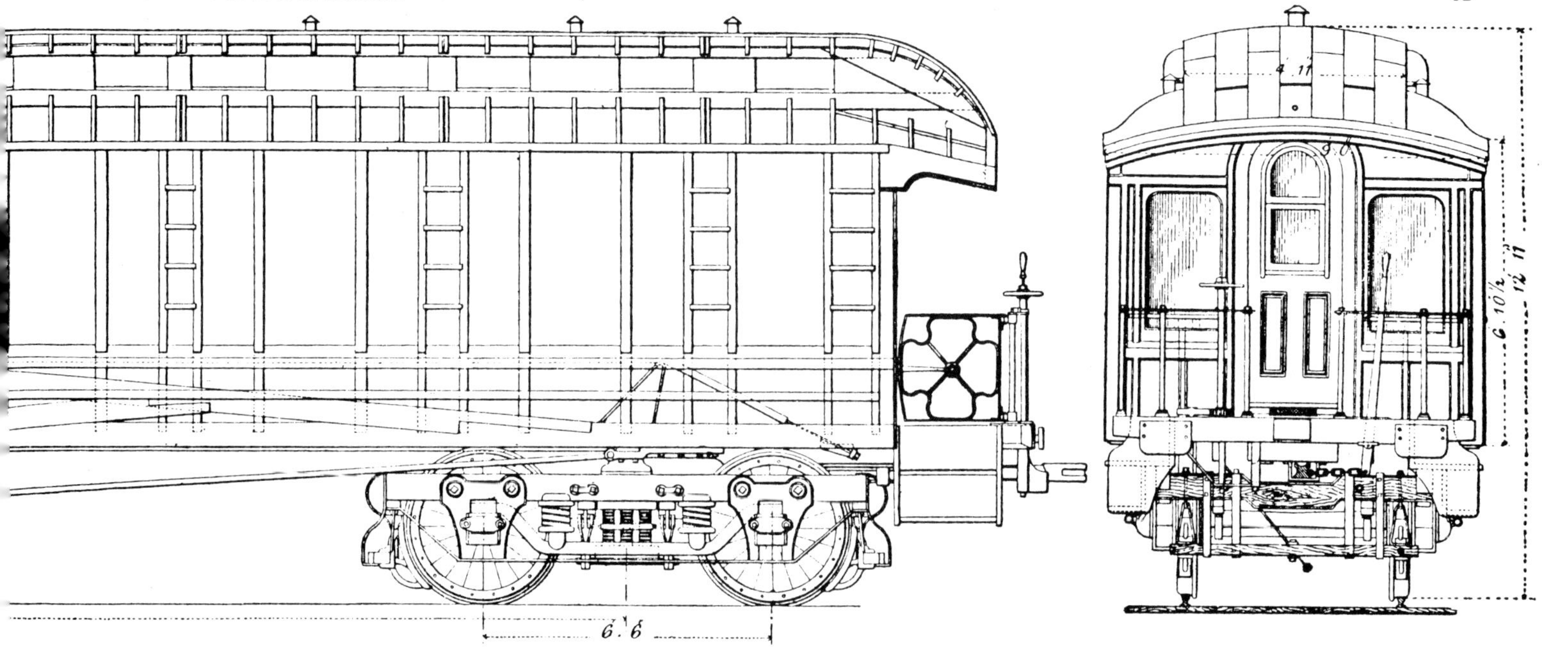

corridor leading to the end platform were separate ladies' and gentlemen's lavatory compartments.

Although they were not Pullman cars, and only outwardly resembled them, it is worthwhile mentioning the other 'day cars' built by Pullman and immediately sold to the Midland Railway, as required by the original agreement, because these cars were included in the original Pullman trains to provide non-supplementary fare accommodation, (ie at no benefit whatever to the enterprising Mr Pullman!). These 'day cars' were of two distinct types, a first and second class composite car, and a third class baggage brake car which were added to the ends of the Pullman ensemble. At first these had open saloons of the American 'day car' layout, but were soon modified, so that the first/second class composite car had accommodation for 32 second class passengers accommodated in four compartments at one end of the car with access from a side corridor which changed sides at the centre of the vehicle to serve a further three compartments each seating six first class passengers (three per side). From the other side, there was a door dividing off first and second class areas in the cross corridor and at each end of the car beyond the compartments was a vestibule and a lavatory compartment. Upholstery and fittings were Midland standard with blue cloth for the first class compartments and green for the second class, the same as ordinary Midland stock of the period. The side corridors once introduced, were a completely new innovation on Midland trains and were hardly to be found anywhere in Britain at the time. The other type of Midland day car of Pullman external appearance was the third class baggage brake which had an open saloon with transverse seats trimmed in red cloth, and seating 32 passengers. Beyond this saloon was a smoking compartment with 18 loose detachable armchairs, 'which could easily take more passengers at times of peak loading'.

The Midland Railway and no doubt the Pullman Co were anxious to demonstrate their new train to the public before placing it in everyday service, and on 21 March 1874 a special demonstration trip was arranged, from St Pancras station to Bedford and back with about 80 invited guests, including notable members of the press. The train consisted of the two sleeping cars *Midland* and *Excelsior* and the parlour car *Victoria*, and it was hauled by Kirtley 2-4-0 No 906 which had a tender adapted to carry the Miller centre-coupling. Quite apart from demonstrating the opulent new 'Palaces on wheels' of Mr Pullman, this train made British railway history, because it was the first British public train on which a meal was served to passengers whilst the train was travelling.

The meal served was a luncheon, provided by the caterers Messrs Spiers & Pond, and was apparently extremely successful. As not one of the three Pullmans had kitchens fitted, it is presumed (precise details not having come to light) that it was a meal of cold meats, salad, etc, prepared beforehand. However it could have been a hot meal if special heaters (or steam boxes) were carried for this unique event, to keep hot pre-cooked food. Certainly hot drinks could have been provided from the Baker heater, although no doubt a fine choice of wines and champagne etc was considered to be more in keeping with this important occasion! One of the contemporary accounts of that historic day was published in *The Times* on 23 March 1874, in which the reporter stated that guests on the train to Bedford were asked to write their names in a special book, and that 'they did so without the smallest difficulty when the train was running at the rate of 50 miles an hour'. This statement is of importance because it shows how well the bogies behaved on the very good MR track, whilst running at a somewhat higher speed than was characteristic of many American Pullman car journeys of the period. It was probably one of the smoothest train rides ever achieved in Britain until that date, aided by the heavy and solid wooden bodies of the Pullman cars, with their centre couplers to hold them very secure at the ends. Another report of this event is of some interest.

'When the first journey of the Pullman train was run, it was from St Pancras to Bedford. "Literally nothing", wrote one who then travelled, "seemed left to desire". Entering the train from one end, you were introduced to the parlour car, a luxurious contrivance for short lines and day-travel only. It was a tastefully and richly decorated saloon, over fifty feet long, light, warm, well ventilated, and exquisitely carpeted, upholstered and furnished. Along each side, and close to the windows, were crimson-cushioned easy chairs, in which, by means of a pivot, you might swing yourself round to converse with your neighbour, or, by means of one of the thousand ingenious contrivances with which the whole train abounded, you might tilt yourself back to the proper angle of enjoyment. The centre is free for passing to and fro. There are various little saloons of the private box order, in which a family party might make themselves happy. Then you came to the drawing-room sleeping car, another long, well-appointed saloon, with fixed seats at the windows like short sofas, two and two, and facing each other. Between them a firm, convenient table could be planted, and upon one of them we were able, while the train ran at over fifty miles an hour, to write without difficulty. The tables removed, the seats lowered to meet each other became an admirable bedstead, while some beautifully ornamented and finished panels overhead, that appeared to be merely part of the sloping roof of the saloon, were unfastened, and in a moment converted into equally comfortable upper berths. By-and-by the saloon was restored to its normal drawing-room aspect, the tables were again put up, waiters entered with snow-white cloths, pantries and anterooms were brought into operation, and there

46
Pullman drawing room car *Victoria* (later to be referred to as a Parlour car) as constructed at Detroit and reassembled at Derby in 1874 for the Midland Railway; showing the constructional details. This was a completely new design and no Amercian Pullman car of this layout had existed before. It had a general saloon with 17 armchairs which swivelled on pivots. Two private rooms (coupés) were flanked by a side corridor, each with two chairs and a sofa.

appeared a dining hall as complete in its requirements as the drawing-room and sleeping-room had been in theirs.' (From *Our Iron Roads*, F. S. Williams 5th edition, London, 1884.)

The reference to 'pantries and anterooms' is rather misleading unless one or more of the state-rooms had been temporarily converted for the preparation of food, which was quite possibly the case. None of the early MR Pullmans were fitted with pantries when built.

The sleeping car *Midland* was non-standard in its internal layout compared to the rest of the Pullman sleeping cars initially ordered for the MR. It had, as already described, 10 sofa sections, of which eight were in the main saloon, whereas *Excelsior*, the second car, and those that followed had eight 'cross-sections' in the main saloon; both types had the standard state-room layout, each with one sofa section and one cross-section. I suspect that the arrangement in *Midland* was done with a view to using this particular vehicle as a roving demonstrator. Rove it certainly did, for on 6 June 1874, only a few days after the Pullman train entered public service on the MR, car *Midland* was shipped across the Channel, and taken to Italy and Austria, as well as France, to demonstrate the superb Pullman design. The story of the outcome of this trip, which secured a foothold for Pullmans in Italy is not within the scope of our present narrative, and interested readers are referred to the very detailed history entitled *Pullman in Europe*, by George Behrend, published by Ian Allan Ltd. Car *Midland* did not return to the MR until June 1877.

On 1 June 1874 the first scheduled British Pullman car train left Bradford at 8.30am, bound for London, St Pancras, where it was due to arrive at 2.5pm. The train was formed of five vehicles of which only two required payment of the special Pullman supplement. All three classes of passenger were carried, and there were lavatory facilities accessible to all on board. Fortunately a record of the formation of this historic train has survived:

Locomotive, with specially adapted tender for Miller coupler.
Car 1, Third Class/Baggage MR, 'day car': 50 seats. No supplement.
Car 2, First and Second Class MR 'day car': 18 firsts and 32 seconds. No supplement.
Car 3, Pullman Sleeping Car: 22 or 26 beds. Supplement charged.
Car 4, Pullman Parlour Car: 27 seats. Supplement charged.
Car 5, identical to Car 1, MR 'day car'. No supplement.

This layout gave accommodation for some 200 passengers, of whom Pullman could cater for approximately 50. His supplementary fare varied according to the distance travelled during daytime — from one shilling to five shillings — but a standard rate of six shillings was charged for the use of a sleeping berth, irrespective of distance. The return train left St Pancras at midnight and the Pullman passengers were allowed to occupy their berths from 10pm onwards, prior to departure, and to remain in them until 8am at Bradford, the train having been timed to arrive there at 5.50am. This arrangement only allowed half an hour for cleaning and servicing the Pullmans before they departed for London once again. Later the same summer a second Pullman sleeping car was added, making a six car train; however it should be noted that a five car train serving 200 passengers already weighed about 100 tons so that an additional Pullman was quite a substantial increase in load for the locomotives of the period.

In his book *American Pullman Cars of the Midland Railway*, J. B. Radford gives a fascinating insight into the engineering, operating, and traffic management aspects of the MR Pullmans, as well as providing a wealth of day-to-day detail extracted from official records. I do not intend to retread that ground in detail, because the book is still readily available.* I want, however, to examine one aspect of this first Pullman venture in Britain which had a very strong influence upon the latterday developments of what then quickly became known as The Pullman Palace Car Co, with activities in both Britain and Europe. This aspect concerns the differing public reactions that greeted Pullman's ideas in Britain, when compared to his unquestionable status and success in America at that time, and it has to be observed that these first Pullmans in Britain were not the overwhelming, immediate success that no doubt both James Allport and George Mortimer Pullman had imagined would be the case.

47

47
A contemporary engraving of the main saloon of the Drawing Room Car *Victoria*, with swivel chairs, and foot hassocks; each side of a wide central aisle. This did not prove to be a very popular layout with British travellers.
Illustrated London News

*Mr Radford's book, published by Ian Allan Ltd in 1984, is also useful in up-dating some details contained in the earlier book by George Behrend entitled *Pullman in Europe*, unfortunately long out-of-print.

CHAPTER 6

A Matter of Class

The American Pullman car was created to serve in a new world where a spirit of adventure and of pioneering made it an outstanding attempt to bring a civilised lifestyle to rail travel. When it was placed, virtually unaltered, on to British rails, it was certainly and undeniably a symbol of wealth and prestige, but there were many conservatively-minded members of the upper classes who would have considered the Pullman's opulence to be both brash and vulgar and — above all — lacking in the privacy to which they felt entitled. Whilst it was true that there were certain refinements such as toilet and washing facilities, almost unheard of in British carriages of the period, they would be forced to share these with other passengers! Whilst it was true that a very courteous and civil Pullman conductor was there to look after one's needs, the habit of the wealthier British travellers was to have their own maid or manservant close at hand, for purely personal service. Because there were no corridors in the typical British carriage, the seating layout was often arranged to have a first class compartment for the wealthy flanked by seconds, or even thirds, on each side, for the servants to travel in. For a lady to travel alone was considered to be a somewhat risky adventure, unless, of course, she was one of the *lower* classes — in which case she was not considered to be a lady!

One British observer of the period, F. S. Williams, wrote about the subject, although it is interesting to note that he assumed that the American traveller in Britain would be using the first class accommodtion of the day:

'How far the Pullman car will be generally preferred in England is a matter of some doubt. Americans themselves when they come to this country, appear well content with the matchless speed of English railway travelling and the comfort of half-filled ordinary first-class carriages. When Lord George Bentinck said on an off-day at Newmarket, only a few owners, trainers, and jockeys being present, "This is what I like: I hate a crowd", he expressed a feeling widely distributed among our population, and one adverse to the adoption of long cars. They are, without question, a great boon to ladies and to solitary travellers. Relieved of the custody of her impedimenta, of all fear of insult, and pleased with the handsome surroundings of the Pullman drawing-room, the lady traveller will appreciate the provision made for her comfort; and the more social varieties of the Englishman if travelling alone, may also take kindly to the long car. A shy and sensitive minority will perhaps prefer the comparative seclusion of the ordinary first-class carriage.' (From *Our Iron Roads*, F. S. Williams, London, 1884.)

The above observation about 'ladies and solitary travellers' is interesting in view of these two contemporary accounts of American journeys, written by ladies. The first describes one of Pullman's rivals of the period, the Silver Palace Sleeping Cars:

'I met with no silver whatever in the Silver Palace sleeping-cars. The fittings, lamps, bolts, hinges, door handles, etc. are of the white metal called pinchbeck or Britannia metal, and the palaces are fitted up in the ordinary hotel style, the floors carpeted and oil-clothed and the seats velvet-covered.

'There are no special cars for ladies and if your opposite neighbour is a gentleman in the day, in all probability you will have him on your shelf at night and it will be well for you if he does not either snore or have nightmares. Although you have blankets, sheets and pillows you have to go to bed in your boots, at least ladies have to, and indeed they cannot undress at all because they cannot shroud themselves behind the curtains without placing themselves in a recumbent position. Besides, what could be done with their clothing? It could only be put on the bed and already stowed away there are hand bags, books, umbrellas, goloshes, wraps, tins of biscuits and possibly a large flask of Bourbon.

'Everyone is obliged to retire to rest at a certain hour and rise ditto, for all the beds must go up and come down together. Then follows waiting one's turn at the silver lavatory, which consists of two basins, a palatial roller-towel and a piece of soap. There is a silver tap whence you procure clean water — and wash the basin too if a score of your neighbours have used it before you. Particular people select the cleanest portion of the towel or carry their own. In the same luxurious manner you are accommodated with comb, brush, nailbrush and toothbrush. I saw a lady take from her pocket a very handsome set of teeth, clean them carefully with the brush provided by the company, and place the teeth in her mouth. "I don't sleep in them," she remarked. "It uses them out so bad and they are right expensive."

'The lavatory is situated at the entrance of the car and cannot be made private either from without or within, and anything like a good wash is out of the question. You cannot even tuck up your hair or roll up your sleeves but some gentleman or conductor is sure to pounce upon you and remark, "Very refreshing to get a good wash." After this vain attempt at ablution you return to your seat, where you can order breakfast, for the bed has disappeared and a small slide is drawn out over which a white cloth is spread. The bill-of-fare is produced and you can breakfast and dine just as you would on a sea voyage. Fresh vegetables may run short but everything that can be pickled, preserved or carried in cans you can order. Bread and cake are baked twice a day in the train and are always hot. With your breakfast comes the morning paper, the *Great Pacific Line Gazette*, printed on board each day, narrating all the items of news of the preceding one, also a love story. The passengers are allowed to contribute and can advertise for lost articles.

'I need not dilate upon the effect of thirty or forty human beings being boxed up together for seven days and nights, crammed close to each other all day, sleeping on shelves at night and in the same atmosphere. It is dangerous to seek relief by going from the overheated cars to the intense cold outside and if an attempt be made to let in a little fresh air by the window some one is sure to have the toothache.

'The insufferable tedium of sitting in one position and even staring at the same persons for so long a time gives you a sort of nervous fidget. You can vary this monotony by a walk down the middle passage but the cars sway so much that if you have not sea-legs you are sure to tumble over some one or some one over you. Worse still, you may fall against the hot stove or the water tank with which every car is provided. These water tanks are the end and aim of most people's promenade and the tin cup is in constant requisition, for there is no stint of iced-water. Gentlemen amuse themselves by walking the full length of the train and jumping from platform to platform of the different cars. There is a smoking car, and cards, etc., form also an amusement, but I think that few people who have spent a week in these palaces on rails would ever wish to spend another.' (From *Teresina in America*, Terese Yelverton, Vicountess Avonmore, London, 1875.)

Her final observation, about the duration of the journey is, of course, also very significant because it was about seven times longer than any journey likely to be made by a British Pullman car! One suspects also that the conditions she described were distinctly inferior to those that would have been the case in one of George Mortimer Pullman's Palace cars of the same period (1875).

This second feminine view, of a somewhat later date (1887) gives a contrast between the merits of the Pullmans for female and male travellers. (The writer also criticises the poor tracks of the line from Niagara to Boston!)

'The terrible jolting and shaking of that journey from Niagara to Boston still lives in my memory in spite of many later and similar experiences. I do not wonder that car sickness is a recognised affliction and upper berths are especially to be avoided on this as well as on many other accounts.

'The Pullman cars are delightful, always premising that they are sufficiently full to be steady. My unhappy experience on more than one occasion has been to form one of a miserable duet of shaking and jerking passengers, the result of an empty parlour-car. Still, under ordinary conditions most of us find

48

48
These fascinating vignettes appeared in *The Graphic* at the time the Midland Railway Pullman sleeping cars were introduced. Remarkably accurate in detail, they give a strong picture of the new way to travel, its pros and cons. The image of the 'quiet luncheon' (bottom right) was only possible if a lunch basket was prepared beforehand, as can be seen on the table, because there were no cooking facilities in the Pullmans. Nothing like these Pullmans had been seen on a British railway before! *Author's collection*

49
In America the Pullmans were accepted as supreme forms of travel by the late 1860s/early 1870s, and because of the long journeys undertaken, which lasted for days on end, every comfort and amenity was given due consideration by George Mortimer Pullman. Hence this Sunday morning scene, on-board a Union Pacific Pullman car in 1869, with hymn singing around a small organ.
Illustrated London News

such carriages extremely comfortable and even luxurious.

'The gentlemen's lavatory in these cars is sufficiently large to accommodate four or five men at once, whereas that set apart for the ladies is so small that you can barely turn around in it and perform your ablutions as best you may, with a rocking, jolting motion that drives you from side to side all the time and the horrible conviction that four or five impatient or reproachful females are standing outside the door ready to take your place the moment you can be induced, by knocking or twisting at the handle, to vacate it.

'Of course, when it is merely a question of one night on board washing becomes a minor consideration and can be supplemented to any extent at home or in your hotel. But the peculiar boast of the American car system is its adaptability to long railway journeys extending over several days. Having received many really well earned compliments from various sleeping-car porters on which I may call Non Lavatory Monopoly, I feel more entitled to speak on the subject than many ladies would be who spend an uncomfortable half hour in an eleborate arrangement of their bangs while some poor wretch is waiting outside for the chance of washing her hands after the long, black night journey.

'In addition to greater space for toilet arrangements men have another advantage on board these cars, namely, that their clothes are much more easily taken off or put on sitting on a berth, with the board of the upper berth within two inches of the top of your head.' (From *A Year in the Great Republic*, E. Catherine Bates, London, 1887.)

Women's fashions of the period, in both America and Britain did not favour ease of travelling, and many a struggle took place between clothing and doorframes on Victorian British carriages (with a door opening slightly less than two feet wide, unchanged since the days of the road stagecoach!). This account of an incident in one woman's American journey was given by a Pullman conductor to a visiting Australian and is certainly amusing (1889):

'He narrated how on one very hot and oppressively close evening a restless lady passenger, resolving to obtain coolness at any cost, violently kicked off her blankets, but in doing so she unfortunately overlooked the fact that her own clothes were stacked near the window that she had thrown open before retiring in order to admit some fresh air. Thus it was that in her struggles to cool herself she inadvertently kicked her wearing apparel out of the window, with the embarrassing result that next morning she was constrained to make her appearance airily attired in a hat, stockings and a blanket.'

The same Pullman conductor also recalled that:

'On another occasion a male passenger of giant proportions, finding his berth too small for the length of his corporeal frame, thrust his feet through the open window as the best and most comfortable solution to the difficulty. When he awoke in the morning he found to his intense disgust that his feet had changed colour during the night. They had been covered by a thick coating of tar by some roadside station humourist who, seeing them dangling outside the window, was unable to resist the temptation.' (From *The Australian in London and America*, James Francis Hogan, London 1889.)

Such behaviour as putting one's feet out of the train window (apparently quite commonplace in America last century) would certainly not have been countenanced by the British upper classes!

It was indeed the class structure which differed so greatly between the two nations at the time that the Pullmans were first introduced to Britain, and this separation of the classes was recognised by the British railway companies to an almost disgraceful degree, as James Allport had observed. An American railroad traveller became one of a large and temporary family on the move, and this was emphasised by the long and open saloon layout of the cars. The typical traveller *talked* to his neighbours, and — believe it or not — even prayed and sang hymns with them, or performed other songs of the times, as witnessed by this 1878 traveller:

'Resuming our journey once more, night overtakes us. We have a little eight-stop organ in our sleeping-car the *Palmyra* and this evening we bring out its tone and our companions in the car contribute a few songs. The instrument has two manuals but will only sound in one and the upper part is devoted to pillow cases and blankets. So for two hours we amuse ourselves with singing and playing, our conductor, who was a bit of a musician in his way, coming and treating us to a few songs. I believe if we had only had room enough we should have got up a dance, but this was entirely out of the question. In this way we spent a very pleasant evening.' (From *Through America*, W. G. Marshall, London, 1882.)

50
A remarkably detailed engraving of the interior of a Canadian Pacific Railroad 'colonists' car, with a layout similar to that created by Pullman, but of a decidedly less sumptuous style in decor and furnishing. It illustrates extremely well the way that travellers had to learn to live together, and share all the facets of human behaviour whilst on the move, perhaps for several days at a time.
Author's collection

If singing was out of the question, there was always conversation, as this British visitor to America observed:

'One of the effects,we will not say advantages, of travelling in a long car may be to promote sociability. "An American," says a St Louis paper, in an article on native politeness, "may not be so elegant at a dinner party, but he will not ride half a day in a railway car without speaking to the fellow-passenger at his elbow, as the Englishman will." "No," remarks an American critic, "indeed he will not: fore George he will not. How often, oh, how often, have we wished that he would! But he won't. He will pounce upon a stranger whom he has never seen before in all his life, and talk him deaf, dumb and blind in fifty miles. Catch an American holding his mouth shut when he has a chance to talk to some man who doesn't want to be talked to.".' (From *Our Iron Roads*, F. S. Williams, London, 1884.)

The author of this present book, a dedicated rail traveller, can recall one journey between London and Brighton in the 1960s when *no one* spoke in the carriage, which was seating some 50-60 people, on their evening return home to the South Coast — between Clapham Junction and the outskirts of Brighton, a duration of nearly one hour!

It must be remembered that when the Midland Railway first introduced the Pullmans, the British traveller was not even accustomed to eating on a train, except in private from a personal basket of provisions brought along for the purpose, and probably not shared with others in the same compartment!

By way of contrast the Americans had a spirit of 'sharing' the journey (which is still to be found if one travels in the company of Americans anywhere in the world today) is well expressed in this writer's view:

'In no part of the world is travel made so comfortable and easy as on the Pacific Railroad. To travellers from the East it is a constant delight and to ladies and families it is accompanied with absolutely no fatigue or discomfort. One lives at home in the palace-car with as much true enjoyment as in the home drawing-room and with the constant change of scenes afforded from the car window it is far more enjoyable than the saloon of a fashionable steamer. For an entire week or more, as the train leisurely crosses the continent, the little section and berth allotted to you, so neat and clean, so nicely furnished and kept, becomes your home. Here you sit and read, play your games, indulge in social conversation and glee, and if fortunate enough to possess friends to join you the overland tour becomes an intense delight.' (From *The Pacific Tourist*, Henry T. Williams, New York, 1976.)

Indeed, he goes on to emphasise how important it was *not* to travel in solitude, because then you would be subjected to unchosen companions! He comments:

'To enjoy palace-car life properly one always needs a good companion, so you can take a section together. From Chicago to Omaha the company in sleeping-cars is usually quiet and refined, but beyond Omaha there is often an indescribable mixture of races in the same car and if you are alone the chance is that your *compagnon du voyage* may not be agreeable. It is impossible to order a section for one person alone and the dictum of the sleeping-car arrangements at Omaha require all who come to take what berths are assigned.

'Fee your porter on the sleeping-car twenty-five cents a day. The porters of both Pacific railroads are esteemed especially excellent, obligating and careful.' (*Ibid*)

There was, if the truth be told, a whole world of difference between British and American railway journeys of the mid-1870s! Whereas a Midland Railway passenger was going to complete his journey either during one day, or by one night (whether in the luxury of a Pullman or otherwise), it was probable that his American counterparts had to resign themselves to several days of travel. The design of the Pullman cars was in reality more suited to the American journeys than to the British. Hence, the same American traveller in 1876 observed:

'The slow rate of speed, which averages but sixteen to twenty miles an hour day and night, produces a peculiarly smooth, gentle and easy motion, most soothing and agreeable. The straight track, which for hundreds of miles is without a curve, avoids all swinging motion of the cars and sidelong bumps are unknown. The cars are connected with the Miller buffer and platform, and make a solid train without the discomforts of jerks and jolts. And the steady, easy jog of the train as it leisurely moves westward gives a feeling of genuine comfort such as no one ever feels or enjoys in any other part of the world.' (*Ibid*)

If a British train of that period, carrying important passengers, had proceeded at such a leisurely pace there would probably have been questions raised in Parliament!

Class distinction in Britain, with its different social behaviour, was perhaps unknown or underestimated by George Mortimer Pullman, when he first brought his cars to Britain, although the design of the parlour car *Victoria* would suggest an attempt by him at meeting British tastes. This design, however, suffered from being basically of the open saloon layout, and one suspects that one of the more popular features of these early MR Pullmans was the limited number of compartment, or coupé style, state-rooms, which gave considerably more privacy to their occupants. Time was to show a public dislike of the parlour cars, but a reasonable acceptance of the sleeping cars. Still in the future was the concept of a British Pullman car as we nowadays think of it!

In Chapter 4 I have drawn comparisons of travelling conditions in Britain and America before the Pullman cars arrived on the Midland Railway, and I observed that the use of Pullmans was just one part of an ambitious plan to change the standards of accommodation offered by the MR, including those for the less wealthy, or 'lower' classes. It was for this reason that Pullman had to supply the 'day cars' from which he would derive no profit, except from the initial sale to the MR. In Sir James Allport's mind there was already another important move, and this was taken on 1 January 1875, after the Pullmans had been introduced. On that New Year's Day the Midland Railway abolished entirely second class travel. This news came as a bombshell to the other British railway companies!

51

51
When this illustration first appeared, in an 1877 illustrated paper, the caption read simply 'Dining at 20mph'. It depicts a Pullman hotel car with the tables set up for dining. Note the large spittoons on the floor by each table, and the upper berths stowed away for daytime travel. In the background an Indian and the train's conductor are engaged in conversation with a passenger. One can but admire the feat of balancing being performed by the servant in the foreground, with no less than five loaded plates upon his arm!
Author's collection

CHAPTER 7

Expansion and Competition

The Midland Railway's historic announcement that it would abolish the second class accommodation on all its trains, was accompanied by a programme of refurbishing its fleet of existing non-corridor carriages. Some poor third class accommodation was scrapped altogether, but the better examples had their wooden seats upholstered; whilst the very best (the mock-Pullman 'day' cars) even had some of the third class accommodation changed to first class, as well as all the seconds. On the majority of existing six-wheeled carriages, the second class became 'improved' third. There was then a reasonably comfortable seat for the third class, upholstered in wool cloth, and together with the superior ride of the six-wheelers it was — as Behrend observes in *Pullman in Europe* — much better than the second class on some lines! This move by the MR inevitably threw the social classes into greater proximity whilst travelling on the same train, and it incidentally made the idea of an exclusive, supplementary fare Pullman car more attractive to those who could afford it, particularly when travelling at night.

The improvement in the existing carriage fleet, and the introduction of new carriages, designed by Thomas Clayton, who had been appointed the Midland Railway Carriage & Wagon Superintendent in 1869, raised a doubt as to the wisdom of running complete rakes, or trains, of Pullmans and 'day cars' of Pullman design. The reasoning was quite simple, and had to do with weight per passenger. There was little to be gained from transporting non-supplementary fare paying passengers in the heavy wooden-bodied bogie 'day cars' when modernised carriages of lighter construction and six-wheel layout would suffice; moreover it would be possible to construct *new* bogie carriages for everyday use that were less lavish in design than the American type, and consequently lighter and cheaper to haul. To take the original Bradford Pullman train as an example (see page 48) the five American cars in this formation weighed about 100 tons, whereas if ordinary MR carriages had been used the same number of passengers could be carried for a weight of only 50 tons. It did not require a genius to observe how much extra coal a steam locomotive had to burn to haul the American cars! The obvious compromise was to use Pullman cars only for supplementary fare passengers (as per the contract with the Pullman Palace Car Co) and to provide normal standards of accommodation, in MR carriages, for the rest of the train's passengers. This could not be done immediately, however, because

52

52
The Pullman drawing room car *Albion*, built in 1876. It is seen here after the addition of buffers and screw couplings and after sale to the MR (see Chapter 10 for details). It has had the name 'Pullman' removed and 'Midland' substituted, and the name *Albion* replaced by No 8. Otherwise it is in substantially the original condition and livery style, with an extraordinary finesse in detail. *National Railway Museum*

53

53
Detail enlargement from the previous photograph, to show the beautiful lining-out in gilt upon every portion of the wooden bodywork. A true work of art, and one which gave much character to all Pullman's cars. Although the finishing touches were by hand, the initial work was done by the means of large metal stencils, in order to repeat the patterns accurately.
National Railway Museum

the American-built cars had non-standard Miller centre-couplers and no buffers, and could not be connected to the existing MR coaches, with their typical British screw couplings and side chains and buffers. Moreover, only specially modified locomotives, with a Miller coupler on the tender, could be used with the American cars.

The decision was in due course taken by the Midland Railway to modify the Pullmans and 'day cars' to have conventional couplings and buffers, making it possible to use them in any train formation, including 'through carriage' workings on to other railway companies' lines. This would widen the future scope for Pullman operations, in particular once the Settle & Carlisle route to Scotland was opened. However, it must be observed with hindsight that the MR would have been better advised to have adopted the Miller centre coupler for all its stock because the 'knuckle' coupler (in American parlance) was a safer and more advanced design for its day. No doubt the cost of conversion would have been very high, but it would have been a progressive step, adding to railway safety in Britain.

When the demonstration run, already described, was made to Bedford and back from St Pancras in March 1874, there were no doubt a number of City of London dignitaries among the invited guests, as well as the members of the Press. We can be certain that the activities of the Pullman Palace Car Co (Europe) were watched with interest by City bankers and merchants, who were in many cases important shareholders in the various railway companies. Bedford was chosen by the MR as a run just long enough to serve a meal on board the train (being some 50 miles from London) and I suspect that amongst those present that day was someone who had strong connections with the London, Brighton & South Coast Railway, whose main line to Brighton was also of some 50 miles. The Pullman Palace Car Co (Europe) soon provided a brand new sleeping car, *Mars*, assembled at Derby in 1875, for the London-Brighton service; with the car's interior arranged permanently for day use. This was obviously a 'demonstrator', for what was destined to be the first of a long-lasting Pullman car fleet on the Brighton line (which of course had no requirement for sleeping cars), and *Mars* paved the way for new parlour cars, of which more anon.

Even before *Mars* went to the Brighton line, John Miller, the English manager of Pullman was very active in attempts to secure contracts with other railways, and apart from the Brighton car *Mars*, and the car *Midland*, which went abroad, in 1875 new construction (or to be precise, reassembly) at Derby had included a parlour car named *Ohio*, and a sleeping car named *Ocean* (fitted up for day use only like *Mars*). These two cars went off to the Great Northern Railway from Derby, and were placed in trial service between London, King's Cross and Manchester, London Road stations, working in what was called the 'Manchester Saloon' train, via Sheffield. Behrend states that the GNR accepted these Pullmans reluctantly, and only had a series of short term agreements to work them, apparently running them chiefly because their rival railways were using them — the Midland in particular, of course.

As more Pullman sleeping cars and parlour cars were reassembled from the crates that arrived at Derby, the Midland Railway expanded its services with them (and provided the cars with their standard screw link couplings and buffers), so that in April 1875 new workings commenced between London St Pancras and Liverpool (both day and night sleeper services), and with the opening of the Settle & Carlisle line in May 1876 Pullman cars worked through from St Pancras to Edinburgh and Glasgow, using both parlour cars and sleepers.

By the beginning of 1876 it was apparent that the Pullman cars had both those in their favour and those opposed to them amongst the MR management. It had to be admitted — for the reasons I have shown in Chapter 6 — that the public had not accepted the parlour cars, or the non-supplement 'day cars' with much enthusiasm (according to one report they were regarded as 'distinctly un-English'!) although the sleeping cars were more popular, for the obvious reason that they greatly improved upon existing night travel facilities, and gave the public excellent value for money. However, Pullman was *not* the first to introduce sleeping cars to Britain (as is frequently mis-claimed in popular histories), and it was not his design of car which the British public was ultimately to find the most acceptable and in due course Pullman's name was to be associated with daytime rather than night-time travel in Britain!

To understand this curious twist of events, it is necessary to look at some rivals of George Mortimer Pullman, and to a lesser extent of the Midland Railway, in the early 1870s.

Prior to the visit of James Allport to America, a young Belgian engineer by the name of Georges Nagelmackers had gone there in 1868, and had been inspired by the Pullman cars he travelled in. Nagelmackers had visions of a European international company to operate Pullman-style services; an idea he published in 1870, entitled: *Project de l'installation de Wagon-lits sur les Chemins de Fer du Continent*. This was the

54

54
Mann boudoir sleeping car No 15 of 1873 in blue livery, and with the engraved initial 'M' on the door windows. Colonel Mann is seated on the running board, and Georges Naglemackers is standing beside him. It is a six-wheeler, without continuous train brakes; screw couplings and side buffers being fitted to standard continental style.
Compagnie Internationale des Wagons-Lits

beginning of the now famous 'Compagnie Internationale des Wagons-Lits'. The history of this company has been extremely well described by George Behrend in several of his books (see Bibliography)* and I see no need to retread the ground here, except to mention the part played by Colonel William d'Alton Mann, because he was another inventor and operator of sleeping cars who was for some time in competition with G. M. Pullman in America; he was also an extraordinary and colourful character!

Colonel William d'Alton Mann was born at Sandusky, Ohio on 27 September 1839. His chosen career was to become a civil engineer, but his studies were interrupted by the outbreak of the American Civil War, and during its first year he joined up as a captain in the First Michigan Cavalry. In 1862 he organised the Fifth Michigan Cavalry, and then later on, the Seventh; whilst in command of the Seventh he was at the battle of Gettysburg. He also displayed a flair for invention, which was not wasted whilst in military service, because he invented, and after the war was over patented, various improvements to the equipment of his soldiers, and made a fortune by selling them to the federal government, with the help of former military colleagues. By then a wealthy member of society, Mann engaged in a number of enterprises in Mobile, Alabama including ownership and editorship of the Mobile *Register*, and he took an active interest in politics. At some stage he became interested in the design of sleeping cars (no doubt aware of the very considerable personal fortunes of both Pullman and Wagner by the late 1860s) and he devised an alternative layout which he patented in 1872; which had compartments instead of open plan sections.

Just as in the case of Pullman, who was not the inventor of the 'section' layout so often ascribed to him (this had been patented by a Mr J. B. Creighton in September 1858) Mann did not *invent* the compartment layout for sleeping. The 'compartment car', as it was called, had first appeared in America for daytime use in 1853, as an attempt to offer an improvement upon the, by then, generally accepted Winans open day car layout. A description of the 'compartment car' was published in the *Scientific American* in August 1853:

'Messrs Eaton and Gilbert of Troy, N.Y., have built a beautiful car for the Hudson River Railroad which is divided into staterooms of 8 feet square. The car is 45 feet long and 9 feet 6 inches wide. Each room is calculated for a party or family and is furnished with a sofa, four chairs, a looking glass and a small centre table. The panels are painted in landscapes, the ceiling hung with silk, and the floor richly carpeted. The rooms are entered by a side passage and each is well lighted and ventilated. There is a wash room in the front of the car. It is the first experiment of its kind in the United States.'

Moreover the Pullman cars of 1865 had several compartments, including staterooms, in their layout for sleeping in, as well as the usual 'sections' of berths and seats. The idea of using only compartments appealed to Mann because they offered much greater privacy whilst sleeping, and in his first design there was no side corridor, and each compartment was entered only by doors in the bodysides, in other words in typical British or European fashion. The compartments each contained a small toilet room, and the seats were designed so as to be convertible into berths, at right angles to the two bodysides. Mann chose to call them 'Boudoir' cars, and he brought his idea to Europe, because he could not at that stage compete with Pullman, Wagner and others who were already well-established in the American sleeping car business.

I think it is reasonable to assume that Mann had the European market in his mind when he produced the patented 1872 compartment layout, because it so closely followed the then normal European non-corridor layout, and ignored the advantages

*Another very good description of the intimate and complex links between the Pullman and Wagons-Lits companies can be found in Chapter 3 of *Pullman* by Julian Morel (David & Charles).

of the typical American car with its open ends and through passageways from one car to the next. Mann and Nagelmackers met each other in London, at a time when Nagelmackers was short of money, and seeking backers for his Compagnie Internationale des Wagons-Lits, which he had set up in Liège in October 1872. Nagelmackers already had some four- and six-wheeled sleeping cars operating on the Continent, which presumably Mann had heard about. As a result of their meeting the two men pooled their resources and the Wagons-Lits name was changed to that of 'Mann's Railway Sleeping Carriage Co Ltd; registered in London instead of Liège. Colonel d'Alton Mann now had an outlet for his ideas, although he apparently left the day-to-day running of the business very much to Nagelmackers (and Behrend records that he even left the old company name and address on the letterheads, alongside the new one). The sleeping cars were painted royal blue and lettered: 'Mann Boudoir Sleeping Car', with a big 'M' engraved on the window panes.

The alliance of Mann and Nagelmackers took place on 4 January 1873, some weeks *before* George Mortimer Pullman's historic meeting with the Midland Railway shareholders, and the following year, on 7 June 1874 to be precise, a Mann Boudoir Car was demonstrated to the press between London, King's Cross and Cambridge. According to Behrend in his book *Pullman in Europe* this Mann Boudoir Car was not the design as patented in 1872, but of the following layout:

'The two four-berth compartments each contained one berth against each wall, which by day had seating all round. For night use, padded tip-up seats on the doors preserved the berth's continuity at the openings. This must have been incredibly draughty for those sleeping across the entrance doors, and embarrassing for the person sleeping across the door to the individual lavatory serving the compartment. The lavatories adjoined, each taking up half the width of the car. None of the compartments communicated. At the end of the car was a second class compartment for servants; its seat backs formed upper berths. The car weighed fourteen tons. It was No 42 in the Mann Company's list.'

The Mann sleeping car afterwards went into service between London King's Cross and Glasgow via the East Coast route on a trial basis. It alternated on this service with a small six-wheeled sleeping car introduced in 1873 by the North British Railway and on both cars there was a supplementary fare of 10 shillings (50p) for their use. The NBR car, which was only 10 tons in weight, had the distinction of being the first sleeping car for public use ever to run in Britain. In 1874 the London & North Western Railway also introduced a sleeping car, to the design of Mr Richard Bore; this worked on the rival West Coast route and also had a supplementary fare of 10 shillings for a berth. Thus there were already rival designs to the Pullmans in Britain, and interested parties from aboard made comparisons of their designs and suitability. One person in particular, Monsieur Banderali of the Northern Railway of France (Nord), produced a report in August 1874 entitled *Note sur les 'voitures à dormir' de divers systèmes en essai sur les chemins de fer anglais*. In M Banderali's opinion the MR Pullman parlour cars were 'absolutely useless' but the Pullman sleeping cars were popular.

The most significant part of M Banderali's report nevertheless concerned the use of compartments for sleeping cars as opposed to the more open Pullman layout. It was clear to the Nord that French travellers would prefer compartments! Banderali observed: 'it is not rare in America to find two in a berth; American families are not embarrassed in a sleeping car in the middle of 20 strangers. Our morals have not attained to that degree of civilisation. French people prefer compartments'. In the present writer's opinion the same observations applied to the vast majority of the British of the period, although the novelty of the Pullman sleeping cars seems to have won some supporters among the more liberal-minded of their day!

The Nord declined offers to run Pullmans, and instead entered into a trial contract with Mann's Railway Sleeping Carriage Co Ltd for a Mann Boudoir car to run on the Paris-Cologne route (the first Wagons-Lits service in France) and also from Paris to Calais. This was a big setback to the Pullman Palace Car Co (Europe) and to George Mortimer Pullman personally, who had visions (to quote the London *Times*) that: 'Pullman Hotel cars would run all over Europe'. Although Pullman did establish a foothold in Italy in 1874, the rival Wagons-Lits (or Mann Boudoir cars as they were called then) proved sufficiently popular on the Continent of Europe to prevent Pullman's dream from ever coming to real fulfilment. What a contrast was taking place in America, however, where Pullman's achievements were on a colossal scale. From the 37 (approx) cars he had in operation in 1867, the fleet had grown to some 700 supplement-earning cars bearing his name, and which operated on more than 150 different railroads by 1873.

Colonel Mann's interest in his company seems to have been of a somewhat dilettante nature, and it was Georges Nagelmackers who did the really hard work, particularly on the Continent. Mann had only two cars running in Britain in 1875; one being the previously mentioned sleeping car No 42 on the East Coast route, the other (his No 43) being a 'Salon' not fitted-up as a sleeper, and running on the London, Chatham & Dover Railway between London Victoria and Dover. The car was apparently to Mann's personal design and both were built and fitted out at the LCDR's Longhedge works in London. Perhaps the most remarkable feature of this particular car, which gives us an insight into Mann's own personality was that it included a compartment specially arranged to suit the needs of 'honeymooners'! It also had a pantry, a drawing room and a smoking room.

In his book *History of Trains De Luxe*, George Behrend relates how Mann became bored with his European sleeping car operations and 'he lounged about in London's Langham Hotel, telling war stories, eating enormous breakfasts of mutton chops and champagne and ignoring Georges Nagelmackers' urgent correspondence'. In the summer of 1875 Mann resigned as chairman and gave Nagelmackers the entire control of the Company. As a result of his determined endeavours, Nagelmackers obtained King Leopold II of the Belgians as his principal shareholder (with permission to use the Royal Belgian Lions in his rolling-stock emblem) and on 4 December 1876 the present 'Compagnie Internationale des Wagons-Lits (CIWL) was founded at Brussels. The CIWL bought out all Mann's interests, including 53 cars. The two cars operating in England were immediately withdrawn by Nagelmackers, and for the time-being Pullman reigned supreme in Britain!

Once back in America, Colonel Mann patented an improved sleeping car layout in 1878, which was complete with side corridor access, and with the toilets and other amenities placed at the ends of the car. It was in fact the European layout, as then being evolved by Wagons-Lits, and it was very modern for its time, in particular because the entrance doors at each end were opened into an enclosed vestibule, unlike the open ends of the Pullman cars. Mann set up the Mann Boudoir-Car Co to build and operate this type of car in the United States, and they were put into service on the Springfield line, between New York and Boston in 1883. In America they were known as 'bed cars' or 'compartment cars'. His 1878 car design and vestibule design was used in the same year for some British-built cars for service in Russia. Strange to relate his Springfield line cars became popular with the travelling public despite the fact that he charged a higher rate than his rivals, because the cars had a very limited carrying capacity, having some double berth compartments and some singles, which were very elaborately furnished and decorated.

The Mann Boudoir-Car Co became a part of the Union Palace-Car Co in 1889, and was absorbed along with it by the Pullman Company later the same year. Colonel William d'Alton Mann then turned to society gossip journalism, purchasing *Town Topics* in 1891 and later founding the *Smart Set*. Behrend relates in *Pullman in Europe* that in the course of these later activities Mann was daily to be found in New York's fashionable Delmonico Restaurant, and he was paid by the leading lights of the theatre world to withhold damaging gossip about them — true or otherwise — including the famous Lillie Langtry. For her, Mann supplied one of his private Boudoir cars! Mann died on 17 May 1920. His contribution to sleeping car design in both Britain and Europe was to prove more significant than that of George Mortimer Pullman, but his name remains unknown to the general public today.

CHAPTER 8

Changing Roles for the Pullmans

In Canada, as we have seen, it had been possible to eat a cooked meal in a Pullman car since 1867 when the first Hotel car had gone into service on the Great Western Railway and in America the first dining car *Delmonico* had commenced running in 1868. In Britain however, the first meal both cooked and served on a train was not until 1878, and then only for a special private party. George Behrend tells us that the party hired a complete train of vehicles including two Pullman cars, in one of which the meal was eaten, to travel all the way from London to Wick, and back. As the Pullmans had no kitchen facilities a MR van was attached at one end, adapted to serve as a kitchen. More remarkable still was a similar van at the other end of the train which was converted to serve as a bathroom. Exactly how the passengers could gain access to this, except when the train was stationary — there being presumably no gangway connections — is not clear, so bathing must have taken place whilst the train was standing at the station!

In the ordinary day-to-day workings of its trains the Midland Railway did not seem to be much interested in the idea of serving food to its travellers, preferring to stop the train at an intermediate station with dining rooms — as had been American practice. We can assume that on the MR the Pullman cars had some hot drinks available, using the Baker heater, and perhaps a limited supply of other beverages (but not intoxicating liquor, for reasons described later) were available upon request, as a part of the special Pullman standards of service. They do not appear to have been advertised, however, in the general promotion of Pullman cars on the MR and may well have been provided at the personal discretion of Pullman's staff on the cars.

If the Midland Railway apparently showed no interest in dining cars, it was characteristic of Pullman that the rival Great Northern Railway should then be offered an opportunity to introduce the idea of serving meals on board the train whilst travelling! To do this the Pullman Palace Car Co (Europe) offered to convert at their own expense an existing Pullman car operating on the GNR, and to let them have it to run as an experiment and without obligation, the Pullman Company to collect the extra receipts from passengers on the journey, whilst the GNR collected the normal fares. This was the usual Pullman arrangement, except that in this case the GNR could give three months notice to end the use of the car, if the trial proved to be unsuccessful. The GNR board accepted the idea of one car, and the standard Pullman parlour car *Ohio*, already in use on their routes, was taken back to the Pullman workshops at Derby, for remodelling as a dining car in 1878.

To understand the reasoning behind the introduction of the converted *Ohio* (which was renamed *Prince of Wales*) for what was to be the first dining car service in Britain, it is pertinent here to take a look at the rest of the British Pullman car fleet as it had developed since 1874. In November 1877 the initial fleet of cars, produced basically as a result of the contract with the MR, had all been delivered to Derby from America, although not all were in service on the MR. J. B. Radford gives the following list, taken from official MR records, and dated 6 November 1877:

55

55
The interior of the Pullman dining saloon in the Great Northern Railway's *Prince of Wales*, the first British dining car. This engraving first appeared in the *Illustrated London News* at the time of the introduction of the *Prince of Wales* into service. *Illustrated London News*

Parlour Cars

Victoria	June 1874
Britannia	June 1874
Leo	June 1874
Jupiter	23 August 1874
Saturn	10 September 1875
Mercury	28 April 1876
Juno	28 April 1876
Venus	3 May 1876
Vesta	28 July 1876
Minerva	28 July 1876
Planet	27 September 1876
Albion	9 October 1876
Comet	31 October 1876
Ariel	31 October 1876
Apollo	19 January 1877
Adonis	19 January 1877
Aurora	1 May 1877
Ceres	1 May 1877
Eclipse	10 July 1877
Alexandra (I)	10 July 1877
Globe	Still in shops, not been in traffic

Sleeping Cars

Excelsior	June 1874
Enterprise	June 1874
St George	28 June 1875
Princess	17 July 1875
Transit	13 October 1875
Saxon	4 February 1876
Castalia	28 April 1876
Scotia	28 April 1876
Norman	2 June 1876
Australia	20 June 1876
India	28 July 1876
Germania	26 August 1876
Midland	June 1877

56
The dining saloon of the Pullman dining car *Prince of Wales*, introduced on the Great Northern Railway in 1879, the first British railway dining car (from an engraving in *The Graphic* of 1 November 1879). An extremely spacious arrangement with only single armchairs each side of the aisle. *Author's collection*

Notes:
Midland went on the Continent July 1874. Returned 20 June 1877.
Mars went to Brighton 26 October 1875. (Arranged for day use only.)
Alexandra (II) went to Brighton 10 October 1877. (Never ran on MR.)
Albert Edward went to Brighton 10 October 1877.

57
The small kitchen of the dining car *Prince of Wales*, showing the chef's assistant (his son) preparing food on the open end of the car, whilst the chimney smoke shows that there is cooking in progress. An interested passenger looks from the car interior. The foreground detail shows the footwalk between the two Pullman cars, and also the handbrakes operated by wheels, a feature of all Pullman cars for many years. The second Pullman car was the sleeping car *Ocean*, arranged for day use. *Author's collection*

Not shown on this list were the two GNR-operated cars, the parlour car *Ohio*, and the sleeping car *Ocean* (which was arranged for day use only like *Mars*.)

This list represents a cornerstone of Pullman's early British history, and deserves a closer study. It made a total of 39 American cars (excluding the eight day cars which had been sold directly to the MR). Of these 39, only 32 were in MR use — the maximum ever operated by that railway — including *Midland* which had returned from the Continent (of which more in a moment). One new car lingered unused in the Derby shops, out of sight since assembly (*Globe*) and three were running on the London Brighton & South Coast Railway (*Mars, Alexandra* (II) and *Albert Edward*); the last two having only just been delivered. Of the entire stock, 23 were parlour cars and only sixteen sleeping cars, of which latter type two were arranged for day use only. We have seen that the sleeping cars had proved to be reasonably popular, and when the pioneer car *Midland* came back to the railway of the same name it was considered worthwhile putting it immediately into the Derby Pullman car shops to be rebuilt to the standard pattern of sleeper, which had by this time been established and which involved changing the sofa sections running along the vehicle sides beneath the windows into the standard cross-section, which comprised a pair of facing seats below each upper berth, having a table between them during the day. For night-time use the table was stored below and the two seat squabs pulled towards each other until touching which, together with the seat back which dropped down, formed a comfortable lower berth. Standard MR buffers and drawgear were also fitted.

The transfer of *Mars* to the Brighton line, already mentioned in Chapter 7, was evidently considered successful, in its role of what was basically a daytime parlour, or club car. Brighton was, and still is, a town with a very large population, of whom quite a number made their living in the City of London. It became known as 'London-on-Sea' because of this, and because it was within easy journey time, being only some 50 miles from the capital city. Some of these 'men of the City' who chose to reside on the South Coast, were extremely wealthy. The train to London each day, and the evening one back to the coast formed quite an important feature of their way of life, and the chance to travel in the comfort of a Pullman, with its added refinements and courteous attendants was agreeable, making the journey less tiresome, and giving it something of the atmosphere of a private club on wheels. *Mars*, which was marshalled into ordinary trains of much humbler aspect and comfort proved so popular, despite its Pullman supplementary fare, that as just mentioned, and as shown on the list, in 1877 the two new parlour cars arrived, *Alexandra (II)* and *Albert Edward*, with the third one, *Globe*, being kept at Derby as a 'spare car' to begin with. Rising receipts encouraged the Pullman manager to introduce *Globe* as the fourth car on the Brighton, in June 1878. Whereas the parlour cars operating on the Midland were often quite lightly loaded, the Brighton cars already had a regular clientele of what are today called commuters. The car *Mars* remained on the Brighton line until February 1883, when it was withdrawn and sent to Italy, where it adopted its designed role as a Pullman sleeping car. In 1886 it was sold to the CIWL.

In August 1878 two MR Pullman sleeping cars, *India* and *Germania* were transferred by the Pullman Company to the Great Northern Railway, who already had two cars the parlour car *Ohio* and the sleeping car *Ocean* which had not seen regular service since 1876, being reserved for specials. Although *Ocean* had been designed as a sleeping car, it was fitted out at Derby for day use only; so in fact *India* and *Germania* were the first Pullman sleeping cars to be used for East Coast night travel, running in the 8.30pm from King's Cross and the 10.20pm from Edinburgh; (both trains also including sleeping cars under direct railway ownership, known as the East Coast Joint Stock). The parlour car *Ohio* was then sent to Derby, as recounted at the start of this chapter, re-emerging in October 1879 as the Pullman dining car *Prince of Wales*, the first British dining car. Thanks to the researches of J. B. Radford we know that on 22 October 1879 Mr Henry Oakley the GNR General Manager wrote to the railway's directors to tell them that the car was ready, and that he proposed that it should be used within the train set working the 10.00am Leeds-King's Cross and the 5.30pm return working (in direct competition with MR services) commencing on 1 November. He also said:

'The car has been fitted with a large general saloon capable of seating 11 persons with revolving chairs and with means of affixing a table between each two after the practice adopted (I am told) in America. A second smaller saloon is also provided which will hold nine, included for those who desire to smoke.

'The kitchen apparatus has been partitioned off and is furnished with a stove capable of cooking all sorts of food for 20 persons.

'By way of experiment it was run down to Peterborough and back on Saturday [note — this was Saturday 18 October 1879] when we invited several representatives of the London press to inspect it. A very good dinner was cooked in the car and served much to the satisfaction of all. Mr Tennant, MP, our Director accompanied the trip down, but left us at Peterboro' where Sir Andrew Fairbairn, who was accidentally coming up, joined us on the return journey.

'It has been agreed that the car should be sent to Leeds and representatives of the local press of Leeds and Bradford and a few of our chief supporters should be invited to take a trip in it, and I shall be glad if the Directors will give instruction on this point.

'The car weighs 22 tons and will of course be a serious addition to the weight of the train. At this season of the year, however, we think that by reducing the number of carriages now run on it by three we shall be able to keep time.'

He did not mention the fact that the seats in the smoking saloon were fixed, nor that the sleeping car *Ocean* (arranged for day use) was coupled to it, to provide more supplementary-fare passenger accommodation.

A very good contemporary account appeared in *The Illustrated London News*, as follows:

'The Pullman Dining Car

'The Great Northern Railway Company has been the first in England to adopt this comfortable American system, which has been daily in operation, since 1 November, in one train each way between Leeds and London. The Pullman Palace Car Company, whose sleeping and drawing-room cars are in use upon several English and Scottish railways, have provided for the Great Northern line a handsome and convenient new carriage, the chief interior apartments of which are shown in our illustrations. It was constructed at Detroit, in the United States* at a cost of £3,000. It is 52ft in length, 8ft in breadth, and runs upon eight wheels. It comprises a dining-saloon, in the middle, a kitchen behind, and a smoking-room in front, with steward's pantry, ladies' dressing room, gentlemen's lavatory, cupboards, and stoves. The dining-saloon has six tables, three on each side, as represented in our illustration; there are 10 easy chairs, large, well-stuffed, and covered with crimson velvet, each revolving on a solid pivot. The ladies' dressing-room, which adjoins this, has a stove, enclosed in a metal-lined wooden cupboard. The sides or walls of the saloon are of American black walnut, decorated with veneer of French walnut. The smoking-room, of which we also give an illustration, has two tables, and seats for nine persons. There are electric bells to summon the attendants, who are the cook, the steward or waiter, and the boy of the smoking-room. The refreshments, including wines, are supplied by Pullman's Company at ordinary hotel dinner charges. Any first-class passenger on the line may use this Pullman Car on payment of half-a-crown over and above the ordinary fare for his journey. He may enter or quit the Pullman car at any station where the train stops, but should give notice beforehand of his intention, lest the car should be full. The up train, to which this car is attached, starts from Leeds at 10am, and arrives at King's Cross, London at 2pm. The down train, by which the car returns, leaves King's Cross at 5.30pm, reaching Leeds at 10.10pm. So that a Leeds man, by this arrangement, may start from home after his usual breakfast, fortify himself with lunch at 1 o'clock, take three clear hours for his business in London, dine comfortably at 6 or 7 o'clock, and get to bed in his own house an hour before midnight. This seems to be just what one should want; and it is to be hoped that Manchester and Liverpool men will soon enjoy the same accommodation. We trust that the London & North-Western, the Midland, and the Great Western Railway companies will not delay to follow the example of the Great Northern. Irregular dining hours, beyond all doubt, have shortened the lives of many prosperous and active men of business who were little past middle age. The Pullman Company therefore deserve our support in this department of their enterprise, as well as in the introduction of sleeping cars, which suit the need of a comparatively restricted number of travellers by night. The principal office of the Pullman Palace Car Company in London is situated in the Midland Railway Station, St Pancras, Euston Road. Mr J. Miller is secretary, Mr H. S. Roberts, manager for England, and Mr A. Rapp mechanical superintendent. We are informed that only one dining-car has yet been constructed for use in this country, and it has been under repairs for a day or two in the present week. It may be expected that this convenience will hereafter be universally applied to our railway passenger traffic.'

The kitchen apparently did not warrant a special description, but we can obtain a good idea of what it was probably like from this contemporary account of an American railroad Pullman car kitchen:

'We were curious as to the working of the culinary department and animated by a noble desire to obtain knowledge we penetrated the sacred precincts of the cook. The kitchen was a perfect gem of a place about eight feet square. A range ran along one side, its dark shining face breaking out into an eruption of knobs, handles and hinges of polished brass or steel. Curious little doors were studded all over it. Pots, steamers and pans were simmering on the top. Every requisite for carrying on the gastronomical operations was there in that tiny space in the neatest and most compact form. Scrupulous cleanliness reigned supreme over all. There was the pantry with its polished silver, glass and china in shining array. The refrigerator, with a plentiful supply of ice, and the larder were side by side. The wine and beer cellar was artfully arranged beneath the car. Thus every inch of space was realised to its utmost extent.

'Toward six o'clock every table was spread with dainty linen and the dinner was exquisitely served

*Note that the press and public were led to believe that the car was a *new* one. The first, but certainly not the last time this ruse appears in British Pullman history!

58

58
Part of the dining car *Prince of Wales* was given over to the gentlemen who wished to smoke, and perhaps converse in more masculine manner than was possible in the dining saloon with ladies present. A large ashtray is on the floor between the feet of the two men in the foreground.
Illustrated London News

according to the previous order of each traveller. The simplest dish as well as the most elaborate was cooked to perfection and everybody fell to with a will.' (From *Through Cities and Prairie Lands*, Lady Duffus Hardy, New York, 1881.)

One interesting feature of the kitchen in the *Prince of Wales* was that, because it was placed at one end of the car, the open platform so characteristic of all the Pullmans running at that time, adjacent to the kitchen, was utilised for storage and food preparation. In charge of the *Prince of Wales* for the opening day and the earlier press day was the Pullman Car Company's own conductor James Bower, who had come over from Chicago where he had served as conductor on the first dining car to run between Chicago and Buffalo late in 1875, and who subsequently made his name in England. James had run away to sea at the age of 14 from his home in Portsmouth, settling later in Chicago. He returned to England in 1877 with Pullman and worked on the Midland Railway Pullman services before his GNR debut.

The change of name from *Ohio* to *Prince of Wales* also heralded a very confusing practice that was developed by the Pullman Palace Car Co, in which the names of cars were often changed during rebuilding or when transferred from one line to another. This appears to have been part of a deliberate policy of making the public believe that they were travelling in a new car! Much detective work by George Behrend and J. T. Howard-Turner in particular has been needed to establish the true identities of some early Pullmans, because the company records were destroyed by German bombing in World War 2 (as will be recounted in Chapter 15).

A particularly confusing choice of name around at this time was *Alexandra* of which there were *three* in Britain at one stage running on the lines of three different railways simultaneously! *Alexandra* (I) was on the MR as a parlour car; *Alexandra* (II) was on the LBSCR Brighton line, as a parlour car, and *Alexandra* (III) was on the London & South Western Railway also as a parlour car. In fact *Alexandra* (III) had commenced life as parlour car *Victoria* on the MR and was transferred to the LSWR in 1880. It was remodelled into an open saloon parlour car with 34 seats, instead of the original 17 armchairs and private state-room, before being sent to the LSWR who put it to work between London (Waterloo) and Exeter. This was the first LSWR Pullman car service, only lasting until 1882 when the car vanished from UK records. We will meet this particular car again later, but meanwhile, no more Pullmans ran on the LSWR until 1890.

Perhaps the most outstanding success of which the British part of the Pullman Palace Car Co (Europe) could boast at the end of the 1870s was that of the dining car *Prince of Wales*, which had proved to be very popular with the travelling public. Success always creates imitators, and in the case of the Pullmans, the imitators were the major private railway companies, who steadily introduced more and more sleeping cars and dining cars of their own between 1880 and the turn of the century, denying Pullman the rights to operate his cars in some cases, and in other cases offering similar services without his special supplementary fares. The role of the Pullman car had already begun to change, and the 1880s witnessed some further exciting developments as George Mortimer Pullman's British management faced the competition, and endeavoured to remain simply 'the best'!

If the *Prince of Wales* was a success story, then the other extreme was true of the MR 'day cars', built by Pullman in 1874 and sold to the MR. These were consistently unpopular, and when an attempt was made once again to use them as a complete train set, together with Pullman parlour cars, on services from St Pancras to Liverpool and Manchester and back (commencing on 11 March 1878) they were so disliked by the public that they were withdrawn after only *one week* in use, despite the fact that they had been overhauled both inside and out. The public apparently still preferred the privacy of the much smaller normal Midland Railway compartment stock, apart from those knowledgeable few who chose to pay the Pullman supplement, and benefit accordingly!

59

59
Although the locomotive dates from 1890, is ahead of our period in this chapter, the train itself is composed of typical carriages of the London Brighton & South Coast Railway. In the midst of which is a stately Pullman; its clerestory roof making a distinctive impression amidst all the pots for oil lamps on the low roofs of the four-wheelers. The engine which is Class 'B1' 0-4-2 No 174 *Fratton*. The train is a Newhaven boat train. *Bluebell Railway Archives*

60

60
A rather superior British main line coach of the late 1870s, which provided a modicum of comfort for third class passengers: a Midland Railway clerestory roofed 40ft brake with padded seating for 50.

CHAPTER 9

Developments in Design

The Pullman workshops at Derby assembled two more sleeping cars in 1880, after a lull of three years when none had come from America. These two were sent directly to the Great Northern Railway, and carried the names *Columba* and *Iona*. Upon arrival on the GNR they replaced *Ocean* and *Germania*; later these two cars were sent to Italy. Prior to the arrival of the two new cars, the pioneer sleeping car *Midland*, suitably renovated, had been on loan to the Great Northern for some weeks between April and July 1879, acting as a demonstrator for the benefit of the GNR directors, who apparently required some convincing as to the relative merits of operating new Pullman sleeping cars, as opposed to using their own jointly-owned stock. We must assume that the rival Midland Railway Pullman sleeping cars were at the time sufficiently popular to be regarded as a threat, and that it was, perhaps, a case of 'if you can't beat them, continue to join them'. The earlier sleeping car, *India*, also remained in GNR service, but was later destroyed in an accident (see Chapter 10).

The American cars had been frequently criticised on account of their weight and size when compared to normal British carriages, and it was certainly true that they were exceptional when first introduced. However, it was also true that they were in many respects superior to their British counterparts, and by the beginning of the 1880s a number of railway engineers were using features first introduced by the American cars, in particular bogies, and the size and weight of British trains increased steadily, requiring new, larger and more powerful steam locomotives to be designed and built. It is therefore a suitable moment to examine some of these developments in British carriage design in general, because they enabled some of the major railway companies to provide better services for their passengers, without contracting the services to the Pullman Palace Car Co (Europe) which in 1882 was renamed Pullman Limited; registered in London.

The 1870s had witnessed a move away from rigid wheelbase four- and six-wheel stock, towards eight- and even 12-wheelers. Some eight-wheelers had rigid wheelbases, some had a degree of flexibility on the outer axles, and some possessed bogies proper. The 12-wheel bogie carriage was pioneered by the Midland and when these appeared they were the longest ordinary passenger carriages in the country, measuring 54ft long over body ends. (The Pullman cars were of course also exceptionally large for their period.) Six-wheelers were commonplace on trains of the 1870s and pot-oil lamps still provided illumination. Trains were frequently composed of a mixture of carriages, with different lengths and wheel arrangements; some possessing bogies, some not, the bogie allowed greater length of carriage to be safely carried around curves and the length of trains increased.

In addition to greater length, the height of carriages was gradually increased. At first this was only by a modest 12in or so, along the centre line of the roof; this was achieved by making the roof an arc-shape. This became possible when luggage was no longer conveyed on the rooftop (a somewhat dangerous practice with hazards such as fire or high winds), and separate luggage compartments, or vans, were provided. The guard also no longer travelled precariously on the rooftop; instead he was positioned in an enclosed elevated 'birdcage' compartment, where he could see over the top of the train. As carriages became yet higher, this 'birdcage' was not always possible within the loading gauge, and an alternative look-out was provided by means of ducket side-windows to the guard's compartment, which was situated within the carriage.

An important feature of the American Pullmans, as we have seen, was the clerestory roof (which had earlier on made a brief appearance in Britain on the broad gauge Great Western Railway 'posting carriages', as already mentioned in Chapter 2) and this now made a comeback on the GWR and the Midland. Besides giving additional headroom, the clerestory roof provided better natural lighting and ventilation. A drawback was the illumination at night. Any type of lamp, suspended within the raised clerestory, tended to cast heavy shadows to each side of the compartment interior in British compartment-style stock. After a while the Midland Carriage Superintendent, Clayton, reverted to arc roofs, but on the GWR, Dean continued the use of the clerestory. A third and more widespread revival of the clerestory roof was to occur in the 1890s.

In Chapter 7 we have observed the introduction of various experimental British sleeping car designs and it seems that to begin with these were sufficiently crude and unsatisfactory to ensure that the most popular early version was the American Pullman saloon layout, with curtains offering a degree of privacy to each berth. These required passengers to undress whilst on the bed — a complicated manoeuvre for Victorian ladies complete with bustled skirt and tight corsets! The British public, however, at least those of the wealthier class who could afford to pay for the privacy of a bed on the train, took even more kindly to the idea of having separate compartments to sleep in, linked by a side-corridor. In 1881 the Great Western Railway produced this arrangement in a broad gauge sleeping car which had six two-berth compartments, no fewer than three lavatories, plus a pantry. This carriage was a 'convertible', with narrow body running on broad gauge wheels, to facilitate ease of change-over to the 4ft $8\frac{1}{2}$in system in due course.

Mention should be made here of the improvements made in lighting, heating and brakes for passenger carriages. The Pullman cars still had their very effective double kerosene lamps with Argand burners (but these were a decided fire risk in a wooden-bodied carriage). They also had oil-fired hot-water heating, a remarkable advance in comfort at a time when the British passenger still froze in cold weather despite his footwarming aids. Lighting on certain railways had changed from the poor pot-oil lamps to coal gas, or compressed oil gas, usually carried in cylinders under the carriage. The compressed oil gas was superior to most systems and it was widely used by British railway companies for many years, despite the serious disadvantage of gas lighting which was the grave fire risk in the event of an accident.

Reverting for a moment to the subject of carriage heating; the footwarmer had been 'improved' by substituting soda-acetate crystals for the use of hot water. The sealed canister of crystals was heated in a mobile vat, and retained the heat for a number of hours. This improvement was first made by the London & North Western Railway, and subsequently adopted elsewhere.

The Pullman dining car *Prince of Wales* was soon to be copied by others and the idea of eating on a train became fashionable, although at first, dining or refreshment cars were the preserve of the first class passenger. The cars did not possess gangways at each end and it was not yet possible to 'stretch one's legs' and head for the diner whenever the fancy occurred. Either you travelled the whole way in the restaurant car, or else changed carriages at a stop *en route*.

Sleeping cars and dining cars were of course still very few and far between (except for the MR Pullman sleepers), and the typical train consisted of non-corridor compartment stock, and the usual British carriage still had hinged doors to each compartment. A carriage produced by the Great Northern Railway in 1881 broke away from the existing non-corridor compartment layout and provided a side corridor to each compartment. As such, this (and the similar arrangement on a Great Western sleeping car, already mentioned) represented the prototype of countless hundreds of carriages built for British railways with this interior layout. As yet there were no gangways between carriages, but on new carriages toilet facilities were accessible to passengers within the carriage.

Brakes for passenger trains had been the subject of considerable debate. There was much interest in providing continuous brakes, following a number of serious accidents, but it was not clear what form these should take. One method utilised chains; but more sophisticated systems using either a vacuum, or compressed air were soon sought after, as the chains were both cumbersome and difficult to apply smoothly. A series of brake trials had taken place on the Newark to Nottingham line of the Midland Railway in 1875, with various railway companies competing; these trials involved automatic air-brakes, hydraulic brakes, vacuum brakes and chain brakes. The results favoured the Westinghouse system of air brakes with which the American cars were fitted from the start, but despite this, many companies chose to go their own way and both vacuum and air brakes were developed. (Could they but have foreseen the events of the 1960s, when British Railways were faced with the reversion to air brakes despite a fairly recent decision to standardise upon the vacuum brake!)

An aspect of braking dear to the passengers' heart was the emergency communication. A cord system between driver and guard had been recommended by a Select Committee of the House of Commons in 1853, but although this was widely adopted the cord was not supposed to be accessible to the passengers. By 1865 recommendations were made to allow passengers access to the cord, and in 1868 under a Regulation of Railways Act, railway companies were obliged to provide an efficient means of passenger alarm communication on all trains making non-stop runs of 20 miles or more. This usually followed the principle of an outside cord, accessible to passengers through the droplights and connected either to the engine whistle or to a gong mounted on the side of the tender. Upon hearing the alarm sounded, the driver applied the brakes. Looking ahead, in 1890 the first move was made towards allowing the passenger to have direct access to the brakes. Approval followed in 1893, and in due course the familiar system of chain running inside a pipe, with the pipe placed inside the carriage, appeared. Pulling the chain (on dire penalty of a fine for improper use) began applying the brakes on the train. And it remains the same in the 1980s. A prominent feature on all the American designed Pullman cars was the handbrake, operated by turning a large wheel mounted on each open platform end. Originally provided for braking the train whilst in motion, when continuous brakes were fitted (as on all the British Pullmans), the handbrake was still required for its other purposes; namely to prevent the car moving on its own when parked, or stabled. Each handbrake was individually linked to the adjacent bogie, and all had to be checked before the train moved.

In 1881 the London Brighton & South Coast Railway (LBSCR) introduced the first complete train of Pullman cars to operate in Britain. It will be recalled that the Midland Railway had insisted upon the inclusion of non-supplementary fare day cars in their trains. On 5 December 1881 the beautiful Stroudley Single No 334 *Petworth* was coupled to four Pullmans, to haul the first 'Pullman Limited Express', which was scheduled to run between London and Brighton twice daily in each direction. What a colourful ensemble this must have been, because William Stroudley, the LBSCR locomotive engineer, painted his passenger engines in a rich golden ochre livery, offset by dark green and claret. The Pullman cars were in the original mahogany, or umber brown, richly embellished with gilt decoration, and of course everything was polished until it positively shone. Stroudley insisted upon cleanliness, and he even had the cab roof on each of his engines painted in white, to ensure that it had to be cleaned regularly. As he was also in charge of the LBSCR carriage stock, he no doubt personally made sure that the Pullmans were well cared for.

The four Pullman cars were not new, but all carried new names, and no doubt the public fondly imagined that they had been specially built for this exciting new train. In fact they were called after the children of the Royal Family and were *Louise* (formerly *Ariel*), *Beatrice* (formerly *Globe*), *Victoria* (II) (formerly *Adonis*) and *Maud* (formerly *Ceres*). It will be recalled that the name *Victoria* had already been carried by the pioneer parlour car running on the Midland. In fact three cars, the former *Ariel, Adonis* and *Ceres* were transferred from the MR, whilst *Globe* had been on the LBSCR beforehand, after being stored when new at Derby, as already recounted. The Derby Pullman workshops had made some alterations to the design of the cars specially for the LBSCR 'Limited', and *Louise* and *Maud* were placed at each end of the four car train, with a sliding door luggage compartment, and a servants' compartment. *Victoria* (II) had a small pantry buffet and was the gentlemen's smoker, and *Beatrice* was called a drawing-room car, with ladies' accommodation for them to withdraw into, as was the manner of the time in polite society. The inclusion of servants' accommodation shows that Pullman had recognised the continuing need for this in Britain.

It was a splendid train for its time, as this contemporary account by F. S. Williams reveals:

'The Pullman limited express on the London and Brighton seem to provide almost the perfection of travelling arrangements. Instead of their being ordinary trains with one or two or more Pullman cars attached, all the carriages are on the Pullman plan and they are so coupled that the officials in charge can walk from one end of the train to the other. "Entering it", says a traveller, "you enter a mansion on wheels. You can roam about from parlour to drawing-room, dining-room, and smoking-room. The ladies have also a boudoir. Servants are at your call by electric bell; but you need not call for light or fire. A pleasant and equable temperature is maintained by hot-water pipes, and lighting up is done before there is time to demand it. No sooner do you enter a tunnel than the bright but soft and equal light on the Edison electric lamp is shed all over the compartments. It noiselessly comes at the moment it is needed, and is as quietly gone when the need for it is over. It beats the good fairy of nursery lore, for it needs no summons. It is with you before you can as much as think of a wishing cap, and it is itself a magic lamp." With this train service Brighton may well come to be more appropriately than ever considered as London-on-Sea.

'At night the whole train is lighted by electricity. "The steady soft white light of the incandescent carbon threads in the little sealed glass lamps — these tiny horseshoes, burning but not consumed — were approved, admired, acclaimed. To read by them was a pleasure. There were thirty lights in the train; none gave a less brilliant light than another; and any one light could be put out without the other being affected. The light came from 34 of Fauré's accumulators in the guard's compartment. These magazines of electricity had been filled at the Strand, but they will hereafter be charged by an engine and dynamo machine at Victoria station." A proposal has been considered for making the train store up electricity for itself by the working of a dynamo machine attached to the axle of the engine.' (From: *Our Iron Roads*, F. S. Williams, 1884.)

The reference to electric lighting was the result of earlier trials, and it is believed that the car *Beatrice* was the first railway carriage in the world to be lit by electricity. (In fact, it would have been more appropriate to have kept its former name of *Globe*!) There were 12 electric Edison lamps, with bamboo-filtered incandescent bulbs to illuminate the car, but the Argand kerosene lamps were also retained, in case of failure. The electric current was supplied by 34 (Behrend states 32) portable Fauré cells (manufactured by the French Fauré Accumulator Co), slung in a shelf beneath the floor of the car. An initial trial trip had already been made for the benefit of the press and other interested parties on 14 October 1881 from London to Brighton, in which the lights were first

61

switched on when entering dark tunnels from broad daylight. Nowadays we are so used to this idea that if the lights do *not* come on we are startled, but of course the contrary was the case in the 1880s! The demonstration was considered to be a complete success, and even better was the evening return journey from Brighton to London when the lights were kept on for the whole trip because it was by then dark. Behrend records that among the invited guests were gas engineers, who calculated that because of the cost of installing the dynamo machine at Victoria station, the cost all told was four times that of the gas supply that would be needed to light the train. In his *History of Trains de Luxe*, Behrend observes that: 'They overlooked the fact that escaping gas turned railway accidents into infernos, by setting the smashed woodwork of the carriages ablaze'. Meanwhile, the Midland Railway had already decided in 1878 to remove the Argand kerosene lamps in all the Pullman cars operating on their territory, and to replace them with 'standard carriage roof lamps' — a retrograde step indeed; although they probably had reason to make the change in their quest for economical standardisation.

The success of the four electrically-lit Pullmans in Britain encouraged George Mortimer Pullman to adopt it for his American cars, which meanwhile had begun to use the Pintsch compressed gas system of lighting. The gas was distilled from crude petroleum, and a brilliant light was obtained; but of course Pullman realised the safety advantage of electricity. Nevertheless when the MR decided in 1885 to fit Pintsch gas lighting to the Pullmans it had, it was no doubt because the majority of their own rolling stock could not accommodate the Fauré cells easily, and the installation of dynamos specially for the Pullmans would have been expensive, at a time when the company was busy providing gas supplies for its trains during the general changeover from oil lamps to gas lamps then taking place. It is interesting to recall that at the time the LBSCR introduced its electrically-lit Pullman train, the average British household lighting was still by

61
A 'G' class Stroudley 'Single' heads the down Newhaven boat train, seen crossing the River Ouse near Southerham Junction, in about 1891. The Pullman car dwarfs the little four-wheelers and six-wheelers that make up the rest of the train, and the sunlight glints upon the oval nameplate on the car side. The identity of the Pullman car, in original condition, with open verendah ends, is not certain.
Bluebell Railway Archives

candles or oil lamps, and in the towns the gas companies were very active in persuading people to accept gas instead! Candles were carried by the four LBSCR Pullmans but only as a decorative feature. On every table along the train there was placed a flower vase-cum-candlestick, some 10in high, with top and base of the same diameter and a pinched-in waist. This vase had a floral pattern, and candles were supplied as requested, to give an added charm to the car's elegant interior.

The 'Pullman Limited' was to meet with an unexpected hostility from the Victorian British public because the LBSCR decided to run it on *Sundays*! On Sunday 11 December 1881, to be precise, it made the runs which gained it the nickname of 'The Sabbath Breaker'; because only the wicked chose to travel on Sundays. As Behrend writes so eloquently:

> 'The Sunday service was a mistake; it lasted a bare two months, but caused a reactionary lack of patronage. Can one not imagine the staid British shyness of this new-fangled, glittering and expensive American train, with its fancy French lighting and lacking, in its foreign way, the privacy of compartments? Would there not be dashing young beaux, lurking in the smoking-room of *Victoria*, who, once the train had started, would sneak across the gangway between the cars to make advances to the chaperon's charges in the drawing-room cars? Just the kind of people with loose morals, who thought nothing of running down to the coast on a Sunday, when they should have been in church!'

The commencement of the 1880s was a particularly busy one in British Pullman affairs, quite apart from the new Brighton line trains, which of course managed to survive the initial hostility! On the Midland, the success of the dining car *Prince of Wales* operated by the rival Great Northern did not go unremarked, and in 1881 consideration was given to converting two of the existing Pullman parlour cars into dining cars, at the expense of the Pullman Company. At the same time the MR was offered two new sleeping cars, 'constructed on the latest improvement'. These were named *St Mungo* and *St Andrew* and were introduced into service on 1 February 1883, replacing two of the older MR cars *Castalia* and *Australia* which were sent to Italy, of which more anon. The new cars were shorter in the body than earlier ones (only 50ft 6in long and 56ft 9in over buffers), and were basically similar to the two cars *Columba* and *Iona* which had been delivered to the GNR in 1880.

To produce the two MR dining cars, the two oldest parlour cars on the MR were taken into the Derby workshops for conversion; they were *Britannia* and *Leo*, which became *Windsor* and *Delmonico* respectively. The name *Delmonico* of course echoing that of the first Pullman dining car in America which was itself named after a very famous New York restaurant of the day. The two cars had six-wheel bogies substituted for the original four-wheel type, to give greater stability of ride; and the interior layout was modernised and included buffet facilities. Once again the press and the public were allowed to believe that these two cars were new, although only the bogies and the interior decor were in fact new. The decor was designed by Mr Gillow of the famous furnishing firm of Waring & Gillow. (Mr Waring having designed the interior furnishings on *Prince of Wales*.) A contemporary account of a trial run of the 'Manchester Palace Dining Car train' states:

> 'A trial trip was made on Friday with the two new Pullman dining room cars named the *Delmonico* and the *Windsor* which this week will commence to run on the Midland line. The *Venus* drawing room car, which has been in use for some time, was also attached to the train. The train left St Pancras shortly after 2 o'clock reaching Leicester at 20 minutes past 4 and returning to the terminus before 7. Among the party on board the cars were Mr M. W. Thompson, Chairman of the Company, Mr John Noble, the General Manager, and Mr F. M. Needham, Superintendent of the Line. The Pullman Car Company was represented by Mr H. S. Roberts, Manager, Mr J. Miller, Secretary, Mr J. Monck, Mechanical Supt, and General H. Porter, Vice-President of the Pullman Car Company of America.
>
> 'The cars, which were built under the supervision of Mr Monck at the Pullman Works at Derby, are 58ft long, and 9ft 10in in width. Each dining car seats 21 passengers while 27 may be accommodated in the drawing room car. In the dining car the tables are arranged at right angles to the windows, with a gangway down the centre. The sides and ceiling are inlaid with mahogany panels decorated with paintings of fruit and leaves, while mirrors are abundant and give a greater sense of space. The seats are upholstered in claret plush and overlaid with hand-worked antimacassars, the floor is covered with Wilton carpet and the blinds are of tapestry. Passengers can summon the waiter by an electric bell without moving from their seats. At one end of the car is a smoking room and lavatory, while at the other end is the pantry and the kitchen. The cooking is done on the journey in a range comprising a grill and which, small as its compass,

can prepare a dinner for 40 persons. The staff for each car will consist of a conductor, page and cook. The drawing room car has its sides and ceiling inlaid with black walnut and it is provided with a number of swing chairs upholstered in crimson plush. Tables are placed at the sides so that dinner can be served here in case of emergency. A smoking room, a private apartment for the use of a family, and lavatory accommodation are attached to this car. The materials for these cars were brought from Detroit and put together at the Derby workshops of the Pullman Company.

'On the trial trip, dinner cooked on the journey was served in the cars with most satisfactory results. The menu, which comprised six courses, beginning with soup and ending with dessert, was as admirably prepared as at a hotel. The trial trip gave promise of the dining cars proving very popular on the Midland line. On and after 10 July 1882 a Pullman dining saloon car will be attached to the express trains leaving London (St Pancras) for Manchester and Liverpool at 5pm and Liverpool (Central) at 4.5pm and Manchester Central at 4.50pm for London. A table d'hote dinner will be served in the car at 6pm on the down and up journeys, the price of which will be 3s 6d, exclusive of wines, etc. Passengers can obtain which wines they require for dinner by giving their orders to the conductor prior to the departure of the train, who will obtain them from the refreshment room.

'Special notice to Manchester passengers:

Manchester passengers by the 4.50pm express are advised that, owing to the limited number of seats in the Pullman dining car, it is desirable to engage their seats as much in advance as possible. This can be done by notifying the stationmaster at the Central Station, Manchester before 3pm, so as to allow time to reserve seats at Liverpool by telegram. Parties of four or five desiring to dine together can be accommodated with special tables by giving notice in advance.'

Prior to the conversion of these two parlour cars into dining cars the Midland had examined the possibility of purchasing the Pullman parlour cars still running on their lines, and had valued them at £1,301 each. Pullman declined to sell them, however, but in 1882 he agreed to the idea of making them more attractive, by allowing the Midland to put buffet facilities in them to sell refreshments, and he took away one more parlour car, *Jupiter*, to use it on the London, Chatham & Dover Railway. The car ran between London (Victoria) and Dover in what was grandly called 'The Dover Continental Pullman Boat Express', and which ran for only two years. The car *Jupiter* then went to the Brighton line in 1884, replacing the car *Mars* which was sent to Italy in February 1883.

Returning to the subject of dining cars, it seems that the MR had some problems with the Inland Revenue over the sale of alcoholic drinks on the Pullman dining cars using St Pancras station. It could not prevent people bringing their own with them, of course, and the MR seems to have got round the problem simply by first of all taking orders, and then purchasing the wines or spirits in the station refreshment rooms, before the train left (as explained in the contemporary account just quoted). However, a more cunning ruse was adopted at some stage, as Behrend explains, whereby the Midland kept a stock of liquor at Bedford, instead of London, and the Pullman conductor took orders as soon as the train had passed Kentish Town. He then threw the order from the Pullman car as the train passed Hendon signalbox, from where the signalman then had to telegraph to the Bedford supplier, who placed the ordered bottles on board the train when it arrived there!

Thanks to the thorough research made into MR history by J. B. Radford we know that the heavy Pullman cars were plagued by overheating of the axleboxes on the bogies (known as a 'hot box' in railway parlance). This trouble commenced when the cars were new, and persisted, despite the use of 'oils of the very best description'. Experiments with different kinds of metal for the bearing surfaces of the journals also failed to make an improvement. I suspect that the higher operating speeds of MR trains when compared to contemporary American running, exaggerated a design weakness that already existed in the axlebox design. Hot boxes were not unknown in America, of course, and seem to have been regarded with a good deal of philosophy as one of the hazards of railroad travel. One American traveller recorded his experience in a Pullman, with a hot box:

'At half past nine in the morning we moved out of Chicago and arrived at St Paul at eight o'clock the following morning, having completed the journey of 529 miles from Chicago in twenty-two and a half hours, including stoppages. Our average pace was under thirty miles an hour, not an excessive rate of speed, but we were delayed by what is technically known as a hot-box.

'Our cars each had a name, such as *Fargo, Minneapolis, Wamduska, Pyramid Park* or *Billings*. I happened to inhabit a section in the coach known as *Billings*, a new and very comfortable vehicle, but as often happens with new cars one of its wheels worked stiffly. The result of this was the overheating of the box and several times during the night the train had to be stopped in order to extinguish the flames.' (From *A Trip to America*, William Hardman, London, 1884.)

The MR management was apparently not so tolerant and sought an improvement. There was also a good deal of trouble with the continuous brakes fitted, and Radford reproduces a report which shows how frequently trains were delayed en route by sticking brakes or burst brake pipes. The fact that the cars were not owned by the MR complicated arrangements when any design alteration was to be made, and careful costings had to be calculated beforehand and agreed.

The year 1882 witnessed the destruction of the Pullman sleeping car *Enterprise* by fire, in a night of high drama, as J. B. Radford also records in his book:

'Shortly before 2am on Sunday 29 October 1882 the 9.15pm down night "Scotch" express left Normanton heading towards its next stop at Skipton, and included in its formation were two Pullman sleeping cars, *Enterprise* on its way to Edinburgh and *Excelsior* bound for Glasgow in the care of Conductors Robert Donaldson and Andrew Baillie respectively. Approaching Hunslet the driver heard the whistle of the safety cord apparatus and, looking back, saw a light above one of the cars and immediately brought the train to a halt near Hunslet South Junction signalbox. Baillie, from one of the vacant berths in his car, also observed a light, reflected in the headboards of his Pullman and, looking out, saw to his horror that the *Enterprise* was on fire. He rushed out on to the end platform of his car but found that the end door on the Edinburgh vehicle was shut, so he pulled the communication cord on his car and then roused all his passengers.

'He assisted in uncoupling the *Enterprise* from the rest of the train and tried with others to extinguish the fire in the vehicle's roof, by now well alight, with buckets of such water as was available.

'Meanwhile his colleague, Donaldson, in the affected car, had been having quite a night already, having had to deal with a Dr Arthur who had unsteadily walked down the platform at St Pancras in an apparent state of intoxication before getting into the front part of the train and who, upon arrival at Leicester, had moved into the Edinburgh car and asked for a brandy and soda. Donaldson had diplomatically said he only had soda water left, whereupon the Doctor had taken a bottle of it and added it to a concoction of his own which he had described as 'Egyptian brandy'. The conductor had then removed the Doctor's shoes and a cigar which he had been attempting to light, and, after asking for his overcoat, Arthur had apparently gone to sleep.

'When the fire had broken out there had been four passengers in the car; a Mr Dove who had left via the end door, a Mr Main who had broken a window in berth No 5 and had been pulled to safety and a Mr Cranston in berth No 8, who had boarded the train at Sheffield and gone to bed with a ready lamp burning. Donaldson had pointed out the danger of such a practice, although not strictly against Pullman rules, and had advised him to be sure to extinguish it before going to sleep. Cranston and Donaldson differed in their accounts of the next series of events, but Cranston escaped.

'In an attempt to douse the flames the car was first moved 876yd down the line to a water crane but the hose would not reach the roof so it was pushed back a few yards to a hose which had been run out from Nicholson's Chemical Works and there the fire was eventually put out.

'A short time later two police officers went into the charred remains of the car and made the grim discovery of the remains of Dr Arthur in berth No 7. Colonel Yolland conducted the official enquiry and concluded that the fire had resulted from Cranston's reading lamp setting fire to the blind in his berth. Dr Arthur probably succumbed to the smoke and fumes whilst in his inebriated state before the flames reached him. Colonel Yolland's main criticism was however levelled at the driver for bringing the train to an immediate stand rather than proceeding to Hunslet station where the fire could have been efficiently tackled and Arthur's life possibly saved.

'The Harrison communication cord, of a pattern not approved by the Board of Trade, also came in for some criticism in that it could only be reached from the platforms of the cars. In this instance the opening of the end doors to get at the cord had so increased the passage of air through the car as to give the fire a complete hold on the vehicle.' (From *American Pullman Cars of the Midland Railway*, J. B. Radford, Ian Allan Ltd.)

The damage to the car *Enterprise* was so severe that Pullman decided to scrap it, rather than attempt to rebuild it, although the MR were obliged to pay costs of nearly £2,500 to Pullman. At that time (March 1883) there were two new cars lying idle in the Derby workshops; shipped over for George M. Pullman's proposed Paris-Deauville Pullman express. These were of 'improved design' and were named *Missouri* and *Michigan*, and *Missouri* was delivered to the MR in February 1884 in lieu of *Enterprise*. The other car *Michigan* remained in store until the MR accepted it in 1886.

The wooden bodies of the Pullman cars were very susceptible to fire damage, if it got out of control. By a strange coincidence another one had caught fire at Leeds earlier in 1882, when on 11 February the Great Northern Railway dining car *Prince of Wales* became alight in the main saloon, to which area the fire was fortunately confined, and the car was repaired at Derby at a cost of £500, and returned to the GNR.

A good example of Pullman enterprise surrounds the transfer of the two MR Pullman sleeping cars *Castalia* and *Australia* to Italy, already mentioned. In clandestine fashion and ignoring all the exclusive rights granted to the rival Wagons-Lits Company for the provision of sleeping cars on the French Nord and PLM Railways, the Pullman Car Co and Thomas Cook & Son Ltd organised a special through train from Calais to Rome, using six Pullman cars, for visitors from London to the 1883 Italian Fine Arts Exhibition, with supplementary fares and a full service of meals on board the train! By doing this the Pullman Co managed to send the six cars with a revenue-earning payload, to cover the expense of moving them to Italy from Derby. More remarkable still was the fact that the trip was advertised in the one direction only, with no details of how the British visitors would be returned! It is said that Georges Nagelmackers of the CIWL was extremely embarrassed and annoyed by this flagrant act of defiance on Pullman's part.

Six Pullmans formed the Calais-Rome special train, five sleeping cars and one parlour car. The five sleepers were *Castalia* and *Australia* (already mentioned) plus *Germania*, *Ocean* and *Mars* and were shipped from England, but the parlour car *Alexandra* (III) was already there in France, having been sent the previous year for a trial service of one year on the Ouest Railway's Paris-Le Havre route running as a 'saloon-dining car' *Alexandra* (III) had started life as the first MR parlour car *Victoria* and had been renamed when sent to the London & South Western Railway (LSWR). It was Pullman's *oldest* parlour car, but the only one spare for the Le Havre service. The outcome was that the Ouest preferred the services offered by Nagelmackers, who had competed in the one-year trial period with a six-wheeled dining car on the Paris-Caen route, and in 1883 *Alexandra* (III) was redundant. Pullman fitters must have travelled to France (with the sleeping cars) and reinstated the chairs that had been in store since LSWR service, in place of the 'saloon-dining car' type seating used for the Ouest trial (see below). George Behrend (in a letter to the author in 1986) believes that this parlour car afterwards became Wagons-Lits No 155, purchased by that Company in order to start a *wagons-salons* service between Amsterdam and Rotterdam for the Netherlands State Railways.

This is an appropriate moment to recall that in 1883 the Wagons-Lits Co and the Pullman Car Co were completely at loggerheads, following an abortive attempt by George Mortimer Pullman to improve his position in Europe, made during a personal visit in 1882, in the course of which he also inspected the electrically-lit Brighton Pullmans. At some stage of the visit he apparently met Nagelmackers and held a long discussion with him in an attempt to form an amalgamation of their respective companies, which he proposed would be called the 'Compagnie des Wagons-Lits et des Wagons-Salons', and which would be registered in London. Moreover he proposed that the supply of *all* new cars would be from America (from Pullman's own workshops, of course). Quite apart from the fact that Nagelmackers must have been suspicious of George Mortimer Pullman's typically flamboyant American attitude — following upon his earlier experiences with Colonel Mann — it was also a move far too audacious on the part of Pullman, who held only a small fleet of seven European cars, in Italy and France (the British cars excepted) at a time when the Wagons-Lits fleet was already 110 strong! Pullman's proposition was on a 50/50 basis, and Nagelmackers refused it, preferring to develop his own services. Meanwhile, Pullman's Italian interest was a considerable handicap to Wagons-Lits expansion, and it was not until 13 June 1886 that Nagelmackers purchased the Pullman rights in Italy, at a high cost, when the Italian railway companies grouped themselves. Thus the American-built and British assembled Pullman cars we have already encountered in this story — transferred to Italy — finally fell into the hands of Pullman's European rival, giving further faithful service under their new ownership, except for *Castalia*, destroyed by fire in 1884.

Meanwhile, in France, where the initial success of Mann and Nagelmackers had prevented Pullman from achieving his aim of operating his cars 'all over Europe', the year 1882 had nonetheless seen Pullman achieve at least one small foothold, because of the trans-Atlantic liner services from America, which were calling at Le Havre. To serve these wealthy American travellers between Le Havre and Paris, Pullman provided a 'saloon-dining car' running over the lines of the Ouest Railway. Pullman then envisaged an all-Pullman service on the Ouest, running between Paris and Trouville, and he had a design prepared as a speculative venture for a type of car more in keeping with contemporary European practice, smaller and lighter than the American sleeping cars he operated in Britain and Italy. Four of these were to be built in the Derby workshops, using six-wheel underframes ordered from the MR carriage works. His lack of success when dealing with Nagelmackers led to the Ouest Railway preventing him operating these cars in France; although he had already selected names for the first two, *St Louis* and *St Denis* — *St Louis* being an American city, whereas *St Denis* was the location of the headquarters of the Ouest Railway in France. Once again fate played a chance card into his hands, providing an opportunity to use these cars in Britain instead (the French presumably having decided against them).

This came about because the Midland Railway was unhappy with the fact that the Great Northern Railway was permitted to run Pullman sleeping cars between London and Scotland alongside sleeping cars of its own joint stock, the latter being not only

62

62
The dramatic events on the night of 29 October 1882, when the Pullman sleeping car *Enterprise* caught fire near Hunslet, Leeds on the Midland Railway, as seen in contemporary engravings, with (bottom right) policemen discovering the remains of Dr Arthur in his sleeping berth after the fire had been put out. The drawing is well detailed with regard to the elaborate Pullman car livery, and places much emphasis upon the wooden clerestory roof being ablaze.
Illustrated London News

less expensive to travel in, but provided extra revenue direct to the railway companies. Pullman then suggested that he should be allowed to supply the Midland Railway with (Radford quotes): 'a somewhat inferior description of sleeping car for the use of which the same fares might be charged as are charged by the Great Northern and London & North Western companies'. He added that: 'four of such cars are now being built in the shops at Derby'. The MR agreed to accept the first two completed, *St Louis* and *St Denis*, of course!

When running initial trials the two lightweight sleeping cars were found to ride badly with their six-wheel non-bogie layout, and short-wheelbase bogies had to be substituted before they went into public service, which was on 1 July 1883, on the through night service from St Pancras to Greenock. Meanwhile, the Great Northern Railway also accepted two cars of this type (thereby finding homes for all four!), and these were named *Balmoral* and *Culross* and were delivered to the GNR on 2 January 1883, to run between King's Cross and Edinburgh. In their curious external style, they no more followed French than British practice and they were completely unlike the American cars; although the bodywork was apparently sent over from America to be mounted upon the MR Derby-built underframes. They had no end platforms or entrance doors, the only doors being a curious outward-opening pair on each bodyside, in the centre. The interior layout comprised a central vestibule 4ft 6in wide, with external doors each side, in half the width of which were housed on one side the standard Baker car heater and on the other a luggage locker and coal box. A curtained-off single Pullman 6ft long cross section sleeping compartment was placed each side of the vestibule with an upper bunk either side and a pair of facing seats below converting to a lower berth for night-time use. Beyond these lay a pair of 3ft 6in wide compartments with a water closet on one side of the gangway and an open lavatory unit on the other. At each end of the vehicle lay a further Pullman cross section sleeping compartment. In service they were not very popular and no more examples were built. The subsequent history of the four cars is however worthy of a brief mention because although we will not encounter them again in our main story, it is not without interest.

The two on the Midland were later used for a while on a through service from London St Pancras to Stranraer, to connect with the steamers to Larne, but neither this, not the previous Greenock service paid its way, and they seem to have been laid aside for some time. It was decided in about 1890 to convert them into 'picnic' saloons (see next chapter). They ended their days as static departmental accommodation, with *St Denis* at Heysham (still on bogies in 1905) and *St Louis* 'grounded' (ie without wheels or bogies) at Cheadle Heath, where it survived until the early 1960s. The two GNR examples apparently ran on that railway still as six-wheelers, despite the poor riding qualities already noted, and it was not until some years after they had been transferred to the Highland Railway (HR) in 1885, that consideration was given to fitting them with bogies. *Balmoral* and *Culross* (which was renamed *Dunrobin* on the HR) were then used on the night service between Inverness and Perth (later extended to Glasgow). Pullman's contract with the HR stipulated that the railway company had to pay the wages of the Pullman staff! Moreover the attendants had to wear the standard chocolate brown uniform of the Pullman Car Co.

The through night service between Inverness and Glasgow (after 1904) was worked by these two oddities of Pullman's sleeping car fleet until 1907, at which date they were the *only* Pullman sleepers still in regular service in Europe (as we shall see in the next chapter). They had a supplementary fare of five shillings and were even operated, at one stage, in mixed trains (a train carrying both passenger and goods rolling stock) which naturally proceeded at a leisurely pace, with occasional stops to shunt a wagon or two, or attach some more. It must have made any attempts to sleep all the more difficult for the passengers!

The Highland continued to operate them as 'spare vehicles' to strengthen their own sleeping car fleet, until about 1918, and then the two cars were purchased by Mr F. Marks, the son of J. S. Marks who had lately been the Pullman Car Co Staff and Rolling Stock Superintendent, having started with the company as a conductor on the first MR Pullmans in 1874. The two cars were then transported all the way from Scotland to Seaford in Sussex, where they were converted into a bungalow where they existed until 1987.

If these four sleeping cars are taken to represent an alternative effort by George Mortimer Pullman to satisfy the different attitudes of the European rail traveller, as compared to those for whom he was so ably catering in America, then they must be regarded as a complete failure. The problem for Pullman was twofold. On the one hand the British public resented paying a supplement to travel in his cars, which ordinary British railway carriages did not demand, except at a less expensive rate for sleeping purposes. On the other hand, the various private railway companies — often in fierce rivalry with each other — saw no reason to patronise an American 'outsider', if they themselves could design and construct suitable and less heavy carriages to match or compete with the Pullmans; and then collect all the revenue earned for themselves.

During the 1870s and early 1880s Pullman showed them the way, and he showed the British public what it was missing, but in too many respects his ideas, and his American cars were ahead of their time at this stage. Only on the Brighton line was there to be found a dedicated following, which was appreciative of the qualities of service and comfort that Pullman considered necessary — and the Brighton line neither possessed, nor indeed needed, any sleeping cars! Whereas in America, George Mortimer Pullman was fast becoming the 'King' of the railroad sleeping car empire, in Britain he was destined to fail in this respect. Nonetheless, his name was becoming associated in the British mind with travelling in style, but by day rather than night.

63

63
In 1923, one of the two Pullman 'short' sleepers that had last worked on the Highland Railway (either *Balmoral* or *Dunrobin*) was caught by the camera while being moved by road from Seaford station to a local building site where it and its sister car were incorproated into the structure of a bungalow. There they were to stay until 1987, when they were nearly destroyed in the course of demolition work. The remains of both cars were moved to Brighton Locomotive Works where, it is hoped, they will form the basis of the reconstruction of one complete car.

CHAPTER 10

Winds of Change and Tragedy

If it appears that George Mortimer Pullman had become somewhat disillusioned by his European experiences by the year 1883 (and it has to be said that the previous 10 years had not witnessed the growth and success for which he had endeavoured, either in Britain or on the Continent) then indeed the true success story was to be found in his homeland of America, where his prestige was at its peak at the commencement of the 1880s, and where he was building a complete 'city' to house his activities and his workers.

Completed in 1881, Pullman City (as it was inevitably named) was the first planned American industrial city, located on the outskirts of Chicago, Illinois on prairie land at a place called Lake Calumet. Covering some 3,600 acres of land, which had to be drained beforehand, Pullman City was designed by the architect Spencer Beman, and included a hotel for visitors, an administrative building (housing also the steam-driven power plant), schools, shopping centre, bank, church, theatre and — most important of all — a huge new workshop, dominated by a clock tower. Pullman City (in reality, of course, only a town of some 12,000 inhabitants) was in its layout a reflection of George Mortimer Pullman's own way of life; a strict one indeed. There were no saloons or bars, and the atmosphere was one of religious hard work, with suitable rewards — hence the position and the size of the houses occupied by the workers was governed by the importance of their jobs. There were splendid parklands and leisure areas; but idling in them was prohibited. Everything that was required to lead a civilised life in a town was provided by the Pullman Co for its employees, and business at the new workshops was thriving, with a production-line assembly of new Pullman cars, as well as other types of railway vehicle, including freight wagons (or box cars as they were called) and street tramcars.

By this time, in Britain, the Midland Railway affairs of George Mortimer Pullman were indeed a minority interest for him, and one which had not expanded as he had hoped; moreover his 15-year contract had not much longer to run, being due to expire in February 1888. Therefore, when the MR again approached him, in 1883, with a view to purchasing the 'day cars' (as the parlour cars had somewhat confusingly become known on the MR) he agreed to sell them all, *plus* the two dining cars that had been produced by rebuilding the parlour cars *Britannia* and *Leo*, which became *Windsor* and *Delmonico* as already described. In so doing, Pullman finally acknowledged the lack of success of these cars (the two dining cars excepted), which was because, as already observed, the British first-class traveller resented paying the supplement required to travel in them. It must be remembered that the Midland Railway had meanwhile been very active in improving the standards of comfort provided in its own first-class carriages. The more popular sleeping cars were still retained by Pullman, but the 'day cars' were sold at £1,600 each to the MR, and at the end of October 1883 this announcement was made in the newspapers:

'**Midland Railway**

'On and from 1 November 1883 Passengers holding First Class Tickets will be allowed to ride in the Pullman Day Cars Without Extra Payment. These Cars run daily between London (St Pancras) and Liverpool, Manchester, Carlisle, and Glasgow, respectively, and are warmed during the winter months. A special Attendant accompanies them, and they are fitted with Lavatory accommodation.

'Dining Cars are attached to the 5.0pm train from London (St Pancras) to Manchester, Liverpool, &c, and to the 5.0pm train from Manchester to London (St Pancras), into which Passengers from Liverpool &c, can change at Derby. No extra charge beyond the sum payable for the dinner will be made.

John Noble,
Derby, October 1883 General Manager'

With the removal of the Pullman supplement for daytime travel, it left only the sleeping cars on the MR to provide Pullman with revenue, under his contract with the railway company. Meanwhile, the MR decided to remodel some of the day cars (the ones originally built for, and sold direct to them) as dining cars, and so a further dining car was produced, named *London* in early 1884, to be followed by two more, without names, Nos 17 and 18. All three had six-wheeled bogies fitted in place of the original four-wheel type.

I have already mentioned in Chapter 3 how Pullman found the dining cars in America to be unprofitable, and opted out of the direct operation of them. One must presume that his attitude to the British cars was the same, because in 1885 he sold the original Pullman dining car *Prince of Wales* to the Great Northern Railway, who

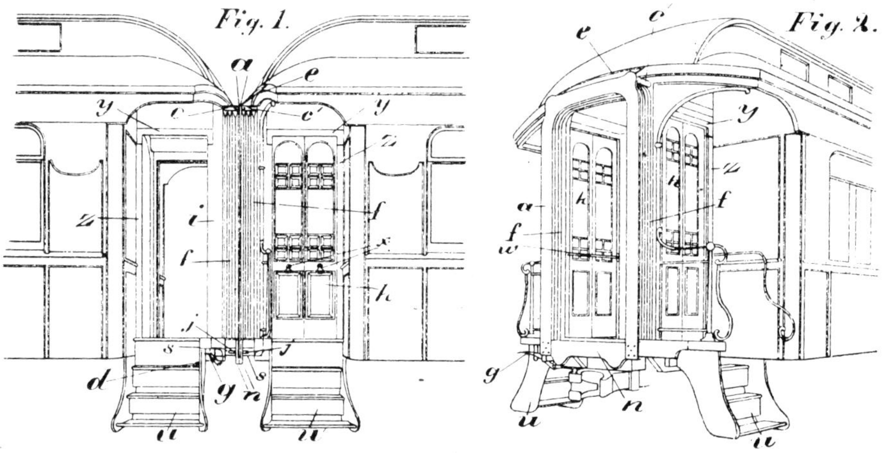

64
The vertical end frame design patented by H. H. Sessions of the Pullman Car Co in November 1887, shown in its original form, with narrow vestibule, small side entrance platforms and inward-opening jack-knife doors.
Author's collection

65

renumbered it as their No 2297 and painted out the name and all references to Pullman. It continued in service until 1901. The MR found *their* dining cars to be popular, and in due course more of the major British railway companies were more or less forced to follow suit, and to construct and operate dining cars but it is debatable if these cars actually made a profit for their owners! In some cases it was simply a matter of 'keeping-up with the Joneses' in order to prevent rival companies from gaining traffic at their expense, because they offered dining car or refreshment facilities. Pullman showed no particular further interest in them during his lifetime, although on the Brighton line, he had the car *Maud* (originally *Ceres*) rebuilt as a dining car in 1884, presumably to satisfy a specific demand from the LBSCR commuters.

In 1886, with only two years to run, the MR viewed the Pullman contract with apparent dislike and in that year the MR Carriage & Wagon Committee was already requesting that alternative designs of sleeping car should be prepared, to be owned by the MR and operated by them, and three different prototypes were authorised for construction in 1887: a 32ft sleeping car costing only £650; a 54ft version costing £1,300, and an intermediate 48ft version costing £900. (It will be recalled that the MR had to pay £1,600 each for the secondhand Pullman parlour cars and diners it purchased in 1883.)

On 18 February 1888 the Pullman contract with the Midland Railway, with respect to the sleeping cars, came to an end. It was not renewed, and (presumably at some inconvenience to passengers) all the Pullman sleepers were withdrawn from service for two weeks, whilst negotiations took place to purchase them outright from Pullman. These negotiations were evidently successful, because Radford reveals the following letter, sent to the MR Carriage & Wagon Committee:

'From John Noble, General Manager.
10 March 1888
Dear Sirs,
We have purchased the 14 Pullman sleeping Cars now running on the Midland line and are under an obligation to take off the Pullman name. I shall be glad if you will do so and substitute the word "Midland" for it, also re-number the cars and removing the names "St Mungo" etc by which they are at present known.
Yours truly,
John Noble'

The Pullman workshops at Derby, which had repaired the cars, as well as initially assembling them, or rebuilding them, were taken over by the MR, who continued to use them for maintenance purposes for many years. When the 15-year contract with the Midland Railway expired, Pullman had not succeeded in interesting the mighty London & North Western Railway or the broad gauge Great Western Railway (to name but two of the many other important pre-Grouping British private railway companies) in his services, and he turned his attention to the one company that did like him — the London, Brighton & South Coast Railway.

Three new Pullman cars were ordered in 1887 for the LBSCR from Pullman City, Illinois, and the LBSCR workshops at Brighton had the task of re-assembling them from their 'knocked-down' condition, because G. M. Pullman did not have a workshop of his own in Britain for the time being. Once again in terms of design, Pullman showed his superiority over the majority of British railway carriage engineers of the day, because the three new cars were not only lit by electricity (supplied by an improved axle-driven method, housed in a six-wheeled van), but they also had enclosed vestibule gangway connections between each car, allowing passengers to walk safely from car to car without exposure to the elements; whereas all previous Pullmans had featured the open platforms, (except the four small sleeping cars mentioned in the previous chapter) which *could* be crossed one to another whilst the train was moving, but at considerable risk to life and limb!

The vestibule, or entrance lobby, and enclosed car end, as opposed to open platform end, was not invented by G. M. Pullman, nor (as is also often claimed), by his superintendent at the American Pullman City workshops — Mr Henry Howard Sessions. It was already patented in America in June 1852 by Charles Waterbury who had been concerned to eliminate dust from entering passenger cars, as much as with making it safer to cross from one to another, and it seems to have been an arrangement of flexible curtains. However, he could not have found at that time a method to keep the two car ends sufficiently tight and close together, and the violent swaying probably wrecked his design (if in fact it was ever put to practical test). We have already seen in a number of contemporary accounts from passengers that the cars sometimes swayed and jolted

65
The Pullman car in the foreground has the original Pullman-Sessions vestibule end, with small entrance platform. It is coupled to a Stroudley Class B1 0-4-2 No 217 *Northcote*; one of the famous 'Gladstones'. *Ian Allan Library*

66
Another LBSCR 'Gladstone' heads a southbound express at Haywards Heath. Behind the engine is the December 1888 Pullman set, with the original narrow vestibules. The first vehicle is the 'Pullman Pup' No 80, followed by the parlour cars *Princess* and *Albert Victor* and the kitchen car *Prince*. Some small LBSCR carriages form the rear of the train. *Bluebell Railway Archives*

66

alarmingly over the indifferent quality American tracks, and the increase in speed made possible, as locomotives were developed which were larger and more powerful, made the end-play between cars a serious hazard. Sessions developed his 'vertical end frame' (or friction plates) to control this swaying motion at speed, and also to act as a buffer between cars (American cars did not have side buffers) in order to cushion the sudden to and fro jerking caused by starting the train, or bringing it to a sudden stop. In his patent, granted in 1887, Sessions disclaimed the vestibule as part of his invention. Moreover, his design was equally suitable for cars with open platform ends.

The first vestibules used with Sessions' vertical end frames (on some new Pullman cars for the Pennsylvania road in 1887) were only the width of the centre car door each end, and did not enclose the entire platform (see drawing). This was the type also at first used on the three new LBSCR cars and on the first of the two six-wheeled vans (already mentioned) which ran with them. This van, which was built in 1888, was styled outwardly to match the Pullman cars, and was designed by William Stroudley of the LBSCR. Given the number 80, it housed his improved method of supplying electricity for the cars: a belt-driven dynamo, with one of the axles giving the drive. The six-wheeler also housed the guard's accommodation, and a luggage area. A second van, No 29, was built in 1895. They became nicknamed the 'Pullman Pups', although the name Pullman was not carried on them, and the company did not own them.

The three new 1888 Pullmans for the LBSCR were joined together, semi-permanently coupled within the set, and officially called the 'Pullman Train No 1', even though it was *not* the first all-Pullman train in Britain. The three cars were named *Prince*, *Princess* and *Albert Victor*, and above their names on the sides they carried the legend 'Vestibule Parlor Car' (note the American, and old-English, spelling for parlour). They were each 58ft long and 8ft 9in wide, and undeniably splendid. A press demonstration was arranged for 10 December 1888, and an enthusiastic *Daily News* reporter afterwards wrote:

> 'The carriages are constructed on what is known as the "vestibule" system, the three waggons of which the train is composed having covered passageways between them. They are fitted with bay windows, and the interiors are finished in a style of unusual splendour. Mirrors are all around; the chairs are upholstered in old gold velvet; the windows are draped with crushed strawberry damask; the floors are covered with Wilton carpets; and the electric light is everywhere.'

In passing, one has to observe with amusement that he manages to refer to both 'carriages' and 'waggons', but not the proper term, 'cars'!

The three cars were mounted upon four-wheel bogies and were the last to be finished in the elaborate original American Pullman chocolate brown livery, with gold ornamental 'arabesque' panels on every available surface. Even the wooden bogie frames had gold lining-out applied. The interiors of the new cars were fully up to Pullman's luxurious ideals and must have made a startling contrast to the run-of-the-mill passenger carriages on the LBSCR! The car *Prince* had a small kitchen/pantry, but no bar. The space occupied by the kitchen/pantry was extremely small, and could only have provided facilities for keeping warm, or heating-up pre-cooked food, plus facilities for washing-up afterwards. It had a serving hatch to the saloon. Furthermore, it is not clear if the meals were prepared beforehand on the train itself, or in the station buffets at the termini and then kept warm in special containers, or perhaps reheated in some way before serving. Presumably the Pullman Co had some arrangement with the station catering contractors, if it was they who provided the food, already cooked. A first class portion had a number of 'cross-sections' (permanently fixed for day use) and next to this compartment was an open saloon with the usual swivel chairs of the parlour cars; finished in old gold velvet. However, these chairs were spaced only three feet apart, giving a density unusual in a Pullman. The car *Princess* was, as its name implied, the ladies drawing-room parlour car, with 19 swivel chairs and eight sofas, upholstered in blue velvet plush. The third car, *Albert Victor*, had a distinctly masculine air, with morocco leather chairs, and this was the gentlemen's smoker, complete with numerous brass spittoons (cuspidors, in polite society), no doubt to accommodate cigar butts.

On the Brighton line, where the Pullmans were popular, the season ticket receipts had grown from £129,000 to £189,000 over the previous 13 years, and they were to grow to £235,000 by the year 1899. But apart from these South Coast dwellers, who worked in the City, there was an enormous potential market in the opposite direction, because Brighton and the surrounding coastal area was a popular holiday resort, especially for day trips from London and back. The LBSCR and George Mortimer Pullman set out to improve the services, and develop the leisure traffic, as well as that of the 'captive' commuters.

The new vestibuled cars were a definitive success and must have done much to restore Pullman's faith in his British venture. Indeed, by 1890 the London & South Western Railway (LSWR) decided to try Pullman cars again, this time between London, Waterloo and Bournemouth, because Bournemouth was being actively developed as a rival coastal resort to Brighton, and a suitable luxury train from London would assist in attracting visitors. Two new cars were assembled at the LBSCR Brighton works and sent to the LSWR, with the names *Duchess of Albany* and *Duchess of Fife*, and these were used singly within normal LSWR carriage formations again making a striking contrast! It is possible that these two cars were intended for the LBSCR, because by 1890 that railway was already in need of more Pullmans, and only one new one *Empress* was introduced in May. The cars for use by the LBSCR and LSWR built in 1890 were finished in a dark bronze green livery, with gold lining: instead of the chocolate brown.*

These pioneering services to Brighton and Bournemouth certainly helped to make the holiday resorts popular, and the added luxury of a Pullman car was a novelty for many, who travelled in them purely for pleasure. On the Brighton line, as we have seen there was, however, also a regular clientele developing, and the Pullmans had taken on a new role, as a 'club on wheels'. From this public response Pullman was able to envisage the type of car and service that would suit the British traveller; and the Pullman car for daytime travel, as we now know it, began to be developed, along with a whole tradition of personal service and attention to the needs of the customer. What had started off as primarily a sleeping car business in Britain was now to focus its attention upon daytime travel, and upon the service of good food and other refreshments, in the most comfortable carriages that could be built.

Whilst Pullman was busy providing for the Brighton line, the London, Chatham & Dover Railway (LCDR) was planning a special London-Dover luxury service, to be run in connection with the 1889 Great International Exhibition, to be held in Paris, which was to be called the 'Paris Limited Mail'. A special steam cross-Channel packet boat was also to be built, called the *Calais-Douvres* and there would be a connecting French luxury train waiting for the passengers at Calais, to whisk them in comfort to Paris. For both the British and the French trains, the cars were supplied by Pullman's European rival, the Wagons-Lits company of Georges Nagelmackers. Four 'wagons-salons' were fitted-out as large saloons (without the more private coupés favoured in France) and the fifth vehicle was a special baggage van-cum smoking compartment car: a 'fourgon fumoir'. There was a form of vestibule, at the car ends, but not of the patented gangwayed Pullman/ Sessions design. Not to be outdone, the rival South Eastern Railway's (SER) Chairman, Sir Edward Watkin, also arranged with Wagons-Lits to provide a train of 'voitures-salons' for what became known as the 'Continental Club Train'. Thus there were *two* Wagons-Lits special trains running in the South of England to connect with only *one* similar luxury train in France. (The LCDR and the SER had to pool their Continental traffic receipts at this time.)

The two trains became popularly known as the 'Club Trains' but the British dislike of paying supplements, plus the fact that two trains was in reality one too many for the traffic available, meant that both were unsuccessful financially. One man who observed this at first hand was an English financier, Mr Davison Dalziel, who actually found himself one day to be the *only passenger* on the service from Paris! Arriving in England he chose the LCDR train to travel to London in. He was no ordinary passenger, however, because at that time he had just become a Vice-President of the Wagons-Lits Company and we will encounter Mr (later Lord) Dalziel again in Part 2 of our story. The 'Club Train' of the LCDR ceased running after 1893 and Behrend records that the Wagons-Lits cars in England were placed in storage for some years, in the hope that some alternative use could be found for them; but without success. They ended their days in France, on the 'Trains-Transatlantiques', the boat trains between Paris and Le Havre and Cherbourg.

Sir Edward Watkin apparently disliked Pullman's activities in Britain, but fancied the idea of running some similar luxury cars of his own, without a supplementary fare going to 'outsiders'. He chose a rival American railroad car-building firm, the Gilbert Car Manufacturing Co, of Troy and in 1891 the new cars were shipped over in knocked-down pieces (just as the Pullmans were) and the SER workshops at Ashford, Kent assembled them. There were six cars and they were put to work after a press demonstration on 2 March 1892, between London, Charing Cross and Hastings, on what became known as the 'Hastings Car Train'. Because of the very tight loading gauge restrictions in tunnels on the Hastings route, the Gilbert cars were oly 8ft 4¾in wide and 51ft 3½in long, and because Pullman and Sessions held the patents for vestibule gangway connections, the rival American firm still adhered to the open ends and platforms; with a primitive iron plate arrangement to permit crossing the gap between one car and the next, if necessary. They had typical British side buffers and screw couplings, unlike the sophisticated centre coupling arrangement that Pullman and Sessions were busy developing (and which will be referred to henceforth in this narrative as the 'buckeye coupling' and the 'Pullman gangway').

The Gilbert cars for the 'Hastings Car Train' were very elaborately furnished and decorated internally, and were clearly influenced by Pullman's designs. They provided an almost incredible standard of luxury at a time when the South Eastern Railway was noted for the low standards of passenger accommodation it provided. Of the six cars delivered, four had a main drawing room 29ft 10in long seating 18, including seven swivel armchairs arranged along the sides. There was a gentlemen's smoking compartment seating seven, and separate toilets for ladies and gentlemen. One of the two other cars was fitted out as a buffet car, with 28 places in a saloon 33ft long, and a kitchen compartment (using oil cooking apparatus, and equipped with a refrigerator). There was also a Baker heater (on each car) and an attendant's cupboard and a lavatory. The sixth car was apparently not fitted out when new and was intended for luggage, but in practice two SER brake vans with luggage space and guard's accommodation had to be attached to provide emergency braking facilities. On the initial press trip it was announced that the cars were capable of being used singly, or in pairs, with ordinary SER stock and would also run on the London-Dover route. The cars were electrically lit and the accumulators had to be re-charged at Charing Cross station, each night. We shall return to these cars later in this chapter.

The early 1890s witnessed, at last, a real effort to improve the ordinary carriages provided by the private British railway companies (no doubt as a result of the Midland Railway's example) and the larger companies began to introduce new rolling stock of various types — dining cars, sleeping cars, day cars, etc — which were better lit and more comfortable to ride in. Apart from the LBSCR Pullmans the use of vestibules did not commence until 1891, mainly using a flexible bellows gangway design together with side buffers and screw couplings. The through-corridor train first made its appearance on the Great Western Railway as a complete set with side corridors and concurrently the Great Eastern Railway produced a set of four six-wheel coaches gangwayed throughout for use on the 'North Country Continental' train from Harwich. One vehicle was a dining saloon and first and second class passengers were allowed to use it without supplement. Although the third class passengers were *not* admitted, one of the third class compartments (with side corridor access) was fitted with a folding table on which meals could be served from the kitchen, via the gangway.

The Great Northern Railway, having experimented with other types, adopted the Pullman gangway, but used retractable side buffers as well, so that their carriages were capable of being coupled to other stock, by the simple expedient of a hinged buckeye coupler, which when dropped revealed a conventional hook behind. In Britain at this time, however, the screw coupling and bellows-type gangway remained more popular (perhaps due to patent charges for use of the Pullman layout) and 50 years were to pass before British Railways finally standardised on the infinitely superior (and safer) Pullman gangway and drophead buckeye couplings; which are still in use today. A Royal Commission had recommended their use on all British railways in 1900, apparently to no avail.

The year 1891 saw a change of management for the British Pullman Limited, which since 1882 had been an off-shoot of the American company and had ceased to use the word Europe in its title. Mr J. S. Marks became General Manager in place of Mr Miller. Marks had been a Pullman Conductor on the MR services from 1874 until 1888. When he took office in 1891 there were only a dozen Pullman parlour cars in service on the LBSCR, plus two on the LSWR, and only five Pullman sleeping cars were still running: three on the East Coast main line,

*This was applied to new cars built for the LBSCR and LSWR until the year 1895. They had gold leaf lining and scroll-work. Roofs were rose-coloured when newly-painted, and bogies and undergear were dark brown, picked out in chrome yellow. It was not until 1900 that in America, Pullmans carried a standard green livery; perhaps influenced by the British initiative.

67

67
A striking view of the South Eastern Railway 'Club Train' the 'Paris Limited Mail', operated by Pullman's rival, the Wagons-Lits Co from 1889-1893, and representing a deliberate move by Georges Nagelmackers to penetrate the British *de luxe* train market once George Mortimer Pullman's Midland Railway contract had expired. The 'Club Train' cars were in an emerald green livery, on both sides of the Channel. The picture shows a Stirling 4-4-0 heading a lightweight load through Tonbridge in 1892, with the top-hatted stationmaster in the foreground. *LPC*

68

68
In 1891 the South Eastern Railway introduced the 'Hastings Car Train' using 'drawing room cars' built by Pullman's American rival, the Gilbert Car Manufacturing Co of Troy. They were shipped over in knocked-down condition and reassembled at the SER's Ashford Works. Illustrated is 'Gilbert' car No 32 which was later to be purchased by the Pullman Car Co and rebuilt as the Pullman car *Carmen*, which was destroyed in the Sevenoaks accident on 24 August 1927 (see Chapter 13). *Doug Lindsay collection*

of the standard type, plus the two smaller ones on the Highland Railway, of the 'inferior' type.

If George Mortimer Pullman's British affairs were at nearly their lowest ebb (and had gained no ground in Europe) then events at home were equally dispiriting for him at this time, and life at Pullman City was not the model one he had envisaged when he built it. Because his company had complete control of the town there were doubts expressed in some quarters about his altruism, and the fact that he would not allow bars, gambling saloons, or brothels within the precincts of Pullman City naturally upset those who controlled them in neighbouring Chicago! Discontent was stirred up by those with vested interests, and to Pullman's dismay the feelings against him grew stronger, until in 1889, as a protest against the rates he charged for gas, water and housing rent, the inhabitants voted in favour of annexation to Chicago. Then in 1893 came the American depression, and Pullman was forced to lay off some 4,000 men from his workshops. His workmen then joined the American Railway Union in large numbers as a result, and in May 1894 a wage dispute between Pullman and the union-led workers resulted in a strike, which then incited an even more serious major US railroad strike, with the banning of all Pullman cars, and mail and freight trains.

At Pullman City the strike became so disruptive that riots broke out and President Grover Cleveland sent in his federal troops to restore law and order, the first instance of this strike-breaking use of armed force in America. The union leader was subsequently jailed for six months, but the shock of this event broke George Mortimer Pullman's health, and he never fully recovered. An inquiry was held into the strike, and this ordered the Pullman Co to dispose of all its property not used, or intended for, industrial purposes, and that Pullman City be forthwith annexed to Chicago, as the inhabitants had previously voted. It was a personal tragedy for George Mortimer Pullman, whose religious ideals, and dreams of a welfare state for his workers, lay shattered by the riots and subsequent events.

The fading fortunes of the British Pullman sleeping cars had suffered a further blow on the night of 2 November 1892, when the sleeping car *India* was badly damaged in a collision at Thirsk, on the North Eastern Railway. The car was included in the second part of the up night sleeper from Edinburgh to King's Cross. At Thirsk, the signalman on duty fell asleep in his signalbox, because of exhaustion. His child had just died, and he was only on duty because the NER had failed to provide another man to replace him. As he dozed, perhaps only for a few fatal minutes, he failed in his duty to protect the sleeping car train, which ran into a goods train. As a result eight people were killed and 39 injured, but none of them were amongst those travelling in the Pullman car, who all managed to escape, although badly shaken. The solidity of the integral wooden Pullman body took the shock of the leading ordinary carriages (which were smashed) and

69

69
The 'Gilbert' cars for the SER Hastings service were extensively rebuilt in 1896, and enclosed vestibules and Stone's system of dynamo-driven electric lighting were added. The windows were altered, and a much more modern-looking train resulted, as seen here headed by Class F 4-4-0 No 183. The livery was crimson lake, elaborately decorated in gold leaf, including the SER four-wheel brakevans at each end. The leading car is a buffet car. *Ian Allan Library*

70

70
The interior of a first class saloon on the 1897 'Folkestone Vestibuled Limited' built by the Metropolitan Carriage & Wagon Co Ltd of Birmingham, England. An almost slavish copy of Pullman's ideas. *Author's collection*

cushioned the blow for its own occupants. Ironically, the gas escaping from the ignited lighting system of the other carriages then set fire to the car *India*, after its passengers (who included the Marquess of Tweeddale, Chairman of the North British Railway) had escaped. One of the chairs from the Pullman was recovered from a lineside field, and according to Behrend was later found again, used in an office at the Great Eastern Railway's Stratford locomotive works. Although not destroyed by the fire, the car was subsequently scrapped, as it was not considered worthwhile to repair it.

With only two Pullman sleeping cars left (*Columba* and *Iona*) on the East Coast main line, after the destruction of *India*, it was bad luck indeed when on 4 October 1894 another accident on the NER at Northallerton (not all that far from Thirsk) involved the car *Iona*. Happily the body of the car was not badly damaged, despite taking — once again — the full blow, and it protected the carriages at the rear. Amongst the shaken but unhurt Pullman passengers were two cabinet ministers and the Deputy Chairman of the Great Northern Railway. *Iona* was repaired and went back into service with *Columba*. Both were sold by Pullman to the joint ownership of the GNR and NER on 1 January 1895, and they continued running on the East Coast main line as part of the East Coast Joint Stock (ECJS). *Iona* became ECJS No 232, and then later GNR No 2992 and was converted into a dining car in 1902. *Columba* became ECJS No 233 and then, also in 1902, it was converted into a 'refreshment car' by the NER who numbered it their 2966. (We will meet these two cars again briefly in Part 2). With these Pullman sales in 1895, only the two HR 'inferior' sleeping cars were left in Pullman's fleet in Britain, all other cars being for day use.

Pullman's South of England activities had meanwhile continued to give more satisfaction, and the two railway companies each received new cars to strengthen their services in 1893. For the LSWR, two cars were assembled from prefabricated sections at their Eastleigh workshops (the only ones ever to be assembled there), and these were named *Princess Margaret* and *Duchess of Connaught*, both were parlour cars. The LBSCR received three new cars, assembled at Brighton at about the same time. These were named *Pavilion, Princess Mary* and *Prince Regent* and were also parlour cars (ie with saloons and no kitchen). In 1895 a further three were assembled at Brighton for the LBSCR and named *Princess of Wales, Duchess of York* and *Her Majesty*. The car *Duchess of York* included a kitchen; the other two were parlour cars.

It was probably the continuing public enthusiasm for Pullmans on the Brighton line that provoked the introduction of some more rival cars on the South Eastern Railway in 1896, and the existing Gilbert cars were also remodelled extensively the same year. Dealing with these first: the changes included making provision for all three classes of passenger (instead of just first class) and provision of two lavatories for each car. The open ends were replaced by enclosed vestibules with gangway connections and the windows were altered along the sides. They were once again exhibited to the press, who were led to believe that they were *new* vehicles! The true new vehicles were British-built, with one exception, and these were intended for an all-car train called the 'Folkestone Vestibuled Limited'.

The Folkestone cars were built in Birmingham by the firm which is nowadays known as Metro-Cammell Ltd, and they were larger than the Gilbert cars, and so could not be allowed on the Hastings route. Eight cars were built, and a centre-coupler arrangement was used for the complete set of cars, which were each 58ft 6in long. In external appearance they were distinctly American, except for the two end cars, which had brake van accommodation; and where the guard was housed the roof had the typical SER raised 'birdcage' to give him a view of the signals ahead. To provide this feature, the clerestory roof needed for the passenger accommodation had to be terminated, and a lower flat portion placed over the luggage space. The bogies were of English design, known as the four-wheel Fox's patent pressed steel type; with four pairs of transverse semi-elliptical springs.

71

71
Stroudley's beautiful 'G' class 2-2-2 'Single' No 329 *Stephenson*, of the London Brighton & South Coast Railway heads a down Pullman train near Tooting, London. The leading 'Pullman Pup' is the second one built, No 29 of 1895 with vestibules to full width, and the three Pullmans are *Her Majesty* (a parlour car), *Duchess of York* (a buffet car), and *Princess of Wales* (a smoking car) furthest from camera. This train was formed in October 1895 and was known as 'Pullman Train No 2'. Ordinary LBSCR stock formed the rest of the train in this picture. *L&GRP*

A curiosity was the ninth car, which appeared in 1896, because it was ordered from America, from Jackson, Sharp of Wilmington, Delaware, who had constructed many Pullman cars, before the advent of Pullman City; it too ran with English four-wheel steel-framed bogies fitted. It was 56ft 5½in long, 8ft 4in wide and seated 20 first class passengers. Because of the restricted width of Hastings line gauge, it had brass bars across the outside of the windows (a feature later also applied to the rebuilt Gilbert cars), to prevent accidents to passengers putting their heads out; a practice which was strictly against the rules and regulations! All the SER cars were painted in their standard crimson lake livery, but embellished with elaborate gold leaf decoration; after the style of Pullmans. The Jackson car has something of a mystery surrounding it, because in 1898 the *Locomotive Magazine* clearly stated that it was running as a Pullman car, when they published the following list:

'SER Car Trains

I. THE AMERICAN CAR TRAIN.—London-Hastings Service

No. 286.	Six-wheeled brake, ordinary S.E.R. build, but painted to match the cars.	
35.	Second-class car.	These cars vestibuled throughout.
33.	Third-class car.	
34.	Third-class car.	
32.	Third-class car.	
36.	First-class car.	
47.	First-class car.	
283.	Six-wheeled brake same as No. 286.	

This train leaves Charing Cross at 3.35 p.m., Cannon Street at 3.45 p.m., each week day but Saturdays, when it leaves Charing Cross at 4.50 p.m., returning from Hastings 8.40 a.m., except on Mondays when it starts at 7.00 a.m.

II. THE VESTIBULED CAR TRAIN.— London-Folkestone and Dover Service.

No. 208.	Third-class car & luggage.	Vestibuled throughout.
207.	Third-class car.	
206.	Third-class car.	
203.	Second-class car.	
205.	First-class car.	
205.	First-class car and buffet.	
204.	Third-class car.	
201.	Third-class car & luggage.	

This train leaves Charing Cross daily at 4.28 p.m., Cannon Street 4.36 p.m., and London Bridge 4.39 p.m., running fast to Ashford. On the up journey it leaves Dover at 8.32 a.m., and Folkestone (Central) at 8.55 a.m.

III. THE TUNBRIDGE WELLS PULLMAN.
No. 171. First-class car only.

Down leaves Charing Cross at 3.44 p.m., up leaves Tunbridge Wells at 9.03 a.m.'

If this 'one-off' American-built car was ordered specifically for the Tunbridge Wells route (being to the necessary restricted loading gauge) then perhaps it was afterwards leased to the British Pullman Limited, who, no doubt, would welcome a foothold on rival SER territory? Or did the SER audaciously steal the use of the name Pullman, soon after the great man himself had died?

The 'Folkestone Vestibuled Limited' made a press demonstration run to Dover on 22 September 1897, and went into regular service to and from Folkestone in October. On the 19th day of that month, in a Chicago street, George Mortimer Pullman had a sudden heart attack, collapsed and died. He was 66 years old.

Many epitaphs could be written about the man who personally brought luxurious comfort and style to the railways, and who created a tradition of quality service to the travelling public in his own lifetime, but no

72

72
Wainwright's Class D 4-4-0 No 733, heading the 'Folkestone Vestibuled Limited' in about 1905. Although American in style, the cars had been built in Britain for the South Eastern Railway in 1897. They passed to Pullman ownership at the end of World War 1.

finer tribute can exist than that represented by all the cars themselves, that have carried his name from 1859 until the present day. A Yorkshireman travelling in America in 1891 had this to say (the italics are mine):

'May 14 1891. I left Philadelphia at twelve noon by the Chicago Limited express train on the Pennsylvania line for Chicago, a distance of 822 miles. This train is timed to reach its destination at 10 a.m. The following day, the local time being one hour later than New York. It is what is called a limited train, which means that it consists of four Pullman sleeping-cars, a smoking-car, a dining-car, an observation-car and a composite-car, the last named having a compartment for luggage, a place for the mail and sleeping berths for the conductors. The smoking-car has a reading-room for the passengers, library, easy chairs, writing and card tables and a barber's shop. The traveller can have his hair cut, curled and shampooed for half a dollar, a shave for a quarter, or a bath at the rate of forty miles an hour for seventy-five cents. *The cars are designated palace-cars and they fully deserve the title, for they are indeed travelling palaces where one can obtain all the comforts of home.*' (From *A Yorkshireman's Trip to the United States and Canada*, William Smith, London, 1892.)

Like the Earl of Sandwich, who gave his name to posterity to describe a simple meal, the name Pullman lives on, as a symbol of excellence in railway travel, and quality in life — of *travelling in style*.

Robert T. Lincoln, son of the late American President, succeeded George Mortimer Pullman as the head of the Pullman Co (the use of the word 'Palace' had been dropped) and he continued to develop the company. Our own story is henceforth essentially about developments in Britain, but it is only fitting to make a brief survey of the Pullman Co in America after Pullman's death, at this point in the book.

At the time of his death, only the Wagner Palace Car Co still rivalled his interests. This was bought out by the Pullman Co in 1899, and American competition disappeared. From then until 1927, when the organisation was changed to become Pullman Incorporated, growth and development was steady. Pullman Incorporated had two subsidiaries, the Pullman Co which operated the cars and the Pullman Standard Car Manufacturing Co to build them. In fact, the company built many other things besides Pullman cars, and there were no less than six big workshops operating as business grew: at Bessemer, Alabama; Worcester, Massachusetts; Michigan City and Hammond, Indiana; Butler, Pennsylvania; and the Pullman Chicago Illinois Workshop. Pullman Standard manufactured all types of railway carriages and freight wagons, tramcars, buses, track maintenance equipment and even boats, and trailers for lorries.

Business continued to flourish without interruption until the outbreak of World War 2, and then in 1940 the company was accused of being a monopoly (in which respect it had fulfilled the late George Mortimer Pullman's ambitions!). The US Government brought a suit against Pullman Incorporated under the Sherman anti-trust law, at the United States District Court in Philadelphia, on 8 May 1944. The company was told it could either continue as a builder, or as an operator, but not both. Pullman Incorporated elected to dispose of the sleeping car business and made an offer for the sale of it to a group made up of the railroads on whose lines the majority of the sleeping cars were running. The offer was accepted by the consortium of 59 railroads, and approved by the court on 18 December 1945. It took some two years to come into operation, and was presided over by Carroll Rede Harding of the Southern Pacific railroad, who had played a leading role in the purchase of the Pullman Co. It continued to operate sleeping cars and other on-board services for 21 more years, until the various railroads took over the services in their own interests, to save money. The last Pullman service in the United States ran on 31 December 1968. In 1971 Amtrak (the National Railroad Passenger Corporation) was formed and now operates these services.

The Pullman Standard Car Co ceased to build railroad passenger cars in 1981, when the last of an order for 284 cars for Amtrak's 'Superliner' services was delivered. The very last car to roll off the assembly line at Pullman, Chicago, Illinois, before it was closed down, was given the name *George M. Pullman*. it marked the 150th anniversary of his birth, and it was, very appropriately, a sleeping car. To this day if you mention the name Pullman to an American he automatically thinks of sleeping car travel; whereas in Britain the name has nowadays become synonymous with comfort, and dining in style.

73
On 19 October 1897 George M. Pullman collapsed in a Chicago street, and he died shortly afterwards. At the time, his Pullman City workshops were busy on the construction of four sets of cars for the 'Pennsylvania Limiteds'. When these appeared in January 1898 they were publicised as the 'newest and most complete Railway Train of this progressive age'. It was indeed a magnificent collection of cars, as exemplified by the sleeping car *London*, a massive 12-wheeler of handsome proportions. The livery was a 'Brewster' green below the waist, with cream at window level and a red fascia board above, with all the lettering in gold leaf, as well as the decorative motifs. The oval window for toilet and pantry accommodation was a feature which was to become characteristic of British Pullman cars for the next half a century; although the Americans soon abandoned it themselves. *Pullman Car Co*

73

CHAPTER 11

A New Door Opens

In Britain, the traditional ceremonies that heralded the New Year in 1900 also marked (although it could not have been realised at the time) the end of the truly golden age of steam railways, when they had reigned supreme as a mode of passenger and freight transport. Henceforth, the embryonic motor vehicle and the powered aeroplane were to be developed as two alternative methods of travel, and developed moreover with quite astonishing speed. Whereas the railway companies were used to rivalry between themselves, they were now to face rivals of a completely new nature, which represented a threat in common to them all.

The steam locomotive had been developed in the latter part of the 19th century into a magnificent machine of considerable power and efficiency, capable of moving several hundred tons of payload over long distances, and at speeds never before achieved on land. The carriages they hauled had became steadily larger, longer, heavier, and more comfortable to travel in. The British railway companies, whilst for the most part having refused to accept the luxurious Pullman cars on their tracks, had nevertheless been influenced by them and some almost slavish copies of his ideas appeared in service, in particular as dining cars and 'boat expresses', around the turn of the century. A random glance through copies of the *Locomotive Magazine* of 1898 produced these brief reports on new carriage progress — in practically every case a development of the pioneering work of George Mortimer Pullman, with features he had introduced to Britain, such as bogies, vestibules, electric lighting and sleeping and dining facilities:

'The Carriage Department

'The sleeping cars of the Great Western Ry. have bodies 56-ft long and 9-ft wide, mounted on steel underframes, running on two four wheeled bogies with wheels placed 10-ft centre to centre. There are eight single berths, each in a separate compartment, placed transversely of the car, and to which access is obtained from a side corridor running from end to end and connecting the central vestibules. There is but one door on each side at the centre of the car, opening into an ordinary compartment, the two seats of which form couches which may be converted into berths at a pinch, bringing the total to ten. Between each compartment is a lavatory with a door on either side to the adjacent berths; there are four lavatories per car, a number which we should think could be reduced by two without inconvenience.

★ ★ ★

'The LC&D Ry have now in service a total of 12 corridor carriages 1st and 2nd class with lavatories, and running on four-wheeled steel framed bogies, they have also four brake vans on bogies.'

Included is this reminder that the poorer class of traveller on Britain's railway was still generally denied the use of the dining car:

'Restaurant cars will be attached to several of the fast main line trains of one of our big railways during the coming summer. All passengers, irrespective of class, will be able to take breakfasts, dinners, teas or other refreshment as they may wish. Such a service is general in America, quite common on the Continent, and we hope it will become an institution here.

★ ★ ★

'Several of the new vestibuled cars for the GC Ry have now been delivered at Gorton. They are painted in the new colours proposed for the line some time back, viz: brown lower panels and greenish grey upper, with gold lines and lettering. Internally they are very handsomely furnished and some of the decorations might almost be termed lavish. In anticipation of the running of these magnificently equipped trains on the through Manchester—London expresss, the L&NW and GN Rys are already looking to the conveniences of their stock on the same services. The former Company's trains are practically all vestibuled cars 1st, 2nd and 3rd class, with dining cars attached, and the latter are gradually coupling the corridors of the "diners" to the adjacent cars. *A little competition soon brings our railways up to the scratch.*'

The italics at the end of the last report are mine.

Meanwhile, the *Locomotive Magazine* did not neglect to keep British readers informed of Pullman activities, and those of others elsewhere:

'Four trains of new Pullman cars have been recently put into service on the Pennsylvania Rd. for their famous "Limited." They are painted in a new style, with the lower portion below the windows dark green, and the upper panels cream colour; the lining and lettering is in gold.

★ ★ ★

'The Western Railway of France run on some of their long distance night trains, 1st class carriages with sleeping berths arranged, very similar to those in the Pullman cars, but placed transversely across the compartment; two of these fold up at an angle by day, and the remaining lower ones form the seats. These carriages are very short, running on four wheels only, and have no attendants or corridor.

74

Still looking fine, three ex-Pullman parlour cars of the Midland Railway are included in this lengthy train of empty stock headed by the Johnson 0-6-0 No 1365, photographed near Hellifield c1900. By this time Pullmans were no longer regarded as front rank MR passenger accommodation and some ended their days as 'picnic saloons', or as third class coaches. *L&GRP*

74

For the larger types of new carriages, the raised clerestory roof was back in fashion, and was made extremely decorative on the inside of the carriages, which were often mounted upon six-wheel bogies, if dining or sleeping cars. Through gangways and corridors now allowed everyone access to toilet and other facilities, but the British still liked the compartment layout, so that the corridor was arranged along the side of the compartments, except for dining purposes.

This is a convenient moment to take a brief look again at the former Pullman cars which had been sold by the company to the Midland, and others; the majority of which were still in existence at this period, although stripped of their former colourful identies. J. B. Radford tells us that the beginning of the end of main line service for the Midland drawing room cars was the great coal strike of 1893, when their heavyweight bodies made them undesirable trailing loads for steam locomotives trying to conserve the fuel in their tenders. They were taken out of service and stored, but concern was then expressed that they would suffer rapid deterioration in storage because of damp. As Thomas Clayton reported to the MR Carriage & Wagon Committee on 5 January 1894, the cabinet work was of 'that description which required to be kept dry and taken care of'.

At that time, one car was 19 years old, eight cars were 18 years old and four cars over 17 years old, and all had been taken great care of, so that the internal cabinet work was still in very good condition. A start was therefore made to convert some of the drawing room cars into picnic saloons. In due course the remaining ex-Pullman sleeping and dining cars were also converted into similar saloons, and they continued to be used for special excursion trains, for private hire, and the picnic specials which were popular family outings, in the days before the private motor car and the charabancs took much of this leisure traffic on to the highways of Britain. The old Pullmans must have been very agreeable to travel in as picnic saloons, giving a great sense of style to the occasion!

75
A former Pullman sleeping car, still in use as such by the MR around the turn of the century, and seen at the head of a train which included old six-wheel carriages and (at the far end) recent clerestory bogie stock. A handsome Johnson 'Single', No 130, makes a fine picture at the head of the train near Mill Hill. *L&GRP*

The two former Pullman sleeping cars *Columba* and *Iona* which had been taken into East Coast Joint Stock in 1895, assumed new roles for their respective owners in 1902, when *Columba* was converted into a 'refreshment car' by the North Eastern Railway and *Iona* was converted to a dining car by the Great Northern Railway, who gave it an odd arrangement, having a new vestibule gangway connection at one end whilst retaining the original open verandah at the other. It had two sections seating 16 and four respectively, and the end of the car which was fitted with a gangway connection was also fitted out with a small kitchen and pantry. The NER refreshment car had a different layout, in which in addition to the main saloon there was a small smoking compartment, a kitchen with a cooking stove and other conveniences, a lavatory and a cupboard for passengers' luggage. Only first class passengers were normally catered for, but the third class passenger was admitted on payment of a supplement of 2s 6d (12½p). The car ran in services between Leeds and Scarborough as a breakfast car in the morning and a tea car on the later afternoon return journeys. It was downgraded to a third class saloon in 1907 and was frequently used by Newcastle United Football Club when travelling to away matches, until its withdrawal from service on 3 November 1910. The GNR car is believed to have ended its days in use as a buffet car on the Nottingham-Skegness services, and it was withdrawn much later, in October 1925. Even then its days were not over, and it rested at Lincoln minus its bogies, as a 'grounded body', serving as a Mutual Improvement Classroom at the engine sheds. It was finally broken up about 1970. A number of the vintage MR and LBSCR wooden Pullmans were eventually 'grounded', and remained in existence for a great many years, as will be described in Chapter 19.

Although the Midland had ended its contract with Pullman, and had continued to run the sleeping cars under its own control, the need for four new cars led them to purchase once again from America,

76

77

where both the Pullman Co and the Wagner Car Co were invited to tender for their design and construction, the latter declining to do so. This move was probably due to an overload of work at the MR Derby drawing offices, and in the workshops; although no doubt the MR was also keen to see at first-hand the latest improvements offered by the Pullman Co, which was constantly refining the design of its cars. The Pullman Co delivered a specification for the construction of either two or four sleeping cars at a price of $15,000 each, with the provision that facilities had to be made available at Derby for their assembly, and that the bogies, running gear and buffers and couplings should be provided by the MR to suit their needs. The offer was accepted for four cars in March 1899, and construction was so swift that by 6 April 1900 the cars were completed and ready to enter traffic.

The new cars were built to the current standard Pullman design, modified to suit the British loading gauge, and each had sleeping accommodation for only 11 passengers; there being no upper berths. Five of these berths were in separate compartments (two doubles and one single) with access from a side corridor. The remaining six berths were to the standard Pullman lower berth section arrangement consisting of a pair of facing seats, convertible into a single berth for night-time use, three of these being placed each side of a central gangway. A gentleman's toilet was provided at one end, together with a small kitchen compartment from which the attendant could supply light refreshments to order. At the other end was a ladies' toilet compartment and a standard Baker car heater enclosure. The vehicle ends were rounded, with a vestibule having side entrance doors, but, initially at least, there were no gangways. Oil gas lighting was fitted, with lamps on either side of the clerestory of the roof. The livery was standard MR crimson lake, and there was of course no mention of Pullman, just the legend 'Midland Sleeping Carriage' and the number (Nos 34-37 in the former MR Pullman car sequence). They went into traffic in May 1900 on the night expresses between St Pancras and Glasgow and Edinburgh, displacing some of the older sleeping cars of Pullman origin.

76
This was formerly the Pullman sleeping car *Iona*, assembled at Derby in 1880 for the Great Northern Railway. When sold it then became part of the East Coast Joint Stock, as No 232 in 1901, and in 1902 was converted to a dining car by the GNR, as seen here. The elaborate Pullman livery detail has been retained and the number 2992 is carried in the oval nameplate. Above the windows is the legend 'GNR Dining Saloon' in the Pullman-style lettering. A curious feature was the addition of a vestibule end and gangway connection at one end only! Note the gas cylinders under the car body. The bodywork lasted until about 1948, grounded at Lincoln and in use as an LNER staff classroom.
National Railway Museum

77
A Pullman sleeping-car, American-designed and built, but never used with a Pullman supplement. This picture shows No 35 of the Midland Railway, one of four sleeping cars supplied in 1900. The six-wheel bogies, buffers and drawgear were of standard MR design and fitted to the cars when re-assembled at Derby. The livery was the standard MR crimson lake, and the legend above the windows read 'Midland Sleeping Carriage' (instead of 'Car'). No end gangway connections were fitted when new, but Pullman style vestibules with inward opening doors were retained. *BR*

Meanwhile the fortunes of the British Pullman Limited (now controlled from America by Robert T. Lincoln) rested mainly upon the success of the London, Brighton & South Coast Railway's operations with them, which continued to flourish. In 1898 the LBSCR had recommenced, on 2 October, a Sundays only all-Pullman train called the 'Sunday Pullman Limited', changed in 1899 to the 'Brighton Limited'. It was first class only, and the use of the word 'Limited' indicated strict accommodation control — ie no standing passengers were allowed, and every seat was sold at a supplementary fare.

Running on Sundays only, it was a train for pleasure-seekers and one can only remark that a great change had taken place in Victorian morality since 1881, when the LBSCR's use of Pullmans on a Sunday had been described as the 'Sabbath Breaker'! The original formation was the Pullman cars *Victoria*, *Beatrice*, *Princess of Wales*, *Duchess of York* and *Her Majesty*, all fitted with vestibules, with the two six-wheeled 'Pullman Pups' at the ends to provide electric lighting as well as brake van and luggage space. The train became so popular that it was necessary to add further cars. Two new cars, *The Arundel* and *The Chichester* were built for it in 1899, with a third one *Devonshire* following in 1900. These had six-wheeled wooden-framed bogies, and their wooden bodies were 64ft long. To haul this impressive load, the new 4-4-0 express passenger locomotives of R. J. Billinton's design were employed (Billinton had succeeded Stroudley at Brighton), and on the 1898 inaugural run the Billinton 4-4-0 *Smeaton* had made history by being the first locomotive to be scheduled to cover the 50.9 miles from London to Brighton in 60 minutes. What a pleasant day out it must have been, whisked along in splendid comfort for just one hour, to reach the bracing sea air of the South Coast, and then to be returned home again in London in time for dinner!

Sad to relate, the Billinton 4-4-0 *Smeaton* made the headlines again in 1899, when on 23 December it ran into the back of the up Newhaven boat train, in dense fog, at Wivelsfield station. The train from Brighton that it was hauling included two Pullman cars, *Maud* and the new car *The Arundel*. The boat train (which was moving slowly when hit) included one Pullman *The Queen*, which was undamaged, and so was the new car *The Arundel*, but *Maud* received severe damage to the buffet end, and the adjoining saloon portion of it was destroyed, as were the two leading LBSCR saloon carriages, marshalled between the locomotive's tender and *Maud*. Three people were killed instantly and three died later from their injuries. To take the injured up to London the car *The Arundel* was used, plus the carriages behind it, which were undamaged. The damaged car *Maud* was subsequently withdrawn by the Pullman Co, but once again the solidity of Pullman's design had probably saved a number of lives, as well as saving the adjacent car, *The Arundel*.

Apart from the Pullmans for the Brighton services and the Waterloo-Bournemouth cars, there were only the two HR small sleeping cars left and the fleet of British

78

79

78
The magnificent 12-wheeler *Devonshire* of 1900 is seen here standing in Brighton station, after a few years of service, to judge from the rather sombre green and gilt livery finish (despite evidence of careful hand-polishing in the reflected light). Perhaps the most elegant Pullmans ever to see service in Britain, these 12-wheelers had remarkably large bodyside windows, for their period and they featured the oval toilet and kitchen windows which were to become synonymous with Pullman cars. The wooden-framed six-wheel bogie design is well portrayed, and the large buffer housing is a noticeable feature. The Pullman car dwarfs the adjoining LBSCR compartment stock!
National Railway Musuem

79
Pullman car *The Arundel* seen in the Lover's Walk carriage sidings, Brighton in the early 1900s, with four of the windows open (upward-opening because of the wooden framework below the waistline in the body construction). Between the massive six-wheel bogies there is a mass of equipment carried for the electric lighting and the gas cooking facilities. From 1900 onwards the Pullman cars were noticeably longer and heavier than their Victorian counterparts, and the 12-wheel arrangement was found to give a very comfortable ride to these majestic vehicles.
National Railway Museum

80
Pride of the British Pullman Co at the turn of the century was the Sundays-only all-Pullman 'Brighton Limited' on the LBSCR. The southbound train is seen near Tooting Bec headed by Class B2 4-4-0 No 202 *Trevithick*. The formation is: LBSCR-owned 'Pullman Pup' dynamo/luggage van No 29 (1895); the 1895 three-car Pullman set, *Her Majesty, Duchess of York* and *Princess of Wales*; one of the 1899/1900 12-wheelers (*The Arundel, The Chichester* or *Devonshire*); *Beatrice* (1877); *Victoria* (II) (1876); and 'Pup' No 80 (1888). The last three vehicles had by this time been modernised with wide vestibules — nevertheless it is noteworthy that cars nearly a quarter-century old were still included in the premier Pullman service. *L&GRP (21175)*

80

Pullmans remained greatly diminished during 1900-1906. There were actually more *ex*-Pullmans in service in Britain, plus the rival cars (SER), than the Pullman Co itself possessed. By contrast, in America the Pullman fleet had grown steadily, and was already covering more than 91,000 route miles, involving 61 railroads, by 1898. The Brighton line remained the star in an otherwise overcast British sky, and in 1906 three more new Pullmans were delivered from America, which will be described anon.*

The Midland Railway found one further use for four of its old parlour cars in 1905, when they were converted into 'push and pull' trailers, coupled to small 4-4-0 tank locomotives borrowed from the Midland & Great Northern Joint Railway (M&GNR). It was partly in a quest for economy, and partly in response to local requests for a better service, in particular on the Derby-Melbourne, Derby-Ripley and Derby-Wirksworth branches, as well as the London area branch to Hemel Hempstead. Conveying third class passengers only, and running in addition to the conventional branch line trains (which still had to be powered by larger locomotives, capable of towing the milk vans and other non-passenger traffic) the 'autocars' ran until 1912. One end of the old wooden Pullman body had to be rebuilt with a driving compartment with front windows for when the car was being propelled by the tank

*It should be mentioned at this point that in addition to the all-Pullman 'limited' trains, the LBSCR also operated Pullman cars singly, or in pairs, within the normal carriage formation, on a number of services.

81

82

81
The Midland Railway found a further lease of life for a few of their old Pullmans, by converting then for use as 'autocar' or 'push-pull' trains. Here is the former parlour car *Minerva* in use as the driving trailing, with a Midland & Great Northern Joint Railway 4-4-0T No 10 as the power, photographed at Wirksworth, Derbyshire, in 1906. The box on the clerestory roof end housed the pulleys and chains which ran down the length of the car's ceiling from the cab of the locomotive to the driving cab built into the far end of the Pullman body, where it was formerly an open verandah. *Real Photos (72047)*

82
The three new American-built Pullman cars *Duchess of Norfolk*, *Princess Ena* and *Princess Patricia* with LBSCR Class H1 Atlantic No 38, and two six-wheel brake vans, in matching livery, photographed at the Crumbles siding, Eastbourne in 1906. This was to demonstrate the new livery of umber brown and cream then introduced by the British Pullman Co Ltd. The cream panels emphasise the two oval windows, and the scheme is noticeably smarter than the previous livery. Mr Davison Dalziel, the prospective new owner of the Pullmans in Britain, was present on this occasion, and another photo exists showing him in the foreground. Note that there are no Pullman coats of arms on the car — these were not introduced to the new livery by Dalziel, until some time after he had completed the purchase of the company. *Charles Long collection*

83
The interior of *Princess Patricia*, one of the last three American Pullmans to be built for service in Britain, and undeniably sumptuous in a decor of greens and gold. Electric lighting pendants with frilly lampshades form the ceiling centrepieces, whilst the clerestory is elaborately disguised (as in

83

84

the 1900 sleeping cars for Midland Railway) and formed into sections. The chairs were pivoted. *Author's collection*

84
Loose, leather-upholstered chairs of more masculine style, were a feature of this car of the final 1906 American-built batch for service on the London, Brighton & South Coast Railway — the last three Pullman cars to be built with clerestory roofs, for service in Britain. *Author's collection*

engine. The system of cables and pulleys which operated the remote control between the driver (when in the leading car), and his fireman (on the locomotive footplate) was housed in the clerestory roof, over the heads of the passengers, and according to one contemporary report produced a 'wondrous rattle' when in use.

In 1905 the Midland Railway rebuilt the four Pullman-built sleeping cars it had purchased in 1900, giving them private berths throughout, with one double berth, seven single berths and one smoking compartment. The British preference for privacy when travelling, especially at night had finally won the day, and British sleeping cars built since then have been to the private compartment plus side corridor layout.

The British affairs of the giant American company were, in truth, insignificant when compared to the revenue it gained in North America and elsewhere in the early 1900s, and if the Brighton line operations had not been successful, one suspects that Robert T. Lincoln would have closed down the London office in due course (where Thomas Powell had succeeded J. S. Marks as General Manager in 1904). One person who viewed the British Pullman Limited with some interest however, was Mr Davison Dalziel (whom we met in Chapter 10) and who later became Lord Dalziel of Wooler. Dalziel had become 'President du Conseil' of the Wagons-Lits Co around this time, at the wish of Georges Nagelmackers, who died in 1905. He saw the possibility of acquiring the British activities of the Pullman Co, and in 1906 he began to negotiate with the trustees of the late George Mortimer Pullman, with a view to purchasing it, to operate it as his own private company.

During his lengthy negotiations with the trustees three new Pullmans arrived in crates from America, and were assembled at Brighton in January 1906. These were named *Duchess of Norfolk*, *Princess Ena* and *Princess Patricia*, and they were the last to be built with clerestory roofs, and the last to be imported into Britain. They also introduced a new livery for Pullmans in Britain, because they were painted in the now famous two-tone umber and cream scheme.* This was in order to match the umber brown livery for LBSCR locomotives and coaches recently introduced by Mr D. E. Marsh, the LBSCR Locomotive Superintendent, who succeeded R. Billinton in 1905, at Brighton.
Mr Davison Dalziel was photographed alongside these new Pullman cars, at Eastbourne, at the time he was negotiating the purchase of the Pullman company. He was soon to be their proud owner, because towards the end of 1907 he successfully negotiated his outright purchase of the British Pullman company, with the goodwill and rights to continue to use the name of Pullman.

*The *official* USA specification, which the author has read, courtesy of Mr Arthur Dubin, clearly states CREAM. Subsequently the colour was sometimes described as Ivory White in railway literature, perhaps the shade was lightened, or perhaps it was just literary licence!

TABLE ONE:

Pullman Cars in Great Britain 1874-1906

List of Pullman Cars and US Pullman-built Cars in service in Great Britain before 1906

To provide this list and the list on pages 151-156, I have compared the researches of Messrs J. T. Howard-Turner, George Behrend, J. B. Radford, Julian Morel and others, who have so painstakingly pieced together the remains of the records that Hitler's bombs so nearly obliterated at London, Victoria station in 1940 when they hit the offices of the Pullman Car Co. It has to be said that there are certain contradictions in these various lists, and in so far as I think possible — or probable — I have combined their researches to make a feasible whole. However there remain some shadowy areas of doubt, and contradiction, which cannot be clarified on existing evidence. Two in this first list in particular must be mentioned:

Julian Morel records the fact that in 1959 a railwayman wrote to the Pullman Car Co seeking some information about an old Pullman Car, off its bogies, together with two others, still in use as sheds at Birmingham, Saltley carriage sidings. There were still traces of brown livery with gold, and engraved windows, a plush ceiling, double flooring, and a name, just discernible — *Cynthia*. No trace of this American-built sleeping car has been discovered in remaining records. It does not feature on the Midland Railway lists as published by J. B. Radford, so for whom and for what service was it built? Or perhaps *Cynthia* was a name applied at some stage to one of the existing cars, and overlooked by the office staff that kept the records? (The other two cars mentioned by the railwayman revealed no traces of identity at all.)

Clerical errors, or false reportage have always plagued historians, and another example concerns the pioneer sleeping car *Midland*, listed by Behrend as becoming MR No 20. When the body was rescued for preservation, as related, subsequent removal of old paint layers in attempts to identify which Pullman car it had been, revealed traces of the name *Midland* and also the number afterwards applied — which was 21. One could then go on to question which car was No 20. Perhaps we shall never know the complete story of these early Pullman cars, but herewith is a list, as complete as I am able to compile it.

The 'Schedule Numbers' shown were applied later to those cars that remained in Pullman ownership after 1915.

Schedule No	*Original name or number*	*Type*	*Date into service*	*Notes*
–	*Midland*	Sleeper	1/1874†	Sold to Midland Railway in 1888 (MR No 21); body preserved by the city of Derby at Midland Railway Trust Butterley. (See notes at head of table.)
–	*Excelsior*	Sleeper	2/1874†	Sold to MR in 1888: MR No 20.
–	*Enterprise*	Sleeper	6/1874	Burnt out at Hunslet, Leeds, in 1882.
–	*Victoria* (I)	Parlour	3/1874†	Used on LSWR 1880, renamed *Alexandra* (III). Went to Ouest Railway, France, 1882 for Paris-Le Havre dining service; later sold to CIWL as their No 155. Visited Italy 1883 as part of special train.
–	*Britannia*	Parlour	6/1874	Rebuilt with 6-wheel bogies as Dining car *Windsor* in 1882; sold to MR in 1883: MR No 15 by 1888.
–	*Leo*	Parlour	6/1874	Rebuilt with 6-wheel bogies as Dining car *Delmonico* in 1882; sold to MR in 1883: MR No 14 by 1888.
–	MR No 1*	Brake Parlour Third	6/1874	Sold immediately to MR as day car, after assembly at Derby. Withdrawn for departmental use c1888.
–	MR No 2*	Brake Parlour Third	6/1874	Sold immediately to MR as day car, after assembly at Derby. Withdrawn for departmental use c1888.
–	MR No 3*	Brake Parlour Third	6/1874	Sold immediately to MR as day car, after assembly at Derby. Withdrawn for departmental use c1888.
–	MR No 4*	Brake Parlour Third	6/1874	Sold immediately to MR as day car, after assembly at Derby. Withdrawn for departmental use c1888.
–	MR No 5*	First/second composite	6/1874	Details as Nos 1-4; then renumbered No 19 by March 1888.
–	MR No 6*	First/second composite	6/1874	Details as Nos 1-4; rebuilt as Dining car No 18 by December 1884.
–	MR No 7*	First/second composite	6/1874	Details as Nos 1-4; rebuilt as Dining car No 17 by October 1884.
–	MR No 8*	First/second composite	6/1874	Details as Nos 1-4; rebuilt as Dining car *London* by July 1884.
–	*Princess* (I)	Sleeper	6/1875	MR No 23 in March 1888.
–	*St George*	Sleeper	6/1875	Sold to MR in 1888: MR No 22.

*Two grounded bodies have survived and are now at Butterley, Derbyshire: identities not established.
†Month completed: entered service 6/1874.

Schedule No	*Original name or number*	*Type*	*Date into service*	*Notes*
72	*Jupiter*	Parlour	8/1875	Used on LCDR in 1882-84; sent to LBSCR in 1884; became Third class No 1 in 1915. (See also Table 2 on page 151.)
–	*Saturn*	Parlour	9/1875	Sold to MR in 1883: MR No 1.
–	*Transit*	Sleeper	10/1875	Sold to MR in 1888: MR No 24.
–	*Ocean*	Sleeper	1875	Ran on GNR in 1875-80 as a day car; then in Italy in 1883; sold to CIWL in 1886.
–	*Saxon*	Sleeper	2/1876	Sold to MR in 1888: MR No 25.
–	*Ohio*	Parlour	1875	Ran on GNR; became Dining car *Prince of Wales* in 1879 (first Dining car in Great Britain); sold to GNR in 1885: GNR No 2297.
–	*Mars*	Sleeper	10/1875	Delivered direct to LBSCR until 1883, used permanently arranged in day position; in Italy as Sleeping car in 1883; sold to CIWL in 1886.
–	*Mercury*	Parlour	4/1876	Sold to MR in 1883; MR No 3.
–	*Juno* (I)	Parlour	4/1876	Sold to MR in 1883: MR No 2.
–	*Australia*	Sleeper	6/1876	On MR until 1883; in Italy in 1883; sold to CIWL in 1886.
–	*Castalia*	Sleeper	4/1876	On MR until 1883; in Italy in 1883: burnt out in 1884.
–	*Germania*	Sleeper	8/1876	On MR in 1876-78, then GNR 1878 80; to Italy in 1883; sold to CIWL in 1886.
–	*Scotia* (I)	Sleeper	4/1876	Sold to MR in 1888: MR No 26.
–	*Venus* (I)	Parlour	5/1876	Sold to MR in 1883; MR No 4.
–	*Vesta* (I)	Parlour	7/1876	Sold to MR in 1883: MR No 6.
–	*Norman*	Sleeper	6/1876	Sold to MR in 1888: MR No 27.
–	*India*	Sleeper	7/1876	On MR in 1876-78; then GNR in 1878; destroyed at Thirsk in 1892 (collision).
–	*Minerva* (I)	Parlour	7/1876	Sold to MR in 1883: MR No 5.
–	*Planet*	Parlour	9/1876	Sold to MR in 1883: MR No 7.
–	*Albion*	Parlour	10/1876	Sold to MR in 1883: MR No 8.
–	*Comet*	Parlour	10/1876	Sold to MR in 1883: MR No 9.
–	*Apollo*	Parlour	1/1877	Sold to MR in 1883: MR No 10.
1	*Ariel*	Parlour	10/1876	On MR in 1876-81; renamed *Louise* in 1881 for LBSCR as Brake Parlour, reverted to full Parlour in 1889; converted to Kitchen car, 1912; withdrawn 1929; became bungalow at Selsey, Sussex. (See also Table 2 on page 151.)
73	*Adonis*	Parlour	1/1877	On MR in 1876-81; renamed *Victoria* (II) in 1881 for LBSCR as Kitchen car; reverted to full Parlour car 1889; later became Third class No 2 in 1915 as Kitchen car. (See Table 2 also on page 151.)
–	*Aurora* (I)	Parlour	5/1877	Sold to MR in 1883: MR No 11.
–	*Ceres*	Parlour	5/1877	On MR in 1877-81; renamed *Maud* in 1881 for LBSCR; became Brake Parlour car in 1884; scrapped in 1899 after accident at Wivelsfield. (See also Table 2 on page 151.)
–	*Eclipse*	Parlour	7/1877	Sold to MR in 1883: MR No 13.
–	*Alexandra* (I)	Parlour	7/1877	Sold to MR in 1883: MR No 12.
74	*Alexandra* (II)	Parlour	10/1877	On LBSCR; became Third class No 3 in 1915. (See also Table 2 on page 151.)
75	*Albert Edward*	Parlour	10/1877	On LBSCR; became Third class No 4 in 1915. (See also Table 2 on page 151.)
2	*Globe*	Parlour	6/1878	To LBSCR 1878; renamed *Beatrice* in 1881 for LBSCR; first Pullman car with electric lighting; derailed at Lovers Walk, Brighton, in 1914; withdrawn in 1918. (See also Table 2 on page 151.)
–	*Columba*	Sleeper	1880	For GNR in 1880; sold to ECJS in 1895, No 233. Became NER refreshment car No 2966 in 1902.
–	*Iona*	Sleeper	1880	For GNR in 1880; sold to ECJS in 1895, No 232. Became GNR dining car No 2992 in 1902 when condemned ended days as a grounded body at Lincoln, in 1925.
–	*St Andrew*	Sleeper	2/1883	Sold to MR in 1888: MR No 28.
–	*St Mungo*	Sleeper	2/1883	Sold to MR in 1888: MR No 29.
3	*Balmoral*	Sleeper	1/1883	3-axle 6-wheel car; delivered direct to GNR; on GNR 1883-85; to Highland Railway 1885-1907. Rebuilt with bogies in 1895, and ended its life running in mixed passenger/goods trains in 1907; stored 1907-18. Became part of a bungalow at Seaford, Sussex. To Brighton Locomotive Works 1987.
4	*Culross*	Sleeper	1/1883	3-axle 6-wheel car; renamed *Dunrobin* in 1885 for service on Highland Railway. Rebuilt with bogies in 1895. Also became part of a bungalow at Seaford, Sussex. Remains to Brighton Locomotive Works 1987.
–	*St Louis*	Sleeper	6/1883	3-axle 6-wheel car; rebuilt with bogies in 1885; sold to MR in 1888: MR No 30.
–	*St Denis*	Sleeper	6/1883	3-axle 6-wheel car; rebuilt with bogies in 1885, sold to MR in 1888: MR No 31.
–	*Missouri*	Sleeper	2/1884	Sold to MR in 1888: MR No 32.
–	*Michigan*	Sleeper	5/1886	In store at Derby until May 1866, from delivery/completion in 1883. Sold to MR in 1888: MR No 33.

Note: All the cars listed above were assembled at the Pullman workshop at Derby, from prior-built and dismantled pieces, sent from America. (Except the wheels and underframes on the 1883 six-wheeled sleeping cars, *St Andrew, St Mungo, Balmoral* and *Culross*, which were supplied by the MR.) The Derby Pullman workshop was sold to the MR in 1888. The assembly of new cars, or the rebuilding of old ones was then undertaken at the LBSCR Brighton workshops except for the cars *Duchess of Connaught* and *Princess Margaret*, which were assembled at Eastleigh by the LSWR. As before, all cars were prior-built in the USA, and shipped over in knocked-down condition, for reassembly. Unless otherwise stated, all the cars listed on the next page, commenced service on the London, Brighton & South Coast Railway.

Schedule No	Original name or number	Type	Date into service	Notes
6 5 7	*Princess* (II) *Prince* *Albert Victor*	Parlour Kitchen Parlour	12/1888 12/1888 12/1888	First cars for England with vestibules and gangways, known as 'Pullman Train No 1'. Built for 'Pullman Limited' service as a 3-car set. Originally *Prince* did not have buffers, since it ran in the centre of the rake. Buffers added in 1915. *Princess* (II) became a Kitchen car in 1920, withdrawn with *Albert Victor* in 1929. Became studio of the railway artist and writer, C. Hamilton Ellis, at Partridge Green, Sussex. *Prince* withdrawn in 1932.
8	*Duchess of Albany*	Parlour	3/1890	On LSWR from 1890 to 1907-12. Became a Kitchen car in 1913 (Pullman Longhedge). On LBSCR 1912-23, SR 1923-29. Withdrawn 1929 to Partridge Green with *Princess*.
9	*Duchess of Fife*	Parlour	3/1890	On LSWR from 1890 to 1907-12. Became a Kitchen car in 1912 (Pullman Longhedge). On LBSCR/SR 1912-29, when withdrawn.
10	*Empress*	Parlour	5/1890	Withdrawn in 1932. Used as bungalow at Lancing, Sussex, by J. S. Marks on retirement as General Manager of Pullman. Dismantled during World War 2.
76	*The Queen*	Parlour	11/1890	Became Third class car No 9 in 1920 with kitchen (Pullman Longhedge).
11	*Pavilion*	Parlour	2/1893	Last Pullman with high windows and clerestory roof to remain in service LBSCR/SR. Became a Kitchen car in 1923 (Pullman Longhedge). Withdrawn in 1934.
12	*Princess Mary*	Parlour	3/1893	Became a Kitchen car in 1909. Scrapped in 1932.
13	*Duchess of Connaught*	Parlour	3/1893	On LSWR from 1893-c1907. Became a Kitchen car in 1914 (Pullman Longhedge). Scrapped in 1932.
14	*Princess Margaret*	Parlour	3/1893	On LSWR from 1893 to 1907, then LBSCR. Became a Kitchen car in 1913 (Pullman Longhedge). Withdrawn in 1932; became bungalow at Lancing, like *Empress*. Dismantled during World War 2.
15	*Prince Regent*	Parlour	3/1893	Became Second class No 3 in 1921 as Kitchen car, for boat train services.
16 17 77	*Princess of Wales* *Duchess of York* *Her Majesty*	Parlour Kitchen Parlour	10/1895 10/1895 10/1895	Became Third class No 19 in 1915 as Kitchen car. Became Third class No 17 in 1915. Became Third class No 10 in 1922 as Kitchen car. These three cars originally formed a permanently coupled set, known as 'Pullman Train No 2', which replaced the 1888 cars in the 'Pullman Limited' Service.
19	*The Arundel*	Parlour	7/1899	6-wheel bogies. Renamed *Majestic* as Kitchen car in 1905, because its name was confused with a destination board. Withdrawn in 1932; became bungalow at Lancing, Sussex. Dismantled during World War 2.
18	*The Chichester*	Parlour	7/1899	6-wheel bogies. Renamed *Waldemar*, as Kitchen car in 1905, because its name was confused with a destination board. Withdrawn in 1932; became a bungalow at Lancing, Sussex. Dismantled during World War 2.
20	*Devonshire*	Kitchen	8/1900	6-wheel bogies. Withdrawn in 1931. Canteen at Preston Park Pullman Works until 1964.

Notes: (i) In addition to *Prince, Duchess of York* and *Devonshire* (known to have some kitchen facilities) it is possible that some of the other cars had a small pantry area where hot drinks, or reheated food were prepared, and glasses and crockery washed up.

(ii) Three more Pullman cars had been ordered from America, for assembly in England, and were delivered after negotiations for the purchase of the company by Davison Dalziel had started. They are therefore listed in Table 2, on page 151. They were *Duchess of Norfolk, Princess Ena* and *Princess Patricia*, and they were destined to be the last American-built/British-assembled Pullmans. Four sleeping cars were built for the Midland Railway in 1900 in America, and shipped to Derby MR workshops for completion, including the provision of MR-style buffers, drawgear and six-wheel bogies. These cars never featured as Pullmans, but they are listed here because they were of Pullman USA design, to MR order and three survived long enough to become LMSR property:

MR No 34		Sleeper	4/1900	Renumbered 2771 (LMSR) 1/1923; renumbered 02771 in 1925.
MR No 35		Sleeper	4/1900	Scrapped c1922.
MR No 36		Sleeper	4/1900	Renumbered 2772 (LMSR) 1/1923; renumbered 02772 in 1925.
MR No 37		Sleeper	4/1900	Renumbered 2773 (LMSR) 1/1923; renumbered 02773 in 1925.

Note: It is believed that the GWR borrowed the three surviving cars for a while in the late 1920s.

85

85
Midland Railway No 13 entered service in 1877 and sold to the MR in 1883.
National Railway Museum

Part Two:

20th Century Growth and Decline

CHAPTER 12

Davison Dalziel's Achievement

Mr Davison Dalziel was a professional financier, and when he purchased the British Pullman Co Ltd, it was with a view to running it at a personal profit, and to operating it only on routes that would pay their way. It was not surprising that he soon withdrew from regular use the two HR 'inferior' sleeping cars, which were put into reserve, and that to begin with he concentrated his energies on the profitable Brighton line. He persuaded the London Brighton & South Coast Railway to let him run the all-Pullman 'Brighton Limited' every day, instead of on Sundays only.

This daily 60min non-stop London-Brighton (and vice versa) run in brand new Pullman cars, complete with a new name, the 'Southern Belle', was the first indication of Dalziel's determination to bring the Pullmans into the limelight once again, and understandably the established London-Brighton service was chosen for what he proudly described as 'the most luxurious train in the world'. To work the prestigious 'Southern Belle' he ordered seven new Pullman cars, the *first* to be ordered from a British firm, the Metropolitan Amalgamated Carriage & Wagon Co of Lancaster, and delivered in 1908.

The new cars had one immediately noticeable design change, in the form of an elliptical, or barrel roof with domed ends, instead of the American clerestory type, giving them a more modern appearance. Finished in the new umber and cream livery, in order to match the LBSCR livery of D. E. Marsh, the cars were mounted upon six-wheel wooden-framed bogies, and were heavier, at 39-42 tons, than any previous cars. The interiors were indeed

86

86
The London & South Western Railway was still operating Pullmans on its Waterloo-Bournemouth services when Davison Dalziel purchased the Pullman Co in Britain, although one car — the *Duchess of Fife* — was transferred to the London Brighton & South Coast Railway in 1906; the remaining three also going to the LBSCR in the period 1907-1912. All four Pullmans were replaced by LSWR-owned and operated dining cars. One of the Pullmans (identity not certain) is the third vehicle in this double-headed express, seen at Swaythling. Two of Dugald Drummond's handsome Class T9 4-4-0 'Greyhounds' powered the train; the nearest one is No 285. The second (number not known) is of interest because it has the wider driving wheel splashers and an eight-wheel 'watercart' tender, whereas No 285 has narrow splashers and a six-wheel tender. The handsome Pullman car positively shines, in its dark green livery.
Ian Allan Library

luxurious, and the publicity brochure that Dalziel had produced for the train, described it also as, 'A chain of vestibuled luxury', and when remarking upon the mahogany panelling inlaid with satinwood, the delicate mouldings and fluted pillars, soft green carpeting with a fleur de lys pattern, damask silk blinds and 'cosy chairs and settees in a restful shade of green morocco', eulogised that: 'A woman's eyes are quick to note these things, and a woman's voice is quick to express delight, "How perfectly lovely! It is as cosy as our drawing room, Dick!" '. (Six of the cars were in fact described as drawing room cars.) It seems that the decor was chosen with the express intention of pleasing the ladies, who would, Dalziel believed, encourage their escorting admirers to pay the necessary first class fares and Pullman supplement that were required!

The 'Southern Belle' was introduced on 1 November 1908, and commenced regular daily running on the 8th of the same month. It was so successful that by June 1909 it was making two return journeys each day, and at a later date it made no less than three trips to Brighton and back on Sundays. This success seems to have provoked some reactions amongst the more staid admirers of Brighton, one of whom wrote:

'The short distance that separates it (Brighton) from London, which was one of the primary causes of its popularity in the coaching days, has, since the advent of the railways, proved itself an element destructive of its exclusiveness. Cheap excursions for "trippers" and half-guinea Pullman Car trains for the well-to-do, have changed the character of the place out of all recognition. Brighton has developed into the Cockney's paradise, the Mecca of the stockbroker and the chorus-girl.' (From *Brighton: Its history, its foibles and its fashions*, Lewis Melville, 1909)

Yet to come was the weekend invasion of the motor cars, motor scooters, buses and coaches that have really robbed Brighton of some of its charm!

The change of roof design for the new Pullmans was in keeping with the start of a general trend in new British carriages, and by 1914 only the Midland Railway was still building new clerestory-roofed main line carriages. The clerestory was undeniably handsome, but very difficult to keep clean. In the early Pullmans, the clerestory ventilators had been in gaudy etched or coloured glass and the two level ceiling panels inside were lined with quilted American cloth, printed with garlands of iris, gladiolus and other flowers. By the 1880s the American cloth had been changed to decorative wood veneers, but now a much more simple 'ceiling' was substituted in the British-built cars, in what could loosely be described as 'Adams style'. The basic construction of the 1908 cars was still to the American integral wooden-bodied style, and the bogies were the American type, but the ventilation, electric lighting and steam heating, were to the latest British standards, and the windows were of modified design to match the changed ventilation arrangements. Davison Dalziel wisely kept the distinctive, large and elongated 'Pullman' lettering on the sides of the cars above the windows, and gave them all names, in the true tradition.

Fortunately a very concise description of the train was published in *The Railway and Travel Monthly* in October 1910, written by G. A. Sekon. The following extracts certainly paint a luxurious picture:

'We have now arrived at the point when the readers . . . will be feeling a pardonable curiosity to learn something of the "make-up" of the *Southern Belle* and of the internal arrangements and fittings of the vehicles that entitle the *toute ensemble* to the appellation of "the most luxurious train in the world.'

'. . . the train consists of seven bogie vehicles of the greatest height and width that the loading gauge permits. The train is vestibuled throughout, a walk from end to end being possible.

87

87
It was the London Brighton & South Coast Railway's services that had really kept the affairs of the Pullman Co going, in the years between the death of George Mortimer Pullman and the purchase of the British part of the business by Davison Dalziel. One of the handsome 12-wheelers of 1899 (either *The Arundel* or *The Chichester*) towers above the low-roofed LBSCR compartment stock in this typical view of a business express; headed by R. J. Billinton's Class B4 4-4-0 No 50 *Tasmania*. By the time this picture was taken the Pullman had been repainted in the new umber brown and cream livery, and would be carrying its new name (either *Waldemar*, formerly *The Chichester*, or *Majestic*, formerly *The Arundel*) because the original names had been too often mistaken for destination boards! *Ian Allan Library*

88
A corner of the interior of *Belgravia*, panelled in pear tree wood with holly inlay, in a style described as 'Pergolesi'. The moveable table lamps had an additional usefulness as a weight if table linen was in use. However, in this charming picture by Fortunino Mantania, the lady in the left-hand foreground has been served with light refreshments upon a tray. This was one of a series of paintings produced for Davison Dalziel by the artist Mantania. *Author's collection*

88

'The train provides seat accommodation for 219 passengers. It consists of a buffet car and six drawing-room cars, two of the latter having the guards' compartments. Each car measures 63ft 10in in length, 8ft 8¾in in width, and a height of 13ft 6in from rail to top of roof, and is fitted with an independent hot-water heating apparatus and a complete installation of electric bells, also with Stone's improved systems of electric lighting and ventilation, which also combines a new and improved device for heating by means of warmed fresh air, freed from all impurities, through ventilating ducts.

'The ceilings in all the cars are treated with a deep coving at the side and ends of each compartment, with a flat panel in the centre, so that the effect is that of a complete symmetrical ceiling. Appropriate decorations in bas-relief have been specially modelled in the plastic material used, and the whole is painted white.

'The usual form of electroliers has been discarded, the ceiling lights being "dished" out of the ceiling, thus retaining to the full extent the feeling of loftiness in the cars. The wall-brackets are charming productions in finely chased ormolu, with moulded crystal globes of flower shape. The basket racks, door mountings, and all other metal work is carried out in the same fine and high-class manner. The tables have all tops of rich curl veneer, and on these are placed the standard lamps of exquisite form, beautifully chased and gilded. The floors are laid with rich pile carpets, whilst the floors of the corridors and lavatories are covered with inlaid rubber of black and white design. All the compartments have mirrors set in the panelling, and these give additional brilliancy to the general effect.

'The ladies' and gentlemen's toilets are furnished with all the latest improvements. The washstands are finished with onyx marble tops and the woodwork is of a fine stripe mahogany.

'A distinctive feature of this train is the variety of panelling adopted.

'We can better appreciate the luxuriousness of the *Southern Belle* by taking a walk through the train, and examining each coach. Commencing at the engine end, we enter the *Verona* car, the front end of which consists of a break compartment, beyond which are the compartments for travellers, seats being provided for 31 people. The *Verona* is finished in wainscot oak, the style of the French Renaissance having been skilfully introduced with highly satisfactory results.

'The oak is inlaid with rich decorative panels, mouldings, pilasters, and friezes of holly-wood slightly shaded to give the effect of bas-relief.

The easy chairs and sofa coverings are of a fine Mohair velvet of a light café colour. The carpets and blinds are in shades of green with self coloured designs.

'The *Princess Helen* is the second car of the train. This coach seats 33 persons.

'The "Adams" style is adopted, and the decoration consists of panelling made from "Plum Pudding" mahogany used in conjunction with kingwood and satinwood. The panels are treated with an inlaid trellis design of kingswood (a striped wood of purple colour) and the spaces are filled in with the stripe veneer running alternately in vertical and horizontal lines, producing an altogether novel and charming effect. The mouldings, pilasters, friezes and cornices are all inlaid with satinwood.

'The pile carpets are in shades of soft green with design of "Fleur de Lys" in self colours. The blinds are of rich damask silk, also in shades of green. The prevailing notes in the scheme of colour are therefore the rich mahogany colour of the woodwork, contrasted with the green carpets, blinds and seat coverings. The seats and easy chairs are upholstered in a dainty green and drab striped moquette tapestry.

'The *Belgravia* is panelled with pear tree wood with holly inlay, and is treated in the "Pergolesi" style, the main features being the beautifully inlaid panels of festoons, draperies, and flowers, the

89

89
A 'camel-back' Bo-Bo electric locomotive of the Metropolitan Railway, built by the Metropolitan Carriage & Wagon Co with British Westinghouse equipment and finished in crimson lake livery, heads a morning up 'Through-to-the City Breakfast Car Express', south of Harrow. Resplendent in umber and cream livery (with white roof) the Pullman car makes a striking contrast to the rest of the passenger carriages. One wonders at the social distinctions obviously then existing, by comparison with present-day commuter travel on the Metropolitan — where all are herded together, without opportunity to eat or drink at leisure whilst on the move!
London Transport

90

general effect being that of ivory on the rich browny-pink wood.

'The sofas and chairs in this car are upholstered in a fine blue Pique velvet.

'The next car, the *Grosvenor*, is one in which the creature comforts of the traveller are catered for, a buffet being provided, from which can be obtained cold luncheons, teas, coffee and other light refreshments, which not only refresh the traveller but afford an additional charm to that feeling of comfort that is so marked a sensation of a journey in the *Southern Belle*.

'The *Grosvenor* (in consequence of the space occupied by the buffet) seats only 25 persons. The "Adams" style is adopted, and the woodwork is of Spanish mahogany, richly inlaid with satinwood.

'The easy chairs and settees are upholstered in a beautiful tone of green morocco, cool and restful to the eye. The carpets, blinds etc are similar to those of the *Princess Helen* already described.

'Continuing our promenade, we next enter the *Cleopatra*, with seats for 33 people. As in the *Belgravia*, the "Pergolesi" style is adopted in this car. The wood is East Indian satinwood, with richly quarterd panels inlaid with grey sycamore, boxwood, greenwood and tulipwood. The panels, pilasters and friezes are profusely decorated with *motifs*, based on the exquisite forms invented by Pergolesi in the latter half of the eighteenth century.

'The easy chairs and settees are upholstered with a closely woven fine Mohair velvet of "Fleur de Lys" design in shades of blue with a delicate tracery of gold showing through the velvet pile. The blinds are of the same colour in rich damask silk and the pile carpets are in shades of deep rose colour.

'The sixth car — the *Bessborough* — is practically a replica of the *Princess Helen* already described, save that stripe mahogany is employed. The seats and easy chairs, however, are upholstered in a fine quality of French drab cloth of a similar tone to that in use in the saloons of the Chemin de Fer du Nord.

We now enter the *Alberta*, which is the last vehicle of the "world's most luxurious train", and therefore contains a break compartment at the rear. In fittings and internal arrangements it resembles the *Verona*.

'Our walk through the *Southern Belle* cannot fail to have advanced our education in the subject of luxurious travel. As already stated the travelling public is largely indebted to Mr Davison Dalziel for the *Southern Belle*, ably seconded by Mr Thos. Powell, the Secretary and Manager of the Pullman Company.

'In addition to the *Southern Belle*, Pullman cars are regularly run on the London, Brighton and South Coast Railway in the express trains between London Bridge, and, or, Victoria and Brighton,

90
Mr Dalziel lost no time in ordering seven new Pullman cars for the LBSCR; the first to be *built* by a British Company — the Metropolitan Amalgamated Carriage & Wagon Co of Lancaster. They were also the first to dispense with the traditional American clerestory roof, although many American features, such as the oval toilet windows and the six-wheel bogies, were retained. The new train was called the 'Southern Belle' and Dalziel proudly proclaimed it to be 'the most luxurious train in the world'. In this publicity photograph the brand new umber and cream cars are posed with D. E. Marsh's beautiful Class I2 4-4-2T No 19, built at Brighton works in July 1908, and painted in the Marsh umber brown LBSCR livery which matched the Pullmans so well. The nearest car is the Brake Parlour *Alberta*, and the second the Parlour car *Bessborough*. This view makes an interesting comparison with picture No 82, taken at the same location some two years earlier.
National Railway Museum

Eastbourne, Worthing Bognor, Hove, and Newhaven (for Paris). Light refreshments are served on all the cars, whilst hot meals can be obtained on several. Suppers are served on the night trains between Victoria and Brighton and *vice versa*.

'Beyond the ordinary first class fare, the charge for the luxury of a Pullman is extremely small, being only 6d or 1s per journey, according to the distance travelled in the car.'

It will be noted that *Grosvenor* is described as capable of serving cold luncheons, etc, whereas later on mention is made of some of the Pullmans on other services being capable of serving hot meals. It seems probable that hot meals were ordered in advance from LBSCR station restaurants and then either kept hot, or reheated in the Pullmans; in which case *Grosvenor* was probably capable of serving hot meals also.

A feature retained was the stylish oval toilet and kitchen windows, first introduced on turn-of-the century American Pullman cars. These consisted of four segments of coloured glass within the oval wooden frame and the lead frames that held them together were joined to form a decorative diamond shape in the centre; within which a circular ventilator could be fitted. One break with tradition was that the chairs in the cars were allowed to stand loose on their legs, instead of being pivoted to the floor (a few loose chairs had previously been used) and the loose armchair was to become another symbol of Pullman luxury in the first class cars.

Having rejuvenated and expanded the Brighton Pullman operations, Dalziel cast his eyes around for further potentially profitable contracts. He was still with the Wagons-Lits company in France, and relationships between the two nations were at a good level due to the 'Entente Cordiale'. It was only natural that Dalziel should consider the boat train services to the Channel ports as suitable material. He had already experienced the 'Club Train', (see Chapter 10) and he approached the South Eastern & Chatham Railway (formed by an alliance in 1899 of the two former rivals, the South Eastern Railway and the London Chatham & Dover Railway). The SECR Managing Committee commenced running Pullmans on 21 March 1910 from London (Charing Cross) to Dover. Six new cars were built, and four more followed very shortly, and all were painted to match the SECR crimson lake livery, although lined out and lettered in gold as Pullmans. They were very handsome cars, on four-wheel bogies, and were built by the Birmingham Railway Carriage & Wagon Co, and consisted of three parlour cars and were usually worked in pairs, a parlour car coupled to a kitchen car, within the normal boat train formation.

Another potential 'plum in the cake' was the Great Central Railway the newest main line to reach London, and one obviously anxious to please its clients in every way possible. Something went wrong for Dalziel in this instance however, because as Behrend tells us, although on 22 October 1909 the GCR Board agreed to accept the offer made by Dalziel to provide their line with his Pullman services for a period of 15 years, no contract was made at the time or ever materialised, because on 28 January 1910 the Board changed their mind and decided themselves to provide, and own, any luxury trains that ran on their lines.

There was an odd spin-off resulting from this lost contract for Dalziel, because the Metropolitan Railway, who operated what are nowadays called commuter services out of London to Buckinghamshire (in part over tracks jointly owned with the GCR) were understandably alarmed that they would lose some custom to the proposed GCR Pullmans from their more wealthy passengers who travelled up to the City each weekday. The Metropolitan therefore decided that they should run Pullman cars, which is more than George Mortimer Pullman ever achieved for an American commuter line of such modest route distance and character! They made a 30 year contract with Dalziel for the provision of two new cars, built by the Birmingham Railway Carriage & Wagon Co Ltd, (BRCWCo) which received the names *Mayflower* and *Galatea*. On 1 June 1910 the remarkable spectacle of these cars running on the underground lines between London's Liverpool Street, Aldgate, and Baker Street stations, and then out into the suburbs and beyond to the pasturelands of what Sir John Betjeman has immortalised as 'Metroland' commenced. Never before had travel on the underground railways of London aspired to such heights as serving morning breakfast, tea and supper, and suitable evening aperitifs such as whisky and sodas! There were no through gangway connections at the car ends (because none of the ordinary Metropolitan carriages had them) and for health reasons the lavatory doors were kept locked whilst the train ran

91

Two months prior to the entry into service of the two Metropolitan Railway Pullmans (on 21 March 1910 to be precise) the South Eastern & Chatham Railway commenced running a 'Continental Express' service from London, Charing Cross to Dover; including Pullman cars. The Birmingham Railway Carriage & Wagon Co built some very handsome four-wheel bogie cars, which usually worked in pairs — a parlour and a kitchen. Illustrated is the Pullman kitchen car *Emerald*, with American-type bogies, and Pullman-type gangway connection complete with retractable side buffers and drophead buckeye coupling. The SECR chose to have their Pullmans in their own standard railway livery of crimson lake, but the gold lettering and lining-out was in typical Pullman style, and the oval toilet and kitchen windows emphasised the Dalziel ownership! An innovation was the experimental circular badge on the waist panels, flanking the car name. This had the words 'Pullman Company Limited' in full around the circumference; within which was a first attempt at a Pullman coat of arms. *National Railway Museum*

91

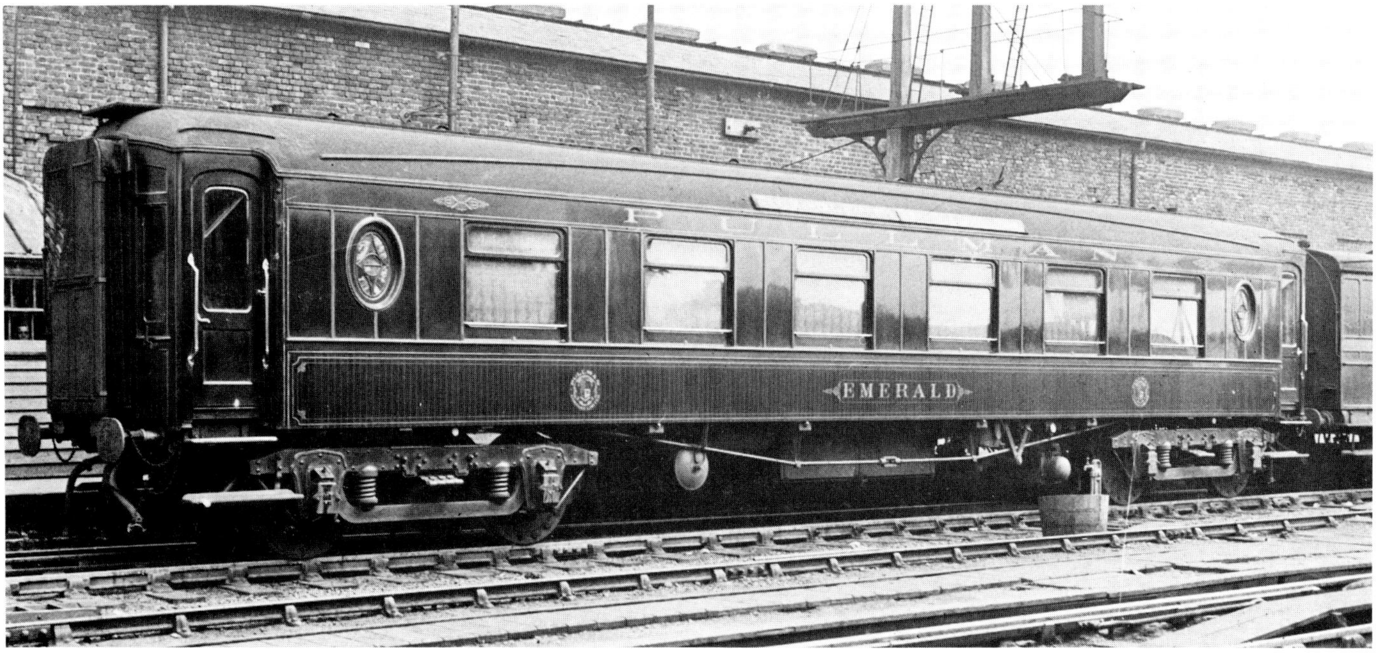

92

through the underground tunnel section. Of pure British design, with interiors by W. S. Laycock Ltd, the two cars nevertheless followed true Pullman traditions, with no expense spared.

Writing in delightful Edwardian prose, G. A. Sekon gave readers of *The Railway and Travel Monthly* of July 1910 a description of the opening ceremony, and of the two Pullman cars themselves.

'In this strenuous age business men find it necessary to spend much time in the Metropolis, and yet yearn for the charm, healthfulness and repose of a country residence. By a country residence we mean one in the heart of the rural districts still to be found from 20 to 40 miles from London, but a home so far from the scene of one's daily business activities is only possible where the train service is good, even then it has many limitations unless the business hours are those extremely curtailed ones, often talked of, but far less frequently found.

'Mr R. H. Selbie, the energetic General Manager of the Metropolitan Railway, whose improvements have done so much to develop that system, appreciated that the charming district through which the Extension Line runs is an ideal one for residential purposes, and he has accordingly provided an ample and convenient express train service, which is greatly appreciated by those who can leave home between 9 and 10am, and commence their return journey before 5pm. But when business necessitates that the hours of departure up are earlier, and those down later than those mentioned, busy men are inclined to say that living in the country "is not worth the candle." At this juncture a good fairy, in the person of the Pullman Company, appears on the scene, explaining that if Pullmans be provided the busy man can breakfast on his up journey, take tea on the down, or if he spent the evening in town can have supper when travelling home by the Theatre Train "after the opera is over", a feature that should be greatly appreciated when ladies form a part of the party. Next the resources of two well-known British firms are called into play, with the result that aided by the art of Messrs. W. S. Laycock Ltd, and the Birmingham Railway Carriage and Wagon Co Ltd, two palatial "Pullmans", the *Galatea* and *Mayflower*, are evolved. So the Gordian knot is cut, and "Pullmans" are running to and from Aylesbury and Chesham on the Metropolitan Extension Railway and the various stations on the Metropolitan Railway between Baker Street and Aldgate.

'Prior to the public service, which was brought into operation on June 1st, an inaugural run was made by a party of guests invited by the Metropolitan Railway and the Pullman Company, on 27th May.

'The special train, consisting of five vehicles, a Metropolitan Railway third class break bogie carriage at each end, the two "Pullmans" and the Metropolitan Railway's own saloon (which by the way did not suffer in comparison with the "Pullmans" between which it was sandwiched)* left Liverpool Street at 11.35am and ran non-stop to Baker Street, where other guests joined. The journey was continued at 11.50, Harrow being reached at 12.12. Here the British Thomson-Houston electric locomotive, of the "milk van" type, the tractive power of which is supplied by four motors, each of 400 horse-power, was detached, and the steam locomotive, No 81 4-4-0T type, substituted. Leaving the scholastic town, smart running was made for the remainder of the journey, the 5 miles 14 chains, Amersham to Great Missenden, occupying but 5min 35sec. Aylesbury was reached at 12.55pm, and the party proceeded to the historic George Hotel, where an excellent luncheon was well served.'

Mr Sekon added a footnote (see asterisk above) as follows:

'As the Pullmans are named, we suggest the Metropolitan Railway bogie saloon be called (unofficially) the "Rothschild", as it is frequently used by this gentleman, whose country residence at Hatton, Wendover, is reached by the Metropolitan Extension Line.'

Describing events at the George luncheon, Mr Sekon recorded that:

'Mr Selbie, the General Manager of the Metropolitan Railway, in the course of a short and interesting speech, said that his Company was endeavouring to meet the public demands in every way, and pointed out that the morning train would enable City men to have another half-hour in bed, and to take their breakfast quietly and comfortably in the train — unless they missed it — as someone had suggested to him. By catching the afternoon train on Saturdays golfers could reach the fine links on the Metropolitan Railway after lunching in peace on the way, and, therefore, starting their game in the best of humours. Finally, theatre-goers, by taking advantage of the *Bakerloo* Railway, would catch the 11.35pm down train with ease, would find a welcome supper awaiting them on the cars, and would not be turned out before they had a chance of doing justice to it.'

92
Working hard past Thornton Heath, a heavy business train on the Brighton line is seen in the charge of R. J. Billinton's 4-4-0 No 201 (as rebuilt to Class B2X by D. E. Marsh with a 'C3' boiler). The picture was taken about 1910, by which time the three leading vehicles in the train, all built in 1895, had been repainted in umber and cream. They are: 'Pullman Pup' van No 29; *Princess of Wales*; and *Duchess of York*. The van was destined to be destroyed by fire in 1913, while both first class cars were converted to thirds in 1915 — among the first to be so treated.
Ian Allan Library

Describing the two cars themselves, Mr Sekon gives a very detailed picture, and one can only marvel at the contrast between these and the 'sit-up-and-beg' primitive compartment carriages that were then the norm on the Metropolitan!

'The Pullman cars have been specially constructed, having regard to the dimensions of the loading gauge of the Metropolitan Railway, the principal "overall" dimensions being:

Length over buffers	59ft 6¾in
Length over platforms	57ft 6in
Length over body corner pillars	51ft

Centres of bogies apart 39ft 3in
Wheel base of bogies 7ft 6in
Extreme width 8ft 5in
Height from rail to top of roof 12ft

'The exteriors of the Cars are finished in the standard "Pullman" colours, namely, umber bodies with the upper panels in white and tastefully lined with gold; the roofs are painted white and the underframes and bogies black.

'The Cars are each divided into two large saloons, one capable of comfortably seating eight passengers and the other seven, and there is also a smaller saloon at one end arranged to seat four; the latter compartment should prove very useful in the case of travellers requiring extra privacy, as it can be reserved and shut off entirely from the rest of the car. All the compartments are provided with tables, fitted with electric standard lamps and bell pushes. A well-fitted lavatory compartment is also provided at one end. The end saloon contains a commodious buffet counter, behind which is a pantry fitted with gas stove,* and cold water, and every convenience for the supply of refreshment.

*George Behrend states that for reasons of safety cooking was actually undertaken by means of a Primus Paraffin stove, because the use of gas was considered too dangerous if an accident occurred in an underground tunnel. The gas stove was probably used for the country part of the route.

'The period of decoration used for the *Galatea* Buffet Car is that of the latter part of the eighteenth century.

'The mural scheme is composed of fine fiddle-back mahogany with the upper panels inlaid with satinwood on a ground of fine quartered and veneered fiddle-back mahogany. There are eight flap tables in the Car which have glass tops, on which are placed electric lamps of chaste design.

'Each compartment has a cored plaster ceiling, with an ornamental band on the flat forming a panel. From this band are suspended the ceiling lights. Wall brackets are also fixed at suitable places on the wall panelling. The bell pushes, switches and number frames are of ormolu, finely chased and gilded.

'The blinds, which are of green silk damask, have been specially woven. Above each blind there is an ormolu baggage rack, with finely chased ornamentation with panels of brass treillage. The floor is covered with a green deep pile carpet, and the chairs are luxuriously upholstered and covered with green morocco.

'The upper panels of the cross partitions are of clear plate glass, thus enabling one to see from end to end of the car.

'The style of the *Mayflower* Buffet Car is that of the Georgian period. The combination of the oak and the holly makes a most pleasing effect. The wall scheme is composed of panels of fine wainscot oak, with the upper panels inlaid with enrichments of the Georgian period on a ground of fine quartered and veneered oak.

'The car is divided into three compartments with cross partitions. An excellent effect is obtained through clear plate-glass having been fixed in the upper panels of these partitions. There are chairs for 19 persons in the car, and patent hinged tables with plate-glass tops are provided.

'Each compartment has a cored plaster ceiling, with an ornamental band on the flat forming a panel.

93

93
Mr Davison Dalziel was President du Conseil of the Wagons-Lits Co when he purchased the British Pullmans and he evidently liked the Wagons-Lits Co coat of arms, which featured the Royal Lions of King Leopold II of the Belgians. (This was by gracious permission of the King, who had been an important backer during the early days when Georges Nagelmackers was busy setting up his company.) Dalziel's Pullman Co coat of arms embodied the heraldry of England, Ireland, Scotland and Wales (despite the fact that Pullmans were unheard of in Ireland and Wales when the design was prepared!) and two imposing lions were prominent in the design. It is thought that Dalziel had hopes of obtaining the British Royal Warrant, in which case the Pullman coat of arms would have been surmounted by a crown. *Author's collection*

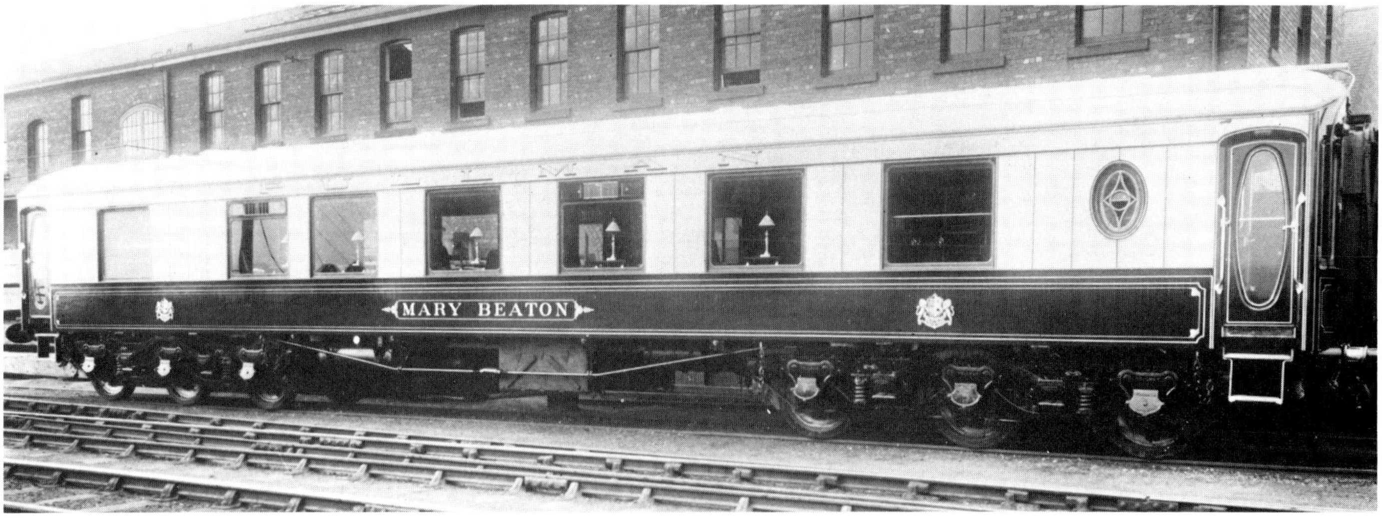

94

94
Scotland was added to Dalziel's list of Pullman operations by the time his cars started to carry the new coat of arms, when he gained a contract with the Caledonian Railway. Cravens of Sheffield built this strikingly elegant 12-wheeler in 1914 for the commencement of CR operations. The new coat of arms added a distinctive finishing touch to the dark umber and cream paintwork. Note the use of bellows-type gangways and long buffer shanks, instead of Pullman-type fittings, due to the fact that these cars had to be compatible with existing CR screw-coupled/bellows gangwayed carriages. *National Railway Museum*

From this band are suspended the ceiling lights. Wall brackets are also fixed at suitable places on the wall panelling. The bell pushes, switches and number frames are of ormolu, finely chased and gilded.

The blinds, which are of green silk damask, have been specially woven and above each blind there is an ormolu baggage rack with finely chased enrichments, with panels of brass treillage.

'The floor is covered with a crimson deep pile carpet, and the chairs are luxuriously upholstered and covered with crimson morocco.

'While the interior woodwork of the two cars varies, some of the decorative features are the same, namely, the electric light fittings, which, in addition to the ceiling lights consists of wall brackets suitably disposed, and table standards of brass gilded by the mercurial process; the door handles, basket racks, ventilators and other fittings of chased and gilded brass, which have been specially designed and modelled to harmonise.'

95

95
Dining services were the Pullman's prime *raison d'etre* in Scotland on the Caledonian Railway, and the interior layout of the cars was arranged to suit, being quite unlike the vast majority of cars used in the South of England, although many fittings such as table lamps, luggage racks and tables and mirrors were virtually identical, as well as the use of wooden panelling with decorative inlays. Leather-upholstered dining chairs provided sufficient comfort for the duration of a meal, as opposed to the usual plush comfort of armchairs. Details to note are the clock above the vestibule door in the background, and the large circular ventilators in the white ceiling. The foreground shows the characteristic table lamps and combined ashtray and matchbox-holder. The further saloon is laid out for dining, with immaculate white table linen. This is the interior of the Cravens-built car *Lass o' Gowrie*, of 1914. *National Railway Museum*

Dalziel also revived the idea of Pullman dining cars, as opposed to parlour, or drawing room cars and buffet-cum-saloon cars which could serve snacks. For a boat train service from London to Queenborough Pier on the Isle of Sheppey, from where the Zeeland Steamship Co's service to Flushing was operating, a Pullman dining car was built in March 1911, once again by the Birmingham company, and named *Shamrock*. It served evening dinners on the way down from London to Queenborough, and breakfast on the return up boat train the next morning. For this service was laid out with 24 upright dining chairs, in green morocco leather upholstery pinned by large gilt nails, and the table tops were of glass, covered of course by immaculate white linen. The Flushing service appears to have been transferred to Folkestone Harbour soon after *Shamrock* was introduced, because the July 1911 issue of the *Railway Magazine* gives it as such. In 1914 it was withdrawn for the duration of the war, and then reinstated, from Folkestone, until 1927 when it was transferred to the LNER via Harwich, Parkeston Quay. The car *Shamrock* was a non-supplementary diner and was not suitable for other SECR Pullman services (which had supplements) and was therefore renamed *Duchess of Gordon* and sent in 1914 to the Caledonian Railway as a dining car on the Glasgow-Aberdeen service (which continued to use Pullmans until 1917).

I am indebted to George Behrend for reminding me of the importance of Dudley Docker at this stage of our story. Docker was virtual owner of the Birmingham Railway Carriage & Wagon Co, and *financed* most of the Pullman cars built. He was on the Board of the Metropolitan Railway and the SECR and later, after 1923 the Southern Railway (SR) until he gave his seat to Sir Herbert Walker in 1937. His London home was located close to the Metropolitan line. At one stage he was also the power behind Metro-Cammell, and he amalgamated Leeds Forge and Midland C&W Co with it. Later there was a row, and he retained the Birmingham Railway Carriage & Wagon Co separately. Prior to this the Company sometimes sub-contracted the building of new Pullmans to Metro-Cammell.

In May 1912, new Pullmans were introduced on the ordinary SECR London-Folkestone trains, built by Cravens of Sheffield (who had delivered two new 12-wheelers to the LBSCR the previous year). The SECR cars were in crimson lake livery, to match the rest of the train; but they were unmistakably Pullman in shape and lettering, carrying the name proudly above the passenger windows, and were complete with oval toilet and kitchen windows at each end. They had four-wheel American-type bogies.

A feature of Pullman car design, in both Britain and America, was the lack of doors along the sides. When the vestibule was first introduced jack-knife entrance doors were recessed close to the gangway connection and reached by a short platform. The 1888 cars and the first of the two 'Pullman Pup'

96

96
Strikingly modern in style for its day was the Pullman observation car *Maid of Morven* built for the Caledonian Railway service between Glasgow and Oban, by Cravens of Sheffield in 1914. Observation cars with an open veranda end were very popular in America and Pullman built a considerable number over the years. Scottish weather and prevailing winds ruled against such a design however, and Dalziel opted for a glazed end, incidentally producing what were almost certainly the largest panes of glass ever used in a British railway carriage, up to and including the present day. American-type four-wheel bogies carried the car, and as can be seen, it had screw couplings plus emergency side chains. An odd feature was to place the doorway at the end, no doubt creating draughts for the end seating, which strangely faced *inwards*.
Author's collection

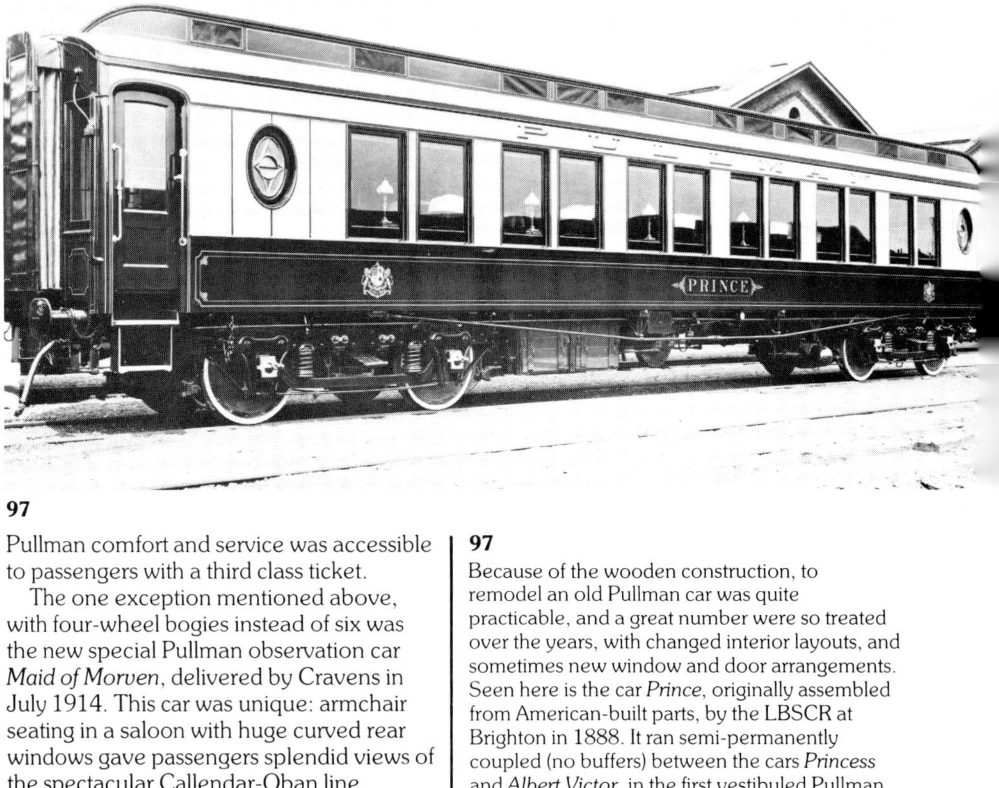

97

97
Because of the wooden construction, to remodel an old Pullman car was quite practicable, and a great number were so treated over the years, with changed interior layouts, and sometimes new window and door arrangements. Seen here is the car *Prince*, originally assembled from American-built parts, by the LBSCR at Brighton in 1888. It ran semi-permanently coupled (no buffers) between the cars *Princess* and *Albert Victor*, in the first vestibuled Pullman train. In 1915 it was remodelled at the Longhedge (London) Pullman workshops, with new standard gangwayed ends, complete with retractable side buffers and drophead buckeye couplers, producing a carriage of very handsome appearance. *National Railway Museum*

six-wheeled vans for the LBSCR had this arrangement, when new. They were later rebuilt to what became the standard layout, with only slightly recessed bodyside inward-opening doors at each end, with a wide transverse vestibule within. The fact that the doors opened inwards avoided loading gauge problems, and was considerably safer than the conventional British carriage door, designed to open outwards. To clear the more restricted loading gauge of the Metropolitan Railway (in tunnels) and parts of the South Eastern & Chatham routes, all the cars for these services had the cantrail (guttering) of the roof cut back over the end doors, which were slightly recessed. All previous vestibuled Pullman cars in Britain had the cantrail continuing over the doorways, but between 1922 and 1951 the cut-back version became standard for loco-hauled cars. A classic feature of Pullman car service was that the passenger was *never* expected either to open or close the door himself.

By June 1911, according to the *Railway Magazine*, there were 49 Pullman cars in service in Britain: with 31 running on the LBSCR; 11 on the SECR; three on the LSWR; two on the Metropolitan and two (the two reserve 'inferior' sleeping cars *Balmoral* and *Dunrobin*) on the Highland Railway.

Scotland next attracted Dalziel's attention, and he was successful in gaining a contract with the Caledonian Railway to provide and operate all their restaurant cars, with the exception of those required to run through over other companies' routes. In 1914 Cravens of Sheffield delivered 10 new Pullman cars to the CR, all carrying Scottish names. There is some evidence to suggest that the CR Pullmans carried a green and cream livery to begin with; although this has not been positively authenticated. With one exception they were mounted upon six-wheel wooden-framed bogies, and weighed around 42½-43 tons apiece. For passengers only wishing to have a meal, and then return to their own seat in the train, no supplement was charged, and to provide dining space for them three composite (first and third class) dining cars were included in the order. So now the Pullman comfort and service was accessible to passengers with a third class ticket.

The one exception mentioned above, with four-wheel bogies instead of six was the new special Pullman observation car *Maid of Morven*, delivered by Cravens in July 1914. This car was unique: armchair seating in a saloon with huge curved rear windows gave passengers splendid views of the spectacular Callendar-Oban line. Scarcely had it arrived on the CR, however, when Britain was plunged into war. While the 'Maid' ran in the late summer of 1914, and again in 1915, it was then withdrawn 'for the duration'. Other CR Pullmans were also pruned.

War did not stop the pleasure-seekers heading to the South Coast however (in fact it may have intensified their urge to do so) and the Pullmans continued in operation to Brighton and Eastbourne, which was by then served by a Sundays only 'Eastbourne Pullman Limited'. The Pullmans were also popular for excursions to the races. Whilst one portion of the British nation prepared to give life and limb, another seems to have carried on regardless! However, on the

98

SECR, busy with wartime traffic to the Channel ports, the 'Hastings Car Train' and the 'Folkestone Vestibuled Limited' — not operated by Dalziel — were both withdrawn, and these Pullman cars went into storage, or special government use.

On 30 September 1915 Dalziel ceased to own the Pullmans privately. A public company was formed to take over all 'the cars, stock, work-in-hand and goodwill'. It was apparently done in great secrecy, and Dalziel became Chairman of the new Pullman Car Co Ltd — to give its correct new title. We now know, thanks to the exhaustive research of George Behrend, that from the start the new company was secretly associated with Wagons-Lits. The new company took over 74 Pullman cars from the private Pullman Co and from the Drawing Room Cars Co, which was one of Dalziel's other financial activities, used to finance the supply of new Pullman cars during his personal ownership, and to retain copyright of the name 'Drawing Room Car'.

An initial move by the new company was the decision to introduce third class Pullmans, with a supplement charged (unlike the CR dining cars where the third class passengers were only seated for service of a meal, and were then obliged to leave the car). There was no longer any second class on British railways, except on boat trains which had to cater for passengers with through second class tickets from the Continent, and so to provide third class Pullman accommodation was not as shocking as it seemed; but it apparently created quite a stir at the time amongst the more conservative of the first class ticket holders. For the new third class passengers, the Pullman Co provided fixed seating of two and two (2+2) each side of the centre aisle. In truth, this was rather cramped — especially if everyone seated at a table for four wanted to eat. (While a 2+1 seating pattern was later adopted for third class cars intended for longer-distance travel, the 2+2 layout was perpetuated in the third class accommodation of the cars built for the Brighton electrification scheme in the 1930s.)

The 'Southern Belle' (curiously American, the use of the word 'Belle' with its New Orleans connotations) was increasingly popular in the early war years and by introducing third class cars the traffic was stimulated still further for a 'day by the sea'; so that by December 1915 there were already six third class services daily to Brighton and back, and third class cars began to run also to Eastbourne. The third class cars were in fact rebuilds of old Brighton line Pullmans (no doubt the public was once again led to believe that they were new however!) and their names were removed. Car *Jupiter* became *Car No 1 Third Class*; car *Victoria* became *Car No 2 Third Class*, etc (see list). This style of numerical description became the standard for all future third class Pullman cars, which were never considered worthy of naming; although the composite cars, with a mix of firsts and thirds did receive names.

Gradually, the realities of war, and of austerity, with staff and food shortages, began to make their mark and at the end of 1916 the 'Southern Belle' was withdrawn from service, and all the normal passenger activities of the Pullman Car Co Ltd were for the time being in abeyance; although some wartime VIP specials were run for members of the royal family, government ministers and military and naval leaders.

This is an appropriate moment to end this chapter, because by 1916 Dalziel had moulded the affairs and policy of the new Pullman Car Co Ltd into the form they were destined to take for many years to come. He had rescued the image of Pullmans as a fine way to travel, but by day rather than by night, and he had retained the long-established traditions of fine personal service, attention to detail, and superb comfort and interior design. The cars themselves were still in the forefront of British designs of the period, and were gradually losing some of their American influence, whilst retaining the same air of luxury and calm. The payment of a supplement continued to be resented by some; whilst others recognised the value it represented. Dalziel had not succeeded in gaining a lot of new contracts when the Great War broke out, but he had built up the fleet of cars, and there were contracts with the LBSCR, the SECR, the CR and the Metropolitan Railway. The London & South Western Railway services had however been replaced by dining cars owned by the railway itself, and their Pullmans had been transferred to the Brighton services.

Most important of all, and known to only a few, Dalziel had allied Pullman to Wagons-Lits. (He did so in another form as well, because his daughter married René Nagelmackers, the son of Georges.) When peace returned to Europe, Dalziel was an important figure in the affairs of both companies.

98
The fourth vehicle in this Caledonian Railway express, is a Pullman dining car, accessible without payment of a supplement to passengers travelling in the ordinary CR carriages, for meal service only. The engine is Pickersgill's beautiful 4-4-0 No 80, in the smart light blue CR livery.
Ian Allan Library

CHAPTER 13

Years of Growth and Competition

Up until now in this book it has been possible to give a step-by-step, and almost a car-by-car, story of the creation of the British Pullman car services. From now on our story is essentially more in terms of general developments, rather than specific cars, because once World War 1 was over, the Pullman Car Co was to experience a considerable period of growth. Many of the prewar cars were reinstated, and the services revived, but there were also new routes and many new cars. It is not possible to give individual attention to all, and I will select the 'highlights' of developments to present the story, although some individual cars are of such importance that they will be featured, sometimes as representative of a whole batch of new cars. It is impossible to say representative of identical new cars, because nearly every Pullman ever built has had some individuality either in decor, or seating layout, or some detail particular to the needs of the service for which it was built. Perhaps it is that which makes Pullman cars so fascinating to so many people. Giving them names was no mere publicity stunt, it was a recognition of this individual personality, of which the Pullman staff were particularly proud.

It may have been something of a surprise for Davison Dalziel when the war ended, to find that there was an immediate demand for a return of the luxury standards of travel his company could supply. The railway manufacturing industry had been deeply involved in the war effort, and such items as new luxury Pullman cars had not figured on their workshop programmes, which had been much more concerned to provide such things as hospital ambulance trains for shipment across the Channel, to care for the wounded soldiers and to transport the dying. Material to build new carriages was also scarce, although somehow or other Pullman's Longhedge Works (which had formerly belonged to the SECR, and which Pullman used when it had become surplus to SECR requirements) managed to build four new third class cars in 1917, which had presumably been ordered earlier because of the success of the Brighton line's activities. When the war ended, materials were still scarce and Dalziel was faced with a dilemma because he soon persuaded the

99

99

The 'Thanet Pullman Limited' express started on 10 July 1921, running on Sundays only from London (Victoria) to Margate non-stop in 90 minutes. The train is seen here passing Bickley in the charge of rebuilt Class E1 4-4-0 No 504 in the plain grey SECR livery. Behind the leading six-wheel brakevan the crimson lake Pullmans, which are all of the four-wheel bogie type, receive a dusting from the locomotive's exhaust! *Ian Allan Library*

Great Eastern Railway to run Pullman cars on its Continental boat train services, in order to match those on the rival South Eastern & Chatham's boat trains. Having obtained the GER contract in 1919 to start in November 1920, he had to find the cars. The solution was typical of the man, because he persuaded the directors of the Pullman Car Co to purchase all the South Eastern & Chatham Hastings and Folkestone Club cars, and to rebuild them as first class Pullmans. They were mostly used on the same routes they had served before, which were now under contract to Pullman. The few new cars he could obtain in time, from Clayton Wagons Limited of Lincoln, could then be used on the GER services.

The Clayton cars were reported at the time, in the *Locomotive Magazine*, as finished in a different livery, being dark olive green below the waistline and ivory white, or cream, above. Presumably the GER did not want to wear the umber brown and cream colours of the LBSCR. This Pullman green and cream may well have been the same as is suggested to have been originally used for the Caledonian Railway Pullmans. On 11 November 1920 the 'Hook Continental' ran with two new Pullmans from London (Liverpool Street) to Parkeston Quay for the first time. These two cars were *Arcadia* and *Corsair*, which were followed by the others over a period of almost 12 months until all 20 were in use. Because they were expected to work also in ordinary trains, the Pullmans had side buffers and standard GER screw couplings, when new. They later received the Pullman buckeye type. The Clayton cars included two second class Pullmans, because of through tickets from the Continent, where second class still operated. The third class cars had 2+2 seating, and all the GER Pullmans had draught-excluding sliding top lights in the main windows (the first so fitted). They ran on six-wheel bogies. The GER cars also ran marshalled in ordinary trains to Cambridge, Norwich and Cromer, Lowestoft, Clacton, Southend, Hunstanton and Harwich Town; but they did not pay their way.

On the Caledonian Railway the Pullmans, which ran in ordinary trains as buffets and diners, were reinstated at the end of the war, although some had to be refurbished beforehand. More cars were delivered in 1922/23 and the services were spread over the system, with the observation car *Maid of Morven* once again on the picturesque Oban route. The two Metropolitan Railway cars also continued to run as before the war. On the South Eastern & Chatham six new cars, intended to run in pairs, were delivered in 1921, all mounted on six-wheel bogies, and finished in crimson lake livery. They were built by the Birmingham Carriage & Wagon Co. Because the Pullman Car Co was now under contract to supply *all* refreshment facilities for SECR trains, the first class kitchen cars had extra large pantries, to serve the whole train. In addition to the new cars, the SECR had the former Club cars now rebuilt as Pullmans. In 1921 the 'Thanet Pullman Limited' was introduced,

100

Lawson Billinton's superbly proportioned Class L 4-6-4T No 333 *Remembrance* — one of the 'Brighton Baltics' — shows her paces at the head of the 'Southern Belle', shortly after the 1923 railway Grouping; when the LBSCR became part of the new Southern Railway (SR). The two four-wheel bogie Pullmans behind the leading bogie brakevan are both third class cars and conversions from former 1914-1918 wartime ambulance train coaches. Some of the axlebox covers on these conversions still carried the initials of their original builder — LYR, LNWR etc. The locomotive carries the new green SR livery, which replaced the LBSCR umber brown. *Lens of Sutton*

101

The newly-created London & North Eastern Railway (LNER) inherited the Pullman contract with the former Great Eastern Railway (GER) and promptly decided to expand Pullman operations on to the main line from King's Cross to the North. The 'Harrogate Pullman' is depicted starting away from King's Cross, in the charge of former Great Central Railway Robinson Class B3 4-6-0 No 6166 *Earl Haig*, with a 12-wheel brake parlour car leading. Note that the cream paintwork was carried over the panelling to the car end, despite the lack of windows.
Ian Allan Library

100

101

102
When the LNER introduced the 'Sheffield Pullman' service in 1924, the Pullman Co issued this publicity picture to the press, depicting, 'the new Japanese Lacquer panels' in the cars. The small seat reservation number, 22, and the art nouveau styled wall light are also of interest.
Ian Allan Library

to run every Sunday during the year from London to Margate, later to Ramsgate and Broadstairs.

The activities of Pullmans on the London, Brighton & South Coast Railway, once reinstated, continued to flourish and the Pullmans began to appear in trains to Worthing, Bognor Regis, Littlehampton, Portsmouth and Eastbourne, as well as Brighton, where the 'Southern Belle' was soon back in service. There were no Pullman cars running on the London & South Western at this time, and Davison Dalziel had still not succeeded in interesting the important London & North Western Railway or the Great Western Railway in operating the Pullman Co cars over their routes.

Except on the GER, wherever the Pullmans were running in the early 1920s they were popular enough to require more cars to be built in a short time, and once again Dalziel showed initiative, by purchasing redundant wartime-built hospital ambulance coaches built by the London & North Western, the Great Western and the Lancashire & Yorkshire Railway during 1914-1918. These had steel underframes and four wheel, steel framed bogies, upon which Pullman-style wooden bodies were provided in place of the originals, and they were of course described as *new*, although the axlebox covers gave away their parentage!

The 1923 grouping of the private British railways into the 'Big Four' did not involve the Pullman Car Co, which remained as

102

before. It did, however, to some extent play fortune into Dalziel's hand, because of the existing contract held with some of the absorbed companies. Three of the four new giant railway companies — the Southern Railway, the London Midland & Scottish Railway and the London & North Eastern Railway — took over services which included Pullman cars. Only the Great Western Railway was now a geographical barrier to the Pullman Car Co in Britain. From this date onwards, the umber and cream livery was standardised by Dalziel for the fleet.

The SR now operated the Pullman services of both the former London Brighton & South Coast and the South Eastern & Chatham — by far the largest part of Pullman's operations — whilst the LMSR took over the Caledonian Railway Pullmans and the LNER inherited the Pullmans of the Great Eastern Railway. The Metropolitan Railway was not affected by the grouping, and continued to operate its two Pullmans regardless.

It was the LNER which inherited the least successful Pullman services, operated by the GER out of Liverpool Street. These, as we have just seen, had commenced at the end of 1920, and they were still not paying their way. The GER's last General Manager had been Mr Henry Thornton, an American by birth, and he seems to have made an error of judgement by introducing so many Pullman cars to East Anglian services, whose clientele did not react as enthusiastically as his American compatriots no doubt would have done! Only the boat train services to Parkeston Quay really warranted the use of Pullmans on a regular basis and the LNER retained some of the first class cars on the 'Hook Continental' and Antwerp continental boat expresses after May 1924, when the other GE section Pullman services ceased to run. Meanwhile, the LNER was anxious to use Pullmans on the more profitable routes it had inherited from the former Great Northern and North Eastern Railways, from London (King's

103
Unlike the LNER, the rival London Midland & Scottish Railway (LMSR) did not see fit to expand its inherited Scottish area Pullman services south of the Border to London, neither over the former LNWR line to Euston, nor the former MR line to St Pancras. It did, however, continue to use them as dining cars in Scotland, because the Pullman Co contract was not due to expire until 1933. One of the former CR 12-wheelers is included (second vehicle) in this nine coach business express, being hauled in fine style by the massive 4-6-4T No 15403, which had been designed by R. H. Whitelegg for the Glasgow & South Western Railway; absorbed by the LMSR in 1923. *M. W. Earley*

103

104

105

104
The Great Eastern section of the LNER lost most of its Pullman cars to the more important East Coast services from 1924 onwards, but a Sundays-only 'Clacton Pullman' was retained — in order to promote Clacton along the lines of Brighton, as an ideal weekend 'day resort'. The down 'Clacton Pullman' is seen here, composed of 12-wheelers in the charge of Class D16/1 4-4-0 No 8783, passing Gidea Park. Pullmans had run to Clacton on Sundays only — as the 'Clacton Pullman' since July 1922, and after the Grouping the stock was supplied by Finsbury Park and worked round empty to the GE Section for the Sunday service. *Ian Allan Library*

105
An SR Eastern Section 'Continental Express' passing Bickley in the capable charge of Maunsell 'King Arthur' class 4-6-0 No E766 *Sir Geraint*. The leading Pullman car is a four-wheel American-bogie, former SECR services example, repainted from crimson lake to umber and cream — one of the 'White Pullman' vehicles that gave the nickname to the service. It is thought that Dalziel decided to standardise the Pullman livery after the Grouping; choosing the former LBSCR colours. *Ian Allan Library*

Cross) to the north-east of England, the West Riding and ultimately to Scotland.

The first service tried out was the 'Harrogate Pullman' which started on 9 July 1923, running non-stop from King's Cross to Leeds Central, where it reversed and continued to Harrogate, Ripon, Darlington and Newcastle-on-Tyne and vice versa. The existing Pullman car fleet was not large enough to form the necessary two rakes of Pullmans that were needed to operate, one each way per day over this long distance. To provide sufficient cars an order was placed for the first time of what was to become known as the 'K' type (see Appendix 1). Not all the new cars were ready in time for the press run and the start of the service, however, and the GE section cars *Ansonia*, *Arcadia*, and Nos 46 and 57 were in the 'Harrogate Pullmans' in 1923, with possibly some others on brief loan during the year.

The 'K'-type cars were considered to be the new standard British Pullman car design, fitted with a new four-wheel standard steel-framed bogie. The first examples built still had wooden bodies upon steel underframes and included four first class cars, *Iolanthe*, *Fortuna*, *Irene* and *Rosemary*, built by the Midland Carriage & Wagon Co Ltd. The third class cars were built by Clayton Wagons of Lincoln. In 1924 most of the GE section cars were transferred away, and those going to the GN section were converted to lower seating capacity (in the case of third class cars) or rebuilt to brake parlours for use on the 'Harrogate' and undermentioned 'Sheffield Pullmans'.

The LNER decided to run a 'Sheffield Pullman', routed via Grantham and Nottingham, but this was poorly loaded, and to gain patronage it was next extended to Manchester in 1925, again without success, and the autumn of 1925 saw it withdrawn completely, when the cars were used for a 'West Riding Pullman' serving Leeds and Bradford. Meanwhile, in July 1925 the 'Harrogate Pullman' had been extended right through to Edinburgh from Newcastle, making it the first (and only) all-Pullman train to run in Scotland. In 1928 it was to be more happily named as the 'Queen of Scots' Pullman train, and was further extended to serve Glasgow, Queen Street, from Edinburgh. (Later still, the 'West Riding Pullman' became the 'Yorkshire Pullman'.)

With the transfer away of all but six of the GE section Pullman cars in 1924, the 1925/26 Newmarket Races specials were worked by Pullman cars borrowed from the SR Eastern Section, and contemporary photographs showed the unusual spectacle of crimson lake-liveried (in some case clerestory-roofed) Pullmans on the LNER. More Pullman cars were returned from the GN section at weekends only, in order to work the 'Clacton Pullman', Clacton evidently considering itself a seaside resort every bit the equal of Brighton! However, before describing the fortunes of the 'Clacton Pullman', and subsequent GE section developments, it is necessary to take a look at another seaside town served by Pullmans, from 1922 onwards. This was Hastings, earlier known for its SER 'Club train' service. Six new Pullman cars were built to introduce the service, and these had to conform to the special route restrictions (see also Chapter 10) being only 8ft 1in instead of 8ft 6in wide. Supplied by the Metropolitan-Cammell Carriage & Wagon Co Ltd, they were originally all first class cars and were built to the new 'K'-type specification (except for width), with wooden bodies on steel underframes, and steel-framed four-wheel bogies. They were named *Barbara*, *Camilla*, *Latona*, *Madeline*, *Pomona* and *Theodora*, and in 1932 they were converted to composite first and third class cars. We will meet them again later.

The Southern Railway was to benefit quite quickly from Davison Dalziel's personal connections with both the Wagons-Lits Co and the Pullman Car Co in the 1920s, because this tireless entrepreneur wanted to forge visible and physical links between his activities in both Britain and France; although they were intended to wear the mantle of 'Pullman' on both sides of the Channel, with only passing references for Wagons-Lits!

Apparently the general travelling conditions on the cross-Channel services were terrible in 1924, and to improve the British portion of the journey, the Southern Railway introduced an all-Pullman boat train between London and Dover on 17 November that year; despite the fact that they were already operating some cars (still in former SECR crimson lake livery) on the regular boat trains between London and Dover. The cars for the new service, called the 'Continental Express' (a name which was actually already in use) were painted in what was now the standard Pullman umber and cream livery, so that outwardly they differed from all the rest of the crimson lake Pullmans running on the former SECR services. Because of this they gained an unofficial nickname amongst railway staff as the 'White Pullmans'. Dalziel wanted also a similar Pullman service on the French side, between Calais and Paris but efforts by Wagons-Lits to persuade the Nord Railway had not been successful in 1923/24. So to begin with the 'White Pullman' was a purely British affair; but Dalziel was not to be defeated so easily.

Meanwhile, the Southern Railway had received new cars by 1925 with lower elliptical roof and four-wheel bogies, for the prospering 'Southern Belle', and some services to Hastings via Eastbourne now included Pullman cars. It was now possible to travel from Hastings to London in a Pullman over what had once been two rival routes. There was also a first class Pullman included in the boat trains from London to Newhaven, to serve the cross-Channel steamers operating to Dieppe. With all the new cars introduced since the end of World War 1, the Pullman fleet was certainly strong on the ground on the SR, because the older American cars were in many cases only just coming under consideration for withdrawal from service. So sturdy was their wooden construction that (accidents apart) it seemed they wanted to live on for ever; only the roof being a weak point. If that deteriorated badly then the car was likely to be a candidate for withdrawal. The Pullman Car Co had a high regard for its older cars,and many were selected for rebuilding, or remodelling, to fit them for further service. The classic example was the British-built 1908 car *Grosvenor*, which was always fitted with a kitchen, and called a buffet car. This car, rebuilt in 1936 with a larger bar, was destined to survive in service until 1960, and even then it was given a further lease of life as a 'camping coach' (see Chapter 16). One way older cars were kept on the Pullman stocklist was by remodelling them for third class accommodation, which was mainly a matter of changing the seats. The old first class 'loose' chairs were then refurbished and used again in many cases in the new cars!

Across the Channel, Dalziel's Wagons-Lits Co was still preparing to introduce the *Wagon-Salon-Pullman*, in order to realise Dalziel's wish to run Pullman cars in Europe. The Nord Railway was, however, still obstinate, and so the first European service was not (as Dalziel would have wished) between Calais and Paris, to link with the British 'White Pullman'. Instead, he commenced with a Milan-Nice service, using 10 British-built Pullman cars, purchased brand new, before they entered service in England, from the Birmingham Railway Carriage & Wagon Co Ltd in 1925. In full British Pullman umber and cream livery these cars, of wooden-bodied construction upon steel underframes and steel four-wheel bogies, were shipped to France via the Harwich-Zeebrugge train ferry in November 1925. At this time the French railway authorities were insisting that new carriage construction should be steel-bodied, and the Wagons-Lits Co was busy planning and introducing a fleet renewal programme of all-steel dining cars and sleeping cars, commenced in 1922. No Pullmans were included in the schedule however, because of the reluctance of the Nord to accept them. Moreover, the 10 British-built cars, with wooden bodies could not be accepted by the French and were perforce based in Italy, although regularly operating into France!

Perhaps Davison Dalziel remembered the story of George Mortimer Pullman's *Pioneer*, which had only been accepted by the American railroads after it had been lent to the widow of the assassinated President Lincoln; because he now found a very clever way to demonstrate Pullman service to the Nord Railway, by running a special press demonstration in what *purported* to be new all-steel Pullmans! It was a masterpiece of strategy, made possible because the 10 British-built Pullmans were fitted with vacuum brakes only and had to be taken into the St Denis (Paris) workshops of Wagons-Lits, to be fitted with standard European air brake equipment. It seems that Dalziel purchased these cars for Wagons-Lits in such a hurry that the work could not be done before they left Britain, and as a result, once in France they had to be hauled as far as Paris like an unbraked goods train at very low speed. Whilst the air brakes were being fitted Dalziel arranged for the chairs and tables to be taken out of some of the British cars and suitably arranged, Pullman-style, inside two brand new all-steel Wagons-Lits *dining* cars at St Denis. These two *mock* Pullman cars, and two brand new all-steel Wagons-Lits sleeping cars were then run from Paris to Calais for the benefit of the Nord as a trial run; with suitable press invitations issued in advance. The Nord was definitely impressed, and the opposition to Dalziel's idea disappeared in the midst of Pullman splendour, cigar smoke and aperitifs! The only problem was that there were no new all-steel Pullmans available to start the service, which had been acclaimed by the press in advance, because the French carriage builders were fully booked-up with orders for other types of new steel cars. A lesser man than Davison Dalziel would have cringed at the possible after-effects of an irate press upon discovering it had been hoodwinked.

Typical of the man was the solution he contrived. The Nord, whilst agreeing to introduce a Pullman car service between Paris and the Channel ports, was insistent upon these new cars being all-steel and French-built. Dalziel agreed, but then observed that there would be an inevitable delay before these could be built and placed in service, and he then proceeded to persuade the Nord to accept in the interim some British-built, all-steel cars, which could be delivered much sooner. The new Pullmans were, however, delayed in delivery by the General Strike in Britain,and it was not until September 1926 that they were available to commence working the French part of what was now called (in France only) the 'Flèche d'Or' ('Golden Arrow') — destined to become one of the most famous named trains in the world. To identify the French Wagons-Lits owned train as a Pullman, Dalziel had the cars painted in the umber and cream of the British company, and even the upholstery and carpets were identical in some cars on both sides of the Channel as was the tableware, which was supplied from Britain. A colourful publicity brochure in English was entitled *The Princely Path to Paris*. It actually described the car's attendants as 'Pullmanic Demigods', and did not overlook the passengers themselves, who were 'fastidious pasengers in a sybaritic age'.

Thus Davison Dalziel — entitled to use the name and goodwill of Pullman — achieved the link between Britain and the continent of Europe which George Mortimer Pullman had personally so much wanted to achieve in his own lifetime. It was indeed ironic that the Wagons-Lits Co, which had been the main obstacle to Pullman's ideals, was now the European operator of the cars bearing his name. Had he still been alive, however, George Mortimer Pullman would surely have approved, because it certainly personified all the ideals to which he had bestowed his name. It was a true train de luxe!

The year 1926 was the golden jubilee year of the Wagons-Lits Co, and — according to George Behrend — it was for that reason that the new train was called the 'Flèche d'Or'. However, the publicity brochure just mentioned, gave another reason for the name, by recalling a historical meeting in the year 1520 between King Henry VIII of England and King Francis I of France, in the 'Field of the Cloth of Gold' near Calais. It did not even mention the 1926 Wagons-Lits golden jubilee. Not until 1929 did the British train receive the title of the 'Golden Arrow', when a new ship was built to link the two trains, the famous SS *Canterbury*, and the through 'Golden Arrow' service was publicised as such.

In the same year, Davison Dalziel commenced running Pullmans in Ireland, for which a subsidiary company, Irish Pullmans Ltd was formed in Dublin, wholly owned by the Pullman Car Co Ltd. Four

106

107

106
Using a powerful magnifying glass it is possible to read the details on the brass plate at the foot of the nearest vestibule entrance door, on the original photograph depicting the new Pullman Parlour car *Minerva*. It reads: 'The Pullman Car Co Ltd/Built by The Birmingham Railway Carriage & Wagon Co Ltd Smethwick'. Complete in every detail (even with menu cards on the tables), this car, and the rest of the same batch of 10 built in 1925 by BRCW never ran in service in Britain in this guise. It was sold immediately to the Wagons-Lits Co (see text for details) and shipped to the Continent. When it did run in Britain, after return in 1928, it entered British service rebuilt as a guard parlour car and renamed *Lady Dalziel*.
National Railway Museum

107
The Birmingham Railway Carriage & Wagon Co also built four third class Pullman kitchen cars for service in Ireland in 1926 — the only Pullmans ever to run in Ireland, where the broad gauge 5ft 3in system necessitated special bogies, which were supplied by the LMSR's Northern Counties Committee (NCC), at Pullman's expense. This picture shows the 'Sligo Pullman' of the Great Southern Railways, Ireland (GSR) headed by a 2-6-0 Class K1 locomotive, one of a class created from parts purchased from the Woolwich Arsenal in England, which had originally manufactured the parts — to the design of R. E. L. Maunsell of the SECR — as a possible new 'national' locomotive. The Pullman car is the second vehicle in the train. *Ian Allan Library*

108
An up Pullman car express on the water troughs near Carstairs, sometime in 1927. The locomotive is former Caledonian Railway McIntosh '903' class 4-6-0 No 14753. The Pullman, evidently recently repainted in umber and cream is one of the 1914 Cravens 12-wheelers, in use for dining purposes. When the Pullman contract expired in 1933, the LMSR did not renew it, and purchased all the cars for further use, gradually repainting them in crimson lake livery and removing their names and Pullman identity. *P. Ransome-Wallis*

108

109

Gresley 'super power' for the down 'Queen of Scots' all-Pullman train, near North Berwick. This was the longest through working by Pullmans in Great Britain, between London King's Cross and Glasgow, Queen Street, via Edinburgh. New all-steel 'K' type four-wheel bogie Pullmans were supplied for it in 1928, when the name 'Queen of Scots' was first used. This is Class A3 Pacific No 2579 *Dick Turpin* heading the eight-car train in fine style. *Ian Allan Library*

109

110

Lord Dalziel unfortunately did not live to see the day when the British Pullman Company finally penetrated the territory of the Great Western Railway (GWR) in 1929, something George Mortimer Pullman had never achieved before him. The GWR operated a Pullman 'Ocean Liner Express' service between London (Paddington) and Plymouth, Millbay Docks (commencing in May 1929) and a 'Torquay Pullman Limited' from Paddington to Torquay and Paignton (commencing in July the same year). The down 'Torquay Pullman Limited' is seen here in Twyford Cutting, headed by 'King' class 4-6-0 No 6015 *King Richard III*. The second and third Pullman cars are new all-steel vehicles. Both services were withdrawn the following year. *Charles Long Collection/LPC*

110

third class Pullman cars were built by the Birmingham Railway Carriage & Wagon Co Ltd and numbered 100-103; they each had a kitchen and two saloons with 46 seats. They had four-wheel bogies supplied by the LMSR, to 5ft 3in gauge; and a small detail difference, allowed by the generous Irish loading gauge was that the cantrails continued over the doorways to the car ends. After 11 years of work, the contract expired and was not renewed. The cars were purchased by the Great Southern Railways (GSR) and they survived in GSR use to be taken over by Coras Iompair Eireann (CIE), who painted them green. One, No 100, was still in service as late as 1962. The Irish office was closed down by liquidation in 1937.

On 14 May 1927 another Anglo-French Pullman service was introduced, known as the 'London-Vichy Pullman', via Folkestone-Boulogne and Paris (it ran until 1930). Just four days after its introduction, on 18 May 1927, in London, Lord Dalziel of Wooler died, aged 82. His estate was valued at the remarkable sum of £2,199,220. A true giant amongst men, he had championed the cause of both the Pullman Car Co *and* the Wagons-Lits Co. Shortly before he died he negotiated the purchase of Thos Cook & Son Ltd, forming the Wagons-Lits/Cook organisation. In his early years he had been involved in newspaper management in America, and with this experience he returned to Britain to found Dalziel's News Agency in London in 1893. Success and fortune in this field led him to a position where, in 1910 he bought the *Standard* and the *Evening Standard* newspapers, in which he kept a controlling interest until 1916. Meanwhile, as we have seen, in 1907 he had purchased the British Pullman Car Co. In the same year he introduced the first motor taxicabs into London operated by the London General Cab Co, of which he was Chairman. After his death the chairmanship of the Pullman Car Co went to Lord Ashfield, the creator of

111

Interior of the first class kitchen parlour car *Phyllis*, built by the Metropolitan-Cammell Carriage & Wagon Co in 1928, a standard steel-bodied 'K' type vehicle on four-wheel bogies. This interior is typical of the period in all respects, showing the draught-excluding sliding toplights in the windows, the clock, ceiling lights and ventilators, small wall lights, moveable table lamps, inlaid panelling and standard small luggage racks. It was included in the 1928 'Queen of Scots' formation. *Ian Allan Library*

London Transport, but Julian Morel, in his book *Pullman* tell us that in reality the Dalziel mantle, and all his directorships passed to his former private secretary, Mr Stanley Adams (who had been General Manager of Thomas Cook), with seats on the Board of both the Pullman and the Wagons-Lits companies. The link between the two, established by Dalziel, was further strengthened by the fact that the Directeur-Générale of Wagons-Lits also had a seat on the Pullman Co Board.

Lord Dalziel's personal byword was 'Pullman and Perfection'. He had been knighted in 1918, as Sir Davison Alexander Dalziel, and had become Lord Dalziel of Wooler in 1927.

Although by 1926 British builders had commenced supplying all-steel Pullmans for use on the Continent (and had already supplied some all-steel sleeping cars for Wagons-Lits as early as 1922), the British fleet of Pullmans was entirely wooden-bodied in the mid-1920s, although the more recent cars had steel underframes and bogies. Wooden bodies were excellent sound absorbers, and gave a comfortable 'feel' to the passengers because they also absorbed some of the shocks and movements of the bogies beneath, but they always represented a fire risk, and there was the danger of sudden disintegration and splintering in a severe accident. The older American wooden cars had been heavier and larger than the ordinary British carriages of the period, but this difference had been progressively diminished, until by the 1920s the 'Big Four' all possessed ordinary bogie carriages of virtually the same dimensions as the latest Pullmans, and of similar forms of construction, with steel underframes. Whereas the old wooden Pullmans had acted either as 'battering rams' or 'shock absorbers' in the event of a collision, the balance was now more even and an old car would not necessarily come off best when struck by adjacent steel-underframed stock. The excellent Pullman gangway and buckeye coupler helped to prevent 'telescoping' of the end of one car into another, but not all the British Pullmans had this feature, because it was only the LNER which had wisely abandoned the screw coupling and side buffer layout in favour of the Pullman layout. Even the SR, which had the largest fleet of Pullman cars on its system, did not use the Pullman gangway and coupler for its own new electric carriages at this time. (It was normally possible to couple the two systems together by virtue of the drophead layout on the buckeye centre coupling — which exposed a conventional hook, when lowered, but there remained the problem of the difference in the gangway connection designs, which meant using an adaptor plate between the two types of gangways.)

It was on the SR, on 24 August 1927, that public attention was vividly drawn to the hazard of wooden-bodied stock in an accident, when the down 5.0pm Pullman car express from London (Cannon Street) to Deal was derailed whilst running at speed at the Shoreham Lane overbridge, near Sevenoaks. The 'River' class 2-6-4T No A800 *River Cray* jumped the rails at about 55mph and embedded itself in an embankment. The carriages behind, including the Pullman car *Carmen* were derailed and flung against the abutments of the bridge and stuffered terrible damage, with 13 people killed and 40 injured. The wooden-bodied Pullman was smashed like matchwood by the abutment. The car *Carmen* was originally built by Gilbert, in America, for the 1891 SER 'Car Train', and remodelled as a Pullman in 1920. It was scrapped as beyond repair after the crash, at which time it had been in service some 35 years.

The Pullman Car Co Ltd wisely decided to adopt all-steel construction for new cars for use in Britain, commencing with new cars for the 'Queen of Scots' Pullman, and deliveries of these commenced in 1928; (although the 1951 'Golden Arrow' cars reverted to a timber-framed, steel-panelled arrangement). The same year, the company opened its new workshops at Preston Park, near Brighton and moved the repair activities from the Longhedge, Battersea location. Battersea still figured in the company's affairs, however, because it was there that the commissary depot was located, handling the supplies for the cars, and the laundering of their linen, etc. Longhedge had first been occupied by the Pullman Car Co in 1912.

Meanwhile, the mighty London Midland & Scottish Railway, the inheritor of the self-styled 'Premier Line' (the London & North Western Railway) and of the Midland Railway — which had first of all welcomed, and then spurned Pullman cars — did not seem to take much interest in them either, and made no attempt to expand their use to England although the Scottish cars, inherited from the Caledonian Railway plus a few more postwar-built, were still under contract to run as Pullmans until 1933. A small Glasgow office, subsidiary to the London office of the Pullman Car Co, looked after Scottish affairs. The LMSR used the cars on a number of new Scottish routes, including between Glasgow (St Enoch) and Ayr (formerly the Glasgow & South Western Railway's territory) and the cars also travelled over former Highland Railway lines to Aviemore, with Carlisle and Perth as other regular locations. The observation car *Maid of Morven* ran between Glasgow and Oban, as usual, but now there were also Pullman restaurant cars used on the route, as well as for Edinburgh and Callander. As there was no

112

112
A change of livery took place from about 1929 onwards, whereby the fascia panels above the windows were repainted in umber brown to match the waist panels. The new scheme was evidently easier to keep clean. In this view the down 'Golden Arrow' express is seen near Petts Wood on 27 March 1931, hauled by R. E. L. Maunsell's classic 4-6-0 No 850 *Lord Nelson*. *Ian Allan Library*

further expansion of the LMSR Pullman fleet by 1933, this is a convenient moment to record that when the contract expired that year, it was not renewed (shades of the MR!), and the LMSR purchased all the cars, which were gradually repainted in its own crimson lake livery, with the name Pullman deleted. Twenty two cars survived in service long enough to become British Railways property in 1948, when they were still in use as restaurant cars in Scotland, finally sporting BR livery.

One must not overlook the tremendous Pullman achievements in America by the late 1920s, and the fabulous era of the transatlantic liners which linked Britain and France and North America. Any self-respecting American visitor to British shores, pampered in great style whilst sailing across the Ocean, would expect to find Pullman standards of comfort for his rail journey from the port to the capital of London. But this quality of service, almost taken for granted in America, was not available! Thus in an effort to please this wealthy, if somewhat exclusive patronage, the Great Western Railway finally accepted the idea of Pullman cars running over its tracks, from Millbay Docks, Plymouth to London. This service commenced in May 1929 and was called the 'Ocean Liner Express', a name already in use for the ordinary boat trains, which were certainly not lacking in comfort, although not with Pullman levels of service to each seat. The problem with boat trains is that they only operate when required, in connection with the arrivals and departures of the ships themselves. The Pullman Car Co therefore sought some additional use for its services on the GWR, to obtain better staff utilisation and to keep their cars fully operative. In July 1929 the 'Torquay Pullman Limited' was therefore placed in service on a trial basis, running from London, Paddington non-stop to Newton Abbot, and then on to Torquay and Paignton; it returned the same day, in order to use only one set of cars.

Any writer who dares to question the activities of the almost sacred Great Western Railway is liable to be 'tarred and feathered' by its immense present-day following! Nonetheless it has to be said that the GWR deliberately played against the Pullman enterprise, which it had grudgingly allowed on its tracks to please the transatlantic voyagers. The Pullmans, both cars and staff, were resented by the GWR staff right from the start, and constant operating difficulties were created by the men on the ground who professed not to understand the additional operating and servicing requirements of the Pullmans. It is said that the hostility even extended to the General Manager, Sir James Milne, who did not get on with the Pullman Co. It was probably mutual, because its own 'Torbay Express' with faster timings and no supplementary fares for its excellent catering facilities, the Pullmans were literally 'run off the road' by the GWR, and the Pullman Car Co retreated in defeat, withdrawing all the cars in 1930, including the boat trains.

The pampered transatlantic clientele evidently did not approve of the withdrawal of the luxurious Pullmans, and the GWR quickly designed and produced its own version specifically for the Plymouth boat train traffic. Of course, they were not called Pullmans (although some people did mistakenly do so) but they carried names on the sides in full Pullman style. Moreover, the GWR chocolate and cream livery was very similar indeed to the umber and cream of the Pullmans. The names chosen were also in Pullman tradition, being members of the Royal Family, such as *King George V* and *Queen Mary*. There were differences, such as bodyside doors that opened outwards instead of inwards (*not* an improvement!) and the famous oval toilet windows were not copied, but the interior was a slavish

113

113
On the LNER, what was known as the 'Pullman Link' was established, using certain drivers and firemen, whose exploits of speed and punctuality records gained them a fine reputation. The 'Pullman Link' favoured the Ivatt Atlantics for these workings for many years, and they got some really astonishing work out of these engines. Here is Class C1 4-4-2 No 4436 at the head of the 'West Riding Pullman' in the early 1930s period. The third vehicle is a LNER bogie brake van, probably separating the Bradford and Halifax portion of the train from the Harrogate and Newcastle portion, the division being made at Wakefield. *Real Photos (T8363)*

114

114
The 'Eastern Belle', which ran 'trips-de-luxe' on the Great Eastern section of the LNER during the 1930s, going to a variety of destinations throughout the week, and as the 'Clacton Pullman' to Clacton on Sundays. Saturday was normally a day of rest for overhaul and maintenance to be undertaken. Pictured here is rebuilt 'Claud Hamilton' 4-6-0 No 8800, seen passing through Chadwell Heath on 30 May 1933 with the 'Eastern Belle' excursion train. The LNER also introduced a 'Northern Belle' tourist train, but not using Pullmans, despite adopting the Pullman 'Belle' brand name. *Ian Allan Library*

copy of contemporary Pullman ideas, with tables and large moveable armchairs and individual table lamps, and there was a supplementary fare for the privilege of using these special 'Super Saloons'. In passing, there was a story — not validated so far as I know — that the GWR actually had the Pullman cars secretly measured up by its Swindon draughtsmen, to base their new design upon, whilst the cars were stored out of service at Old Oak Common, London in 1930.

The Pullman Car Co was faced with not only a financial flop, but also a redundant fleet of cars and staff. The Pullman Car Co therefore looked around for a possible new use for them and the Southern Railway, which appreciated its services, soon came up with the solution, which ironically placed the cars in competition with the GWR to some extent, because the SR was keen to capture a big part of the transatlantic boat train traffic. To do this it was developing the railway-owned Southampton docks as an ocean liner seaport, capable of taking the latest and largest liners, which were becoming too large for Plymouth. To cater for the new and highly discriminating clientele that could afford to travel in such style, a service of Pullman cars was required, and the company was able to provide these immediately. Once again, Pullman cars commenced running over the tracks of the former London & South Western Railway, from London's Waterloo station. At about the same time a few Pullman cars began operating as dining cars on Scottish area services of the LNER, marshalled into ordinary trains, as opposed to the all-Pullman 'Queen of Scots' service, and in some cases the LNER Pullmans ran in competition with the LMSR Pullman cars, in particular between Edinburgh and Glasgow.

From this period also dates the establishment of perhaps the most extraordinary of all Pullman trains in Britain. This was the 'Eastern Belle Pullman Limited' launched in the summer of 1929. I have briefly mentioned already the promotion of Clacton as a seaside resort. This was by the operation of a Sundays-only 'Clacton Pullman'. Now the cars (which were the wooden-bodied cars displaced by the new all-steel type on the GN section) were to be employed on weekdays as well, for the summer season only. Each day throughout the week the 'Eastern Belle' ran to a different East Anglian holiday resort. For example on Mondays it might go to Cromer, or Felixstowe Town; on Tuesdays to Skegness, etc. These 'trips de luxe', with full Pullman service became very popular indeed and continued until the outbreak of World War 2. Ordinary fare paying passengers were allowed on board, if there were spare seats, but they had to pay the supplement. Of course, some people booked a whole week or more of journeys in advance, enjoying the comfort and dining facilities of the train, plus a 'change of scene' each day. Clacton remained the Sunday destination, and the train ran as the 'Clacton Pullman' at a higher fare than the weekday trips, even if the 'Eastern Belle' went again to Clacton on a weekday! Another Pullman car service which had commenced running out of Liverpool Street station in 1927, was known as the 'Flushing Continental' and gave a connection at Harwich, Parkeston Quay to the Zeeland Steamship Co service to Flushing; where the Wagons-Lits Co had a connecting Pullman car in the boat train to Amsterdam. I have already referred to this train in Chapter 12.

Apart from the Great Western débacle, the fortunes of the Pullman Car Co had greatly improved by the commencement of the 1930s, offering standards of comfort and service which were still the envy of the emergent competitors — the road motor coach and the aeroplane — as well as some of the motor car manufacturers. The name Pullman was synonymous with stylish travel, but the names of Ford, Morris, Austin, Lanchester and others, were becoming equally recognised as synonymous of personal freedom, and the ability to travel at will. The car was a threat not only to the services of the Pullman Car Co but to the 'Big Four' railways. The 1930s were to witness some remarkable developments, to combat the growing competition on the roads and in the air, and to cater for the changing social tastes and needs in Britain.

CHAPTER 14

'Belles' and Streamliners

The Pullman Car Co had since the end of World War 1 steadfastly clung to its traditions in respect of both service and car design, so that by 1930 the typical Pullman car was still instantly recognisable as such by its furnishings and decor, as well as by its external style and livery. But the same could not be said for the typical railway carriages of the Big Four. Technical advancement there was a-plenty, with bogies, brakes, steam heating and electric lighting vastly superior to those of the turn of the century, but the interior design had become both plain and dreary, apparently as a reaction to the sumptuous Edwardian phase when the private railways had almost outpaced Pullman in providing lavish decor and luxurious accommodation, in particular in dining cars.

The increasing use of side corridors, the change from the clerestory roof to the elliptical form, and the preference for the compartment layout had altered the physical dimensions of British carriages considerably, but their furnishings were dull in the extreme, with bleak white ceilings, fussy little lampshades, cloth-upholstered seating and dreary wooden panelling devoid of inlaid decoration. It was all excused away as 'neat simplicity' whereas 'lack of imagination' would have been more appropriate. If one searches for any real improvements during this period, it was without question to be found in the vastly improved toilet facilities offered, and in the aforementioned heating and lighting developments. Only the Pullman cars resisted this ghastly fashion for austerity, and perhaps that made them all the more popular. A first sign of reaction, was the introduction on the LNER in 1928 of a dining carriage for the 'Flying Scotsman' decorated in the 'style of Louis XV'. Luckily it did not catch on! Instead a trend towards jazz patterns for furnishings — often in contrasting and gaudy colours — began to develop. It had much in common with the furnishings of the new suburban homes, and of the upholstery of the road motor coaches and buses of the period. It only succeeded in making the Pullman cars seem even more exclusive and refined by comparison, although at a later date

115

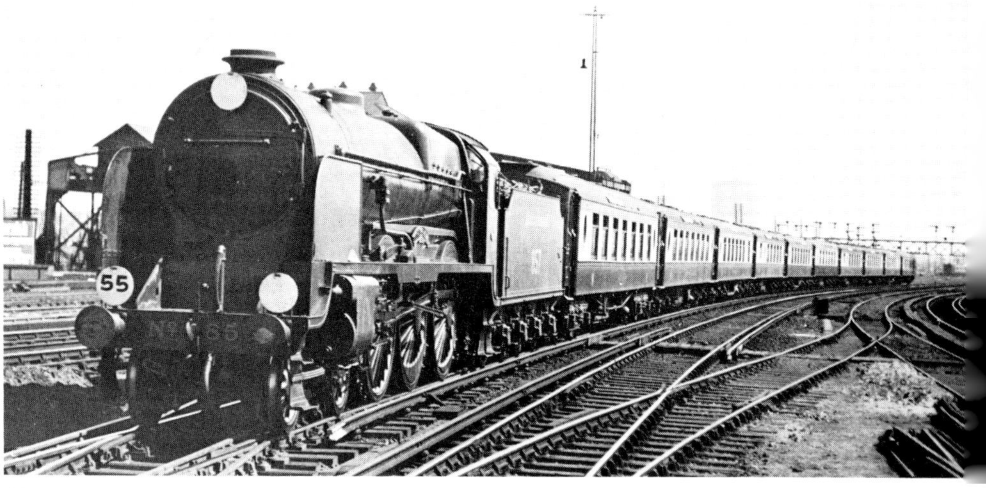

116

115
Another 'Belle' had gone into service on 5 July 1931, between London (Waterloo) and Bournemouth, aptly named the 'Bournemouth Belle', and running in the summer season only (until 1936 when it became a daily service). A variety of existing four-wheel and six-wheel bogie Pullmans were used, and no vehicles were specially built for this train when introduced. In this view, taken at Clapham Junction on 4 June 1933, the down 'Bournemouth Belle' is headed by 'Lord Nelson' class 4-6-0 No 857 *Lord Howe*, and the leading car is an ex-SECR example, on American-type four wheel bogies.
H. C. Casserley

116
An 11-coach business train on the Brighton line, where commuter services had become so heavily patronised during the 1920s and early 1930s, that the SR wanted to extend its third rail electrification from the London suburbs all the way to the South Coast towns, in order to cater for the growth. Remarkable in this picture is the presence of no less than three American clerestory-roof wooden-bodied Pullman cars (the leading vehicle and the fourth and fifth) out of the four Pullmans included in the train's formation. Two American cars lasted in service on the Brighton line until 1933 *Pavilion* and *Princess Patricia*, whilst the last day of steam haulage between London and Brighton, the 31 December 1932 had seen *Duchess of Fife* and car No 4 (previously *Albert Victor*) still in use. At the head of the train, in this picture, is 4-6-4T No B328, one of Lawson Billinton's majestic Baltic tank engines. *Ian Allan Library*

117

117
New Year's Day 1933 witnessed the start of electric train working between London and Brighton. It also heralded a new generation of Pullman cars, specially designed and built for the SR electrification. Included in the order, placed with the Metropolitan-Cammell Carriage & Wagon Co, were three five-car all-steel multiple-unit sets to operate the 'Southern Belle' (renamed the 'Brighton Belle' in 1934). The livery was the standard umber and cream, complete with the Pullman Co coat of arms on the front and sides. A 10-car formation is seen here at Merstham, after the units had been renumbered; leading is No 3053, originally No 2053. The SR classified them as 5-BEL units. *Ian Allan Library*

(mid-1930s) the Pullman Car Co had a brief flirtation with jazz-pattern upholstery!

The livery of the Pullman cars underwent a modification from 1929 onwards, as cars fell due for repainting. The umber was retained for below the waist, but replaced the cream above the windows on the fascia board carrying the name Pullman, effectively sandwiching the cream portion. Henceforth all new cars were delivered with this scheme, and it remained the standard livery of Pullmans in Britain until 1939-45. Exceptions were the SECR Car Train cars now running as Pullmans, in crimson lake livery; these ended their days in the crimson colour but the more recent Pullman cars for the ex-SECR lines were repainted in due course to umber and cream. The two Metropolitan Railway Pullman cars were also a crimson lake livery; no doubt because the original cream and umber was difficult to keep clean on stock running through lengthy tunnels all its working life.

A new 'Belle' was added to the Pullman fleet in the summer of 1931, mostly made up from existing cars. This ran between London (Waterloo) and Bournemouth, and was to begin with a summertime only working, commenced on 5 July 1931. The 'Bournemouth Belle' was to prove very popular, and became a daily service in 1936. To improve services to the port of Southampton, already operating with Pullmans in the special Ocean Liner expresses, the 'Bournemouth Belle' was scheduled to make a Southampton stop in each direction; which no doubt added to its popularity amongst businessmen in particular.

A change in social patterns had developed during the latter part of the 1920s, and it was also changing the landscape through which the stately Pullman cars ran by the early 1930s. Aided and abetted by the electric tramways, by buses and coaches, and by recent suburban railway electrification, the urban population was migrating to the newly-built suburbs around London. The Metropolitan Railway's trains, and those of the SR to the South Coast, now ran through a considerable area of housing development before reaching green fields and hillsides. On the Southern Railway this growth justified electrification schemes to transport the increasing number of suburban dwellers back into their City work areas each weekday. Moreover, as the gaucherie of the suburbs spread fanwise bringing row upon row of 'mock Tudor' and other horrors, the previously remote landowners of the Surrey and Sussex, Kent, Hampshire and Buckinghamshire county towns and villages felt pressurised and many of them sold off land to the developers, meanwhile establishing new homes in more 'refined' areas. A considerable growth of new housing took place along the South Coast, because it was in such easy reach of London, but just far enough way to avoid becoming part of 'suburbia'.

The Southern Railway was anxious to keep its main line services to and from the South Coast competitive, at a time when the roads were being developed and improved, and more and more motor vehicles were running upon them. It was steadily pushing the surburban electrification farther out of London to supply the needs of what we now call the 'commuter', and the SR realised the logical step of extending the electric power supply all the way to the South Coast, enabling main line trains to be formed of electrically-powered rolling stock. (There had been an early 1900s proposal for an independently-run London-Brighton electric limited-stop express railway, to compete with the LBSCR; this would have used Pullman cars, but the scheme never materialised.)

118

119

118
A first class parlour car of one of the three 5-BEL units, as first placed in service in 1933. Notable was the improved ceiling with enclosed lighting behind peach coloured glass, frosted with rising suns. The very elaborate mirrors in the partition gave an added individuality to the very high quality wood panelling and furnishings. Once again the luggage racks could be criticised as being of remarkably small capacity for a train which could expect heavy patronage. These racks, and the door furniture and light fittings were all silver-plated. The pictorial images in the marquetry panelling differed from car to car, and to some extent so did the upholstery, although carpets were uniform. *BR*

119
The third class accommodation of the 5-BEL units (and of the SR electric composite cars) retained the 2+2 layout of earlier short-distance locomotive-hauled cars built for service on the LBSCR and GER. Jazz-pattern upholstery was evidently considered to be appropriate to third class travellers' tastes of the day! A new system of pressure ventilation was used in the electric cars, as well as the standard sliding toplights in all passenger area windows. Large wall lights gave added illumination to each seating bay. *BR*

On New Year's Day 1933 the SR introduced the first main line electric railway service in Britain, over the existing tracks between London and Brighton. The Pullman Car Co had co-operated from the outset, and 38 new all-steel Pullman cars were supplied by the Metropolitan-Cammell Carriage & Wagon Co Ltd for the scheme. They were the only Pullmans ever built for multiple-unit electric services. The SR chose not to replace steam locomotives by electic locomotives, but instead to place traction motors — or 'power bogies' — under the leading and rear cars of set formations. These sets, or units, were arranged to be able to work in multiple, hence the term. Fifteen of the new cars were for the 'Southern Belle', and the remainder were individual cars, which were coupled within ordinary SR main line multiple-units with gangway connections, for use as refreshment cars with supplement.

The 15 'Southern Belle' cars were divided into three complete Pullman units, and as a rule two were in use with the third as standby. The end cars, known in SR parlance as 'motor cars' had a driving compartment at one end, with no gangway connections, and the bogies on these cars were powered by four BTH 225hp traction motors, all the other cars being unpowered trailers. The livery remained the standard umber and cream, complete with the Pullman coat of arms on the front ends, which were umber, with a black bufferbeam. A design change as subtle as it was attractive was that the sides of the cars were sloped inwards slightly from waist to cantrail level; whereas Pullmans had previously been noted for their flat sides, quite unlike the curved 'tumblehome' characteristic of British carriage design. For the three 'Southern Belle' sets this feature made an attractive ensemble, but the individual Pullman composite trailer cars did not match the contour of the rest of the trains of ordinary stock; which were also different in being painted in green livery. It helped to boost the prestigious Pullman image however, because they looked so different externally.

The interiors were superb, and clearly reflected a real effort to produce the *best*. For the 15 'Southern Belle' cars each had an individual interior, if only in small details such as the wood veneers and pictorial marquetry which panelled the walls. The first class cars carried names in true tradition, including all the composite first and third class trailer cars. An innovation was the use of electricity for cooking purposes — quite naturally — instead of the gas system, which had first been introduced in 1908 on the steam-hauled 'Southern Belle'. The composite individual Pullmans were marshalled into the middle of trains composed of five ordinary carriages — the end two being motor cars — and formed what were known as the 6-PUL and 6-CIT sets, the latter having more first class seating, specifically for London Bridge services. The 6-PUL and 6-CIT sets were used on services to Brighton, Worthing, and in 1935 to Eastbourne, Hastings and Ore, when electrification was extended there. Haywards Heath was now very well served by Pullmans, into the bargain. The Pullman Co held an exclusive contract to provide all train catering services on the SR's Central and Eastern Sections. After 1933 some cars were replaced by Pullman-staffed, railway-owned catering vehicles, which gained the description of ROVs in Pullman parlance.

Eighteen months after the commencement of electric Pullman services to Brighton, the 'Southern Belle' was renamed as the 'Brighton Belle', with a suitable christening ceremony at Brighton station on 29 June 1934, performed by Miss M. Hardy, then the Lady Mayor of Brighton. The 'Brighton Belle' was a name destined to become very famous.

The introduction of the new all-steel cars for the SR electrification released a lot of the vintage wooden-bodied cars from the stalwart service they had given on the Brighton line, although by 1933 only two American-built cars remained, *Pavilion* and *Princess Patricia*. Not all were scrapped however, and a number of the withdrawn cars were purchased for conversion into seaside bungalows, whilst two ended their days in rural pasturage at Partridge Green, Sussex, grounded without bogies. Later they were used for some time as a studio by the railway artist and writer C. Hamilton Ellis (see list for details). About this time some older Pullmans were converted into what were called 'supply cars', or 'kitchen cars' which did not carry passengers, but which were marshalled into excursion trains, or boat trains, to supply refreshments to adjacent saloons. These supply cars were painted in an overall umber brown livery, and it seems that they were produced as an economical answer to a need for more kitchen and storage facilities at peak periods.

On the LNER in the early 1930s, where the new all-steel Pullman cars were giving very satisfactory service, the railway's Chief Mechanical Engineer Sir Nigel Gresley and his assistant Oliver Bulleid were concerned to improve the comfort and amenities of the carriages owned by the railway company. They were also concerned to improve the speed and schedules of the main line express trains, to counter the competition of the West Coast route operated by the LMSR. This was the era of spectacular 'non-stop' running, a sort of battle of superiority to gain public accliam, and to

120

120
The Pullman Car Co indirectly benefited from the introduction of the Gresley streamliners, in as much as they shared the superb new Pacifics of the 'A4' class (that were specially designed for the services) on numerous occasions once there were sufficient examples of the class in service. In terms of power and speed the 'A4s' were superlative, and could whisk along the stately Pullmans in fine style. Alas neither the Pullman standard bogie, nor the earlier 12-wheel bogie, had been designed for the sustained 90-100mph running (or more) that the Gresley Pacifics could provide, and drivers had to give due attention to the passengers' comfort — not to mention the need for the Pullman attendants to remain on their feet whilst serving to the tables! Here is the 'Yorkshire Pullman' of the 1938/39 period, headed by double-chimney Class A4 No 4903 *Peregrine*, with the locomotive in striking Garter blue livery. *E. D. Bruton*

publicise the services in the face of ever-increasing road competition. Better carriages were just one part of the activity, but it is significant that neither the LMSR nor the LNER opted to use Pullmans; choosing instead to improve their own rolling stock. The LNER honoured its existing contract for Pullman car services, but was more and more inclined to offer modern alternatives of its own, without the Pullman supplement. In 1930 the carriages of the principal morning London, King's Cross-Leeds service, and the afternoon return working were fitted with headphone sockets at every seat, and passengers could enjoy radio programmes. When passing through long tunnels however, the radio reception was blotted out, so a record player was necessary as well to fill the gaps! It was a free service, with headphones available upon demand. This idea was next extended to Anglo-Scottish trains, and the LNER also produced a 'cinema coach' with hour-long programmes of newsreels, documentaries and cartoons; operated by the Pathé Co within a converted passenger brake van. It attracted a good deal of patronage from those who preferred not to look out of the windows, or at their fellow travellers. For night travel, Gresley and Bulleid produced some superb first class sleeping cars, and both the LMSR and the LNER had introduced third class sleeping cars in 1928, with the GWR copying the idea the following year. For the prestige 'Flying Scotsman' — which had cooking by electricity, and electric refrigeration before these appeared on British Pullmans — Gresley really went to town, and there was a hairdressing salon (for both sexes) and a ladies retiring room. All these ideas had been pioneered by the American Pullman Co, but for some reason the British company was slow off the ground, preferring to retain its recognised role.

The GWR had placed itself in the limelight with 'The World's Fastest Train' — the 'Cheltenham Flyer' — until 1933 when the German diesel train, the 'Flying Hamburger' seized the title. The battleground was now one of *speed*, and to assist in producing faster trains, the idea of streamlining the exterior was gaining

121
A Southern Railway London-Hastings train at Grove Park in 1938, headed by Maunsell's 'Schools' class 4-4-0 No 938 *St Olaves*. The second vehicle is the 1926-built Pullman car *Barbara*, one of six built to the special narrow width of only 8ft 1in for the route. In original form it had seating for 20 first class passengers, plus toilet and kitchen facilities, but in 1932 these cars had been converted to 'composite' first and third class. After the war (1946) they were converted again, three becoming first class cars once more, for the 'Ocean Liner' boat trains to Southampton (*Camilla*, *Latona* and *Theodora*) and the other three (*Barbara*, *Madeline* and *Pomona* as non-supplement buffet, or refreshment cars (Pullman-staffed) on the Hastings Line. *Barbara* survives in use today on the Kent & East Sussex Railway, together with *Theodora* (see Chapter 20). *Lens of Sutton*

121

SOUTHERN
938
81
No 938

122
The 'Bournemouth Belle' awaiting departure time at Waterloo station on 21 May 1938, with 'Lord Nelson' class 4-6-0 No 864 *Sir Martin Frobisher* at the head. Next to the locomotive's tender is a 12-wheel 'Car No 40 Third Class' dating from 1921. This car was destined to be destroyed in an air raid at Preston Park and was condemned in 1943. Nearest the camera is 'Car No 36 Third Class': a standard 'K'-type parlour car of 1926, with the standard four-wheel Pullman single bolster bogie. Note that the umber brown paintwork was extended to cover the guard and luggage accommodation area of the 12-wheeler; the cream paint stopping short at the passenger portion. *National Railway Museum*

122

popularity — in particular in America and Germany. Diesel traction was emerging as an alternative to steam to power these high speed 'streamliners' and the publicity surrounding these new developments was a useful weapon in the increasingly severe competition from pneumatic-tyred road vehicles. In Britain, with ample coal supplies, the steam locomotive remained in favour, and the mid-1930s were to witness its zenith of development in a high speed streamlined form — basically for aerodynamic reasons, but also to gain publicity.

First off the mark was the LNER, when Sir Nigel Gresley designed the Class A4 Pacific for the London-Newcastle 'Silver Jubilee', introduced in the autumn of 1935. A complete new train was built, the flush-sided carriages having a silver rexine finish. The silver grey livery was extended to the locomotive itself and the appearance of this ultra-modern ensemble caused a sensation only equalled by its performance, when a maximum speed of 112½mph was achieved on a demonstration run. The interior of the new carriages was designed by Sir Charles Allom, whose clientele included the Royal Family and Cunard, and it broke away from established style. London to Newcastle was run in four hours, and a supplementary fare was charged by the LNER for the use of the 'Silver Jubilee' service.

The streamlined shape of the train, its new livery and style, succeeded in establishing a new fashion for railways, in keeping with the tastes of the 1930s, and it prevented the steam locomotive from becoming thought of as obsolescent in the public eye a time when the aircraft and the motor car were rapidly developing 'modern' styles of their own, which caught the public's attention by their suggestions of 'progress'.

The year 1937 witnessed two new streamlined trains of even more impact, one of the LNER and one of the rival LMSR. Both celebrated the fact it was Coronation year in Britain, and carried the names 'Coronation' and 'Coronation Scot'

123

124

123
At first glance this could be mistaken for a typical LNER East Coast main line Pullman working; complete with Ivatt superheated 'Atlantic' at the head. However, the American-type bogies on the leading Pullman car, and the two kitchen windows on the second car show that these are from the Southern Railway South Eastern Section. In fact the whole train of eight Pullmans had been sent from the SR to the LNER to work a Race Special from London (King's Cross) to Newmarket — the sort of charter trip that the Pullman Car Co was always happy to arrange. The locomotive is Class C1 4-4-2 No 3295; the time probably sometime in the mid-1930s. *Ian Allan Library*

124
London Transport continued to operate the two Pullman cars on their Metropolitan line, until the outbreak of war in 1939. Pictured here is the *Mayflower* at Aylesbury on 2 May 1936, in latterday crimson lake livery. No gangway connections were carried by these cars. Beneath the oval toilet and kitchen windows wooden plaques carried the message: 'PULLMAN CAR. Available to holders of First Class Tickets on payment of a Supplement. Refreshments Available'. *H. C. Casserley*

125
The Royal train conveying Their Majesties The King and Queen, en route from London to Portsmouth Docks, at the commencement of their visit to Canada and the USA, on 6 May 1939. The train was formed of 'K'-type Pullman cars, hauled by Class T9 4-4-0 No 718, resplendent with the traditional four head lamps (white discs in daytime) on the SR that denoted a royal train in Britain in steam days.
Ian Allan Library

respectively. New speed records were established, and both trains had striking new liveries, and specially fitted-out carriages. The LNER train included a special 'beaver-tail' observation car at the rear. Neither train had any involvement with the Pullman Car Co, but both demanded a special supplementary fare or seat reservation price, and both provided *de luxe* amenities. The Pullman cars were in danger of being regarded as passé. However, the Company realised the competition, and designs were prepared in 1938 for new cars for the LNER services. To match Gresley's achievements it was agreed that these should feature his double-bolster design and underframe construction (allowing higher speeds and more comfort). This meant a return to a wooden-framed, steel-bodied construction. The first trains due to receive them were the 'Queen of Scots' and the 'Yorkshire Pullman' and orders were placed for an initial seven cars in 1939, with the Birmingham Railway Carriage & Wagon Co. The outbreak of war put a stop to them, and they were not finally constructed until 1951, becoming some of the new cars for the SR 'Golden Arrow' service (see page 114) instead.

The previous year had seen the introduction of through sleeping cars between Paris and London, using specially-built train-ferry vessels for the Channel crossing. Known as the 'Night Ferry' service, the sleeping cars were owned by Wagons-Lits, but built to the British loading gauge. The service began on 14 October 1936, and was very successful, running until the outbreak of World War 2, when the cars were returned to France. In England two Pullman cars were attached to the train to serve meals*, and as such the train represented the joint enterprise of the two companies — once such bitter rivals! George Mortimer Pullman had dreamt of such a through service, as part of his plan to operate Pullman cars 'all over Europe', but Mann and Nagelmackers had thwarted his efforts, as recounted in Chapter 7. The Wagons-Lits contract with the SR in England was until 1976, and the service was resumed after the war, starting again on 4 December 1947, as related in Chapter 18.

Whilst these exciting developments were taking place on the rails, there were significant events in the skies above. The 'Big Four' railways were aware of the threat to their long-distance services posed by the civil airliners of the developing private air network and they wanted to stifle the competition. They were empowered by Act of Parliament to operate and own aircraft of their own, and the GWR had commenced air services in 1933. These were later merged into what became known as Railway Air Services Ltd (RAS) which had developed considerably by the time the streamlined trains were being introduced. RAS services were operated principally on routes that were poorly-served by rail. The capacity of the de Havilland planes was very limited, but it was regarded as a prestigious way to travel, and it posed yet another potential threat to the long-established services of the Pullman Car Co. More serious was the threat to established international services — London-Paris, etc — and the 'Golden Arrow' was demoted to a mixed Pullman/ordinary stock formation in the 1930s. Audaciously, the passenger accommodation of the Handley Page HP42 airliner was described as 'Pullman-style'!

Events on an international scale were however destined to interrupt *all* these developments in a most dramatic manner; with the declaration of World War 2 in September 1939.

*In a recent, excellent book *Night Ferry* by George Behrend and Gary Buchanan, Jersey Artists Ltd, details are given of the Pullmans used. A very interesting feature was that one Pullman served the sleeping cars and the other, the 'foot' passengers, and the gangway between the two Pullmans was kept strictly locked to passengers and supervised by the Customs Officer who patrolled the train. The gangway end of the Pullman adjacent to the French sleeping cars had to have a specially adapted connection to 'marry' it to the continental type. Third Class Pullmans Nos 17 and 19 were the two cars so adapted (one as spare vehicle). Soon after the service was reinstated after the war, ordinary SR catering cars were used, but staffed by Pullman men.

126
Attention to detail was a Pullman tradition, carefully nurtured over the years. Passengers were greeted on the platform and shown to their seats. Each car had a metal plaque fixed on the doorway, in this case showing it was car D in the seating plan of the 1939 'Bournemouth Belle'. When the train was ready to leave the plaque was removed and the door was closed by the attendant or the Conductor; in this case Conductor Werry, pictured at Waterloo station on 10 June 1939. *National Railway Museum*

CHAPTER 15

Interregnum and Postwar Restoration

The outbreak of war did not take the railways by surprise, because as early as 1937 their managements had held talks with the British government, to plan their actions if the worst came to the worst, and war was declared. On 1 September 1939 the government took formal control of the 'Big Four' railways and London Transport and established a Railway Executive Committee (REC). The REC promptly withdrew from service all the streamlined trains and the Pullman cars. No Pullmans ran between 11 and 17 September 1939, but then it was realised that on the Southern Railway the electric cars were useful to the war effort and the 'Brighton Belle' and some single cars were restored to service and these provided 'limited catering facilities' to begin with. There was however no Sunday service run with the Pullmans, some of which had the cream parts of their livery repainted umber in due course to make them less conspicuous to enemy aircraft.

127

127
The wartime NAAFI catering organisation took over six SR South Eastern section Pullman cars for war use as buffet cars for military trains. They were repainted in a plain green livery and the NAAFI numbers appeared on the sides above the solebar level adjacent to the entrance doors. This is NAAFI Car No 29 at Preston Park workshops after return to the Pullman Car Co for reconversion to a Pullman. The exact identity is uncertain, but it is evidently a former SECR Continental services car, complete with American-type four-wheel bogies.
National Railway Museum

128
Photographs of Pullman cars actually in use during World War 2 are extremely rare, due to the conditions of secrecy that surrounded much of their activities. However, the Southern Railway issued this publicity photograph in July 1941, depicting the second Bulleid 'Merchant Navy' class Pacific No 21C2 *Union Castle* making a special VIP demonstration run from London to Brighton and back (pictured at Merstham en route to Brighton). The invited guests travelled in the two Pullmans (second and third vehicles) which no doubt offered suitable refreshments despite the wartime restrictions! Both Pullmans are in drab all-over umber paintwork, to be less conspicuous to enemy aircraft. The 'Merchant Navy' class had been proposed prior to the war by Bulleid, for the heavy 'Continental Express' services; wartime interrupted the plan, but it was the first of these locomotives, No 21C1 *Channel Packet*, that hauled the first postwar 'Golden Arrow' train. *Ian Allan Library*

129
Back in business! Using refurbished prewar Pullmans, the 'Golden Arrow' started running again in April 1946 — and all the Pullman trimmings were back in place, making a superb presentation after the dreary wartime years. Note that the name of the train, plus a large golden yellow arrow was actually painted on each end of the umber waist, necessitating the Pullman coat of arms to be moved to the extremity of the waistband. The Pullman attendant awaits the 'right-away' for the first peacetime departure from Victoria. When the whistle blows he will remove the metal plaque for seat reservations in car P, and firmly close the entrance door. The vestibule doors opened inwards and were held in place when open by a large brass handrail — known as a commode rail — which passengers could use when boarding or alighting. The attendants ensured that all the handrails were clean at every station stop, before passengers used them. *Ian Allan Library*

128

On the LNER the all-Pullman trains were withdrawn from public service, and some cars were stored, but 30 were subsequently painted brown and loaned as ordinary first class carriages, complete with LNER running numbers. A few Pullmans were selected for special wartime use, including car *Joan* which was used by the King and Queen, and by Winston Churchill. Later in the war *Joan* was in a special train attached to an armour-plated LNER sleeping car, used by the Chiefs of Transportation of the US Army. It became known as 'General Ross's car' and was based at Cheltenham, working all over Britain. In 1962 it was specially groomed and sent to the London Midland Region to work attached to the 'Thames-Clyde Express' from St Pancras to Scotland, conveying the ex-US President Eisenhower during a British visit. On this occasion it was staffed by BR Catering Services, not by Pullman staff. In recognition of the important role it had played in wartime the car received a special service award, but unfortunately, *Joan* was later scrapped, despite its wartime importance. Other locomotive-hauled Pullmans were used by wartime VIPs in conditions of great secrecy throughout the war. These provided an 'Airways Special' between London, Victoria and Christchurch (for Hurn Airport) and Bournemouth West (for the Poole Harbour flying boats) which connected with trans-Atlantic flights, and ran as required. However, the majority of the locomotive-hauled cars spent the war years in storage, their luxurious appointments definitely out-of-step with such austere times. The two railway companies which still had contracts with Pullman, the LNER and SR, paid an annual compensation to the Pullman Car Co to keep the cars intact in storage, ready to be used again once the war was over. Six of the Pullman cars normally used on the London-Folkestone and Dover services were taken over by the NAAFI organisation

129

130
The 1917 12-wheel Car No 5 was remodelled as the *Trianon Bar* for the reintroduction of the 'Golden Arrow' service. Unfortunately it developed an axlebox fault on a press demonstration run, and was not able to commence the new passenger service; a hurried substitute being found (see next picture). The old car's interior had been dramatically modernised in pink, grey and cream coloured plastic finishes, as seen in this view of a corner of the bar saloon. It was intended as a showpiece of 'Things to Come' by Pullman Chairman Mr Stanley Adams, who had vested interests in the plastics industry. *Ian Allan Library*

131
Because the axle failure on the *Trianon Bar* (see previous picture) could not be rectified in time, the bar car *Diamond* was hastily renamed to become *Trianon Bar* for the opening of the 'Golden Arrow' service. The interior, which featured many traditional Pullman fittings was in marked contrast to the 12-wheeler's plastic finishes. This car later became *Diamond* once again and was then called the *Daffodil Bar* for use on the WR 'South Wales Pullman'. However, in the meantime it had also run with the names *One Hundred Bar* and *New Century Bar* on the SR, once the 'plastic' version of the *Trianon Bar* was back in service! *Southern Railway*

to run as buffet cars in military trains, for which they were painted in a plain green livery. The stored cars were painted all over in grey to render them less conspicuous.

During the night of 9/10 October 1940 whilst the Blitz was raging upon London, a bomb hit 'Brighton Belle' unit No 3052 parked inside Victoria Station. One of its motor cars was seriously damaged, and the unit had to be withdrawn from service for storage as a 'casualty'. The other two 'Brighton Belle' sets continued running until 14 May 1942, when they too were withdrawn 'for the duration', along with the composite electric cars in the 6-PUL and 6-CIT units. It appears that the 'Brighton Belle' cars were stored in Devon for most of the rest of the war, although there were also reported sightings at various times of Pullmans at Epsom goods yard; Glynde, Leatherhead, and Crystal Palace high level station (although it is possible that only the composite single cars were stored in these locations). The two Metropolitan Railway (London Transport from 1933) Pullmans were also withdrawn at the outbreak of war, and never reinstated, the contract being terminated mutually when the war was over.

Forced to bide its time apart from government war contracts, the Pullman Car Co had the misfortune to be bombed itself during the Blitz. The offices were located at Victoria station and on four occasions they were damaged by fire and high explosives. As a result many valuable records of Pullman history were destroyed. This alone accounts for some of the confusions that have arisen in various published works produced since that date. At the Preston Park workshops four Pullman cars were damaged beyond repair and the works building was hit on 25 May 1943, in an air raid, during which five bombs were dropped on railway property in the Brighton area by German fighter-bombers; twelve other Pullmans suffered some damage and no fewer than 43 Southern Railway electric coaches were damaged in the adjacent Lovers Walk sidings. Although 120 Pullman cars had suffered some damage of varying severity during the war years (60% of the roughly 200-strong 1939 fleet including the 30 non-supplementary-fare SR vehicles known as ROVs, or Railway Owned Vehicles), the fact that most of them had been stored meant that, in general, they were in better shape than the mass of ordinary railway vehicles, which had been in continuous intensive use with the minimum of maintenance for five long years, in many cases, flogged to the limit to keep wartime services rolling. When the war ended at last, the railways emerged in an exhausted and run-down state, and in poor financial shape. There was no question of restoring the brilliant prewar streamlined express trains, but it was feasible to reintroduce Pullman services quite quickly.

On the Southern Railway which had lost some 450 carriages during the war years due to bomb damage or sheer wear and tear and lack of proper maintenance, the prospect of reintroducing the Pullmans, for which the contract still existed, was an attractive one. For one thing, the resources to build new carriages of their own were very limited due to material shortages, and there was a big backlog of repair work to catch up with on the existing fleet. Meanwhile — just at the time when the railways could have done without it — there was a great surge of patronage. This had commenced whilst the war was still on, in the summer of 1944, when the skies overhead were menaced day and night by the German flying bombs — the 'doodlebugs' — and when despite a publicity campaign to the contrary, as mentioned below, the government had reluctantly agreed to allow civilians to travel by rail for holidays away from bomb-threatened areas, and to evacuate children. A war-weary populace, unable to use motor cars because of petrol rationing, had created chaos by thronging to the main line termini in order to escape for a few days by the seaside, or deep in the country. Until then the government had actively discouraged civilian journeys by rail, wanting to keep the carriage seats available for military use and other essential passenger traffic, and had issued posters saying: 'Is your journey really necessary?'.

After the British nation had celebrated the return of peace, it was still faced with austerity, and there was still fuel rationing. But people were determined to enjoy

132

133

134

132
Pullman services were restored one by one to both the SR and LNER, to some extent dictated by the availability of suitably reconditioned Pullman cars. A strike by workshop staff delayed the 'Queen of Scots' cars, and this service did not commence running again until July 1948 by which time Britain's railways had been nationalised. The tail end of the 11.30am 'Queen of Scots' Pullman is seen leaving King's Cross on 6 July 1948. A pleasing touch was the addition of train name boards on the rear of the train, as well as on the front of the locomotive. In the left-hand background a Gresley LNER Class V2 2-6-2 No 915 awaits its turn of duty — still painted in plain wartime black, in striking contrast to the gleaming Pullmans. *C. C. B. Herbert*

133
The year 1947 saw a new 'Belle' added to the Pullman list. This was the 'Devon Belle'; very much a service for holidaymakers. A special feature was the use of an observation car at the tail end of each of the two rakes of Pullmans required. These were fashioned upon the frames and bogies of vintage 1914-18 former LNWR ambulance train vehicles, which had already been rebuilt as Pullmans after World War 1. In style they had the austere modernity of the late 1940s/early 1950s period in Britain, although the livery was quite striking, with an abundance of cream stripes and large lettering. It was necessary to turn them at the end of each journey, on a locomotive turntable, in order to be facing the correct way at the tail end of the train, an inconvenience to operating staff. One of the two cars is seen here being pushed on to the turntable at Ilfracombe on 28 May 1949 ready to be turned for the next up train. The locomotive is Bulleid 'West Country' light pacific No 34041 *Wilton* in malachite green livery, with red and yellow 'Devon Belle' name boards on the smoke deflectors. *C. A. Todhunter*

134
'Devon Belle' observation car interior, showing the arrangement of settees and pivoting arm chairs. The seats were deliberately made a little hard, to persuade passengers to vacate their seat for others, after enjoying the changing scene and some light refreshments for a while, and to return to their own seat in the train.
National Railway Museum

themselves after years of menace and deprivation. The populace turned to the railways to transport them, only to find them in an appalling state because of the dedicated service to the nation they had performed from 1939-45. To make matters worse the 'Big Four' railways lost their RAS air services and aircraft. By 1939 the RAS, and associated companies, was responsible for some 80% of internal air mileage and business was booming. On the outbreak of war, the services had already been requisitioned for the war effort. As the war progressed some passenger air services were allowed to recommence, but it was not until 1944 that civil aircraft were allowed over the London area. In 1947 the RAS and other private international and internal air services were merged by the government to form British European Airways (BEA), and all railway involvement ceased. What had once been a developing asset of railway operations was now a positive rival to their interests, including the cross-Channel and Irish Sea railway-owned steamship services. This creation of BEA (now part of British Airways) was to have serious effects upon the boat train services of the Pullman Car Co as time progressed and the civilian way of life was restored to Britain.

However, the immediate situation presented to the Pullman Car Co when the war was over was a very promising one, because of all the care taken with the storage of their cars; although, as we have seen, they did not escape damage. If they could be reinstated swiftly, the company would be providing attractive trains, at a time when the railway's own carriages were run down, due to war use. The Pullman Car Co evidently realised the strength of its position, and it quickly set in motion a programme of repair and refurbishment and, in some cases rebuilding, for its fleet of cars. As a result Pullman cars began running again in 1946.

On 15 April 1946 the 'Golden Arrow' was once again standing in London's Victoria station, an immaculate rake of umber and cream cars, with a malachite green air-smoothed Bulleid SR 'Merchant Navy' Pacific locomotive at the head, decorated with large golden arrows and the flags of Britain and France. A few days earlier a special press demonstration had been run, to gain suitable publicity for the return of the Pullmans. Included in the train formation was a 12-wheel car which had started life in 1917 as third class kitchen car No 5, built at Longhedge. Earlier in 1946 it had been taken into Preston Park workshops for an internal refit, making the whole car into a bar. It was then renamed the *Trianon Bar*. The interior design was completely modern, using pink, grey and cream coloured plastic laminates, supplied by Warerite. The Pullman Car Co Ltd was an early and enthusiastic user of plastic laminates, particularly in the various bar cars. Mr Stanley Adams was a director of the Bakelite/Warerite Company, whilst Chairman & Managing Director of Pullman. This car was intended as a showpiece of 'things to come' and it was the first British Pullman car to break away completely from the traditional wood finishes and marquetry panels. No supplementary fare seating was installed in the car, which was intended for use solely as a bar, to which people 'stretched their legs' during the journey. On the press trip the 1917-era bogies betrayed their age, when a hot box developed, and the car had to be taken out of service, an unfortunate event in full view of the press.

135

135
A postwar development was the use of 'cigarette proof' Formica table tops, with a pattern of buff linette and with the Pullman coat of arms interlaminate printed. The effect was a little out of keeping with the pleasing 1920s wood marquetry panelling on the wall behind, in this LNER Pullman (photographed c1948) but the scheme was very practical, nothing being more unsightly than cigarette burns on wood or fabrics in elegant surroundings. *Ian Allan Library*

When the regular 'Golden Arrow' service commenced on 15 April, the *Trianon Bar* was still 'hors de combat'. A quick substitute was found, by renaming the car *Diamond* as *Trianon*. In this car there was a mixture of seating, with 12 first class supplement seats, and a bar occupying half the length of the car. The 12-wheel *Trianon Bar* was then renamed *New Century Bar* and put to use on the Victoria-Dover-Ostend boat trains. However, as it was always intended that this bar car should feature in the 'Golden Arrow' service, it was reinstated in July 1946 and renamed once again as the *Trianon Bar*! To add confusion upon confusion, the other car *Trianon* was then given the name *One Hundred* and later the name *New Century Bar* and put to work on the Ostend boat trains from Victoria and Dover. Thus the two cars had exchanged both names and roles over a very short period of time, suitably confusing latterday historians.

The spring of 1946 also saw the return to the SR of the 'Brighton Belle' (minus one five car unit, still to be repaired after the bomb damage). Next to reappear were the 'Bournemouth Belle' and the 'Yorkshire Pullman', the latter restoring the LNER link with the Pullman Car Co. The public responded to such splendid comfort and service in fine style, and the popularity of the Pullmans was re-established after a break of some six years; although only two of the 'Big Four' railway companies were running them, and in the case of the LNER there was a further delay in re-introducing the 'Queen of Scots' Pullmans because of a strike by the workshop staff refitting the cars. This service was restored in July 1948. A new 'Tees-Tyne Pullman' service was introduced in September of that year, between London, King's Cross and Newcastle, via York and Darlington. It could not match the prewar 'Silver Jubilee' in terms of speed (due to the deterioration of the tracks and signalling during the war) but it could certainly match it in comfort and service. The 'Tees-Tyne Pullman' was Newcastle-based, and included the excellent 'Hadrian Bar'.

Whilst the Pullman Car Co was steadily getting back into fine form, the 'Big Four' railway companies had another view of life, because of the threat of nationalisation that had arisen from the 1945 election of a Labour government, with a large House of Commons majority. The Labour party had always advocated the takeover by the state of the 'Big Four' railways, even during the war, and the Party was now in a position to implement this, as part of its overall objective which was to have state control of, and co-ordination of, all public transport services. They envisaged this utopia as finally ending 'the destructive war between road and rail' (to quote Herbert Morrison in a 1938 political pamphlet entitled *British Transport at Britain's Service*). The 1947 Transport Act, creating the British Transport Commission (BTC) came into effect on 1 January 1948. The Pullman Car Co was not included in the list of private companies that were taken into state control at that date; and its existing contacts with the SR and LNER were subsequently honoured by the BTC, whose railway affairs were handled by a subordinate Railway Executive.

Whilst the politicians were busy putting the final touches to their 1947 Nationalisation Act, the Pullman Car Co was invited to add another 'Belle' to their list, by Sir Eustace Missenden, the General Manager of the Southern Railway. The 'Devon Belle' commenced running on 20 June 1947, and operated a long weekend, Fridays-Mondays, service to suit the needs of holidaymakers, running only in the summer months. The 'Devon Belle' — 'Pullman's best to the Glorious West' — was run between London, Waterloo and Ilfracombe, with a portion going to Plymouth detached at Exeter Central. It had more third class than first class accommodation, and it ran with up to 14 Pullmans in the formation. Two complete trains were required to operate the service, and attached to the rear of both was a modern-styled Pullman 'observation car', to provide views of the countryside. None of the Pullman cars were new, however, and even the apparently 'new' observation cars

136

136
One of the three electric multiple-unit SR 'Brighton Belle' 5-BEL Units No 3052, in full Pullman umber and cream livery passes Clapham Junction in September 1963. *J. G. Dewing/Colour-Rail*

137
The most famous of all Sir Nigel Gresley's classic 'A3' class Pacifics, No 60103 *Flying Scotsman* passes Hadley Wood in August 1960 with the down 'Yorkshire Pullman'. *J. F. Aylard/Colour-Rail*

137

138
The first BR Standard locomotive designs (intended for use on all Regions) appeared in 1951, commencing with the two-cylinder Pacifics of the 'Britannia' class. Examples were used on important named trains to exhibit them during 'Festival of Britain' year (1951). The SR placed them on two Pullman workings, the 'Golden Arrow' and the 'Bournemouth Belle'. Pictured here is No 70009 *Alfred the Great* on the down 12.30pm 'Bournemouth Belle' passing Battledown Flyover on 4 August 1951.
E. D. Bruton

139
Teething troubles with the new 'Britannias' arose during the summer of 1951, and for a period the Bulleid Pacifics once again ran the 'Golden Arrow' services. Seen here is 'Merchant Navy' class 4-6-2 No 35028 *Clan Line* in the short lived BR blue livery, passing Folkestone on 16 June 1951. The train is composed of the new 1951-built Pullman cars, except the leading second class car which is a modernised prewar vehicle. The fourth Pullman is the new *Pegasus* containing the *Trianon Bar*. The new train had entered service only five days earlier.
A. C. Cawston

138

139

were no exception; but of course the public were led to believe otherwise, in true Pullman tradition!

The two observation cars were 27-seat third class Nos 13 and 14, and they had been rebuilt at Preston Park with new bodies upon old underframes and bogies, taken from 'J'-type Pullman cars; which had in 1921 themselves been reconstructed by Clayton Wagons, Lincoln, from former London & North Western Railway ambulance coaches of 1918 vintage (see Chapter 13). At the end of each trip the observation car had to be detached from the rear of the train and hauled to a locomotive turntable, to be turned to face the opposite direction for the return journey. The two cars differed in external appearance from all other Pullmans, by having windows that went right up to cantrail level (including the observation end windows) and a very modernistic version of the Pullman livery. The interior treatment used Warerite plastic finishes and the seating was an arrangement of pivoting arm chairs and fixed settees. The chairs could be swung at any angle. These seats were deliberately to a less than usually comfortable Pullman standard, to dissuade passengers from staying in the car for the whole journey, the idea being that the observation car was available to all. Refreshments were served at the seats, as an added amenity.

The summer-only London Victoria-Ramsgate 'Thanet Belle' was inaugurated in May 1948. In 1951, for one season only, the train included a Canterbury portion; it was renamed the 'Kentish Belle' — a title subsequently retained.

With the creation in 1948 of Regions rather than railways, by the BTC, the Pullmans were henceforth operated over four Regions, the Southern, Eastern, North Eastern and Scottish. Neither the LNER nor the Eastern Region of British Railways decided to re-introduce the Pullmans on the Liverpool Street-East Anglia services after World War 2. No Pullmans ran on the London Midland or the Western Regions, neither of which declared any interest in them. The new Railway Executive was set the task of standardising the passenger carriage fleet, by taking the best features of the existing designs of the 'Big Four' railways and moulding them into a new design, to be mass produced in order to replace older wooden-bodied stock as soon as possible. There were still some gas-lit carriages running on British Railways at this time, and a considerable quantity of pre-grouping designs, including clerestory roof examples. The privately-owned Pullman Car fleet did not come under review by the RE, but two important features which Pullman had pioneered in Britain were selected for the new 1951 standard BR main line carriage which was of all-steel construction. These were the Pullman gangway and drop-head buckeye coupling, which were adopted as standard for main line stock (only the LNER and the SR were using them beforehand). The interiors of the new BR carriages echoed current LMSR design, and were in plain wood finishes with boring patterned moquette upholstery; they were distinctly utilitarian, and succeeded in making the Pullman car interiors even more inviting by comparison!

1951 was 'Festival of Britain' year and the nation was anxious to show the world its potential as a marketplace of manufacture, invention and culture. To coincide with this important event, when a great surge of foreign visitors was expected, the Pullman Car Co introduced a new set of Pullman cars for the 'Golden Arrow' service. These were the cars designed before the war, and for which much material, including the marquetry panels had already been made, and which had been safely stored throughout the war years. By this time the London-Paris services were menaced by the rival air services, of British European Airways in particular. Wartime development of fighters and bombers had given way to a new generation of civil airliners, vastly improved in comparison with prewar days. Air travel was being actively promoted as *the* way to the Continent of Europe, making the train and ship combination look old-fashioned and slow, by clever publicity.

The aircraft which were all heavily subsidised by government funds, had female hostesses instead of male staff, to serve during the flights, and a great emphasis was placed upon modernity in styling the cabins. Yet to come was the further threat of the private car and the car-ferry services — private motoring still being restricted by fuel shortages. The new 'Golden Arrow' cars presented an opportunity to fight back against air competition. On Monday 11 June 1951, seven new Pullmans went into traffic, built specially for it: *Aquila*, *Carina*, *Cygnus*, *Hercules*, *Orion*, *Pegasus* (in which there was a *Trianon Bar*) and *Perseus*. Two parlour brakes were refurbished for use, *Minerva* and No 208, and one parlour car, No 35. In appearance, these straight-sided, timber-framed, steel-panelled Pullmans clung to prewar tradition and livery, except

140

141

142

140
Built by the Birmingham Railway Carriage & Wagon Co Ltd to a design prepared by the Pullman Co in 1938 (delayed by the war) the 'Golden Arrow' cars were on underframes and bogies of Gresley LNER pattern, and the bodies were of strong timber and galvanised steel construction. A feature later to pose problems was the use of sprayed 'Limpet' asbestos to improve insulation, for the interior of the roof and side panels. A break with tradition was the rectangular toilet window; otherwise the general livery and styling followed established Pullman traditions. Pictured is the first class kitchen car *Orion*. *Graham Warburton collection*

141
Interior of the 1951 'Golden Arrow' car *Orion*, as first delivered to traffic. Each car had a different scheme of interior panelling and finish, with carpets, chairs and fittings to match, although the marquetry panels of the cars, as well as the ceiling and wall lights, were of restrained style compared to earlier Pullmans. The main windows were double-glazed and the sliding toplights above them were considerably larger than before, to improve the ventilation of the cars. There was a new method of fan-operated air extraction installed in the roof of each car. The luggage racks had greater capacity than prewar cars and could accommodate suitcases safely. A public address system, worked by the train's Conductor, was fitted to each car, with speakers in a central position above saloon doorways and partitions. *Graham Warburton collection*

142
Interior of the 1951 'Golden Arrow' car *Pegasus*, showing the *Trianon Bar* area. Instead of individual bar stools, 'stand-up' seats were provided on the wall facing the bar counter, together with handrails. The counter had two bay portions to give more space (one in foreground) and the portion of the ceiling over the bar was moulded so that strong recessed lighting could be fitted. The principal decorations of the new bar area were two large aluminium plaques, taken from the previous *Trianon Bar* car. Calor gas was used for cooking on all the 1951 Pullman kitchen cars, with the gas cylinders safely mounted in steel boxes beneath the underframes. *Charles Long collection*

143

143
Rebuilt 'Merchant Navy' class 4-6-2 No 35030 *Elder Dempster Lines* leaves Southampton with the down 'Bournemouth Belle' in August 1964.
M. Chapman/Colour-Rail

144
Bulleid SR 'Merchant Navy' class Pacific No 35027 *Port Line* passes Brixton in August 1954 with the down 'Golden Arrow'. *Colour-Rail*

144

145

146

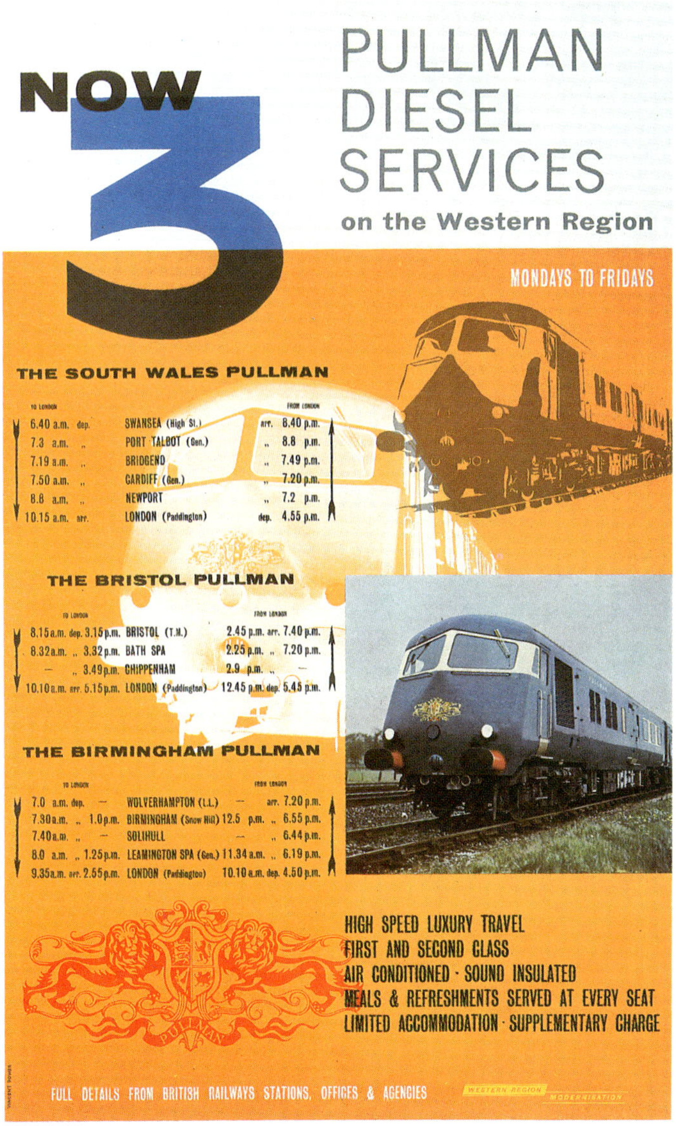

147

145
One of the two London Midland Region diesel-electric six car sets, in original Nanking blue and grey livery, with revised coat of arms is seen at St Albans in March 1966. *H. D. Ramsey/Colour-Rail*

146
A Western Region publicity poster announcing the fact that the region is now operating three Pullman diesel services using the new diesel-electric 'Blue Pullman' sets. *Crown Copyright National Railway Museum, York*

147
The Nanking blue livery of the Pullmans was replaced by a version of the standard blue and grey scheme. An unidentified set is seen here at Bristol Temple Meads in 1971 is in grey with blue window surrounds, along with full yellow ends. The looks of it have been spoilt by the addition of jumper cables. *J. Palk/Colour-Rail*

148

148
The interior of the second class parlour car No 35 of the 1951 'Golden Arrow' train, a 1926-built third class car remodelled for second class use. The original oval toilet windows were at this date replaced by the new rectangular type. Plenty of luggage is visible in this posed publicity picture! *BR*

that the famous oval toilet windows were replaced by a rectangular version. One improvement on the new cars was the use of LNER Gresley design of double-bolster bogies which gave a superior quality of ride to the cars than was possible with the Pullman steel-framed standard bogie. Regular travellers on the LNER East Coast main line could compare the two types easily; the Gresley design won without question at speed. Three more cars followed in 1952 for the service. These were Car No 303, *Aries* and *Phoenix*. The last did not have Gresley bogies, being in fact a new body on the prewar chassis and bogies of car *Rainbow* which was damaged by fire in 1937.

The interiors of the new cars which were beautifully finished, also clung to tradition, and offered the typical high levels of comfort and cleanliness associated with the name Pullman; whilst of course the service was superb. For existing clientele the 1951 cars were reassuring signs of Pullman's private enterprise in the midst of nationalisation and standardisation. However, it must be observed that in certain respects the 1951 (1938 designed) 'Golden Arrow' represented a lost opportunity to demonstrate the railways' ability to match progress in the air. It was not a question of inferiority in design and amenity, so much as one of failing to recognise changing public attitudes and tastes. One must recall that this was the commencement of the British postwar recovery, when the emphasis was very much upon change and of exciting new developments, of which the civil airliner was a spearhead symbol. The Pullmans seemed to belong to the 'old school' by comparison. On the other hand it may be argued that the Pullman Car Co was wise to resist change for its own sake, and to keep its standards for those who recognised their true qualities; even if this represented a minority in a potentially growing market.

The fickle British electorate, having banished Churchill and the Conservative Party Government after the war, promptly put it back in power late in 1951. The Tory Party loathed nationalisation, and was pledged to a policy of decentralisation. Many people thought that they would quickly reinstate the 'Big Four' railway companies, but this was going too far. Instead, in a 1953 Transport Act they abolished the Railway Executive, and gave the British Transport Commission direct dealings, through a Railway Division, with each of the Regions, whose powers were greatly enhanced. At this stage the standard BR carriage livery was replaced by some

149

150

149
Pullman Pride of the East Coast Main line! 3,300hp 'Deltic' No D9006, (later named *The Fife and Forfar Yeomanry* in December 1964), crosses the River Ure at Ripon with the down 'Queen of Scots' Pullman. A prewar brake third car is next to the locomotive, followed by new Metro-Cammell 1960 built all-steel cars.
Colour-Rail

150
Class 86/4 No 86439 heads the 'Liverpool Pullman' on 12 March 1987. The locomotive and rolling stock are all in the present day InterCity grey white and red livery including the Pullman Cars. *Chris Morrison*

151
Class 86/3 Bo-Bo electric locomotive No 86317 passes Standon Bridge south of Crewe with the up 'Manchester Pullman' on 19 July 1982, the day services were resumed after the ASLEF strike. The train was usually formed entirely of Mk 2 Pullmans in grey and blue livery at this time. *Hugh Ballantyne*

Regional variations, such as chocolate and cream on the Western Region and green on the Southern Region, reflecting the decentralisation of the system. In 1953 the Pullman Car Co, unaffected by nationalisation, gave a 15% dividend to its ordinary shareholders, and the company was still allied to Thomas Cook and Wagons-Lits. Mr Stanley Adams was Chairman and Managing Director of Pullman, and Chairman of Thomas Cook, whilst Wagons-Lits were represented on the Pullman directorate by Monsieur René Margot-Noblemaine, the *Directeur-Generale* of the Compagnie Internationale des Wagons-Lits. It was a powerful alliance of interests, but there was one problem which they had not resolved, namely the continuation of the contract to operate Pullman cars on British Railways. In 1953 the BTC brought together the existing LNER and SR contracts under a new one, but it refused to consider extending the contract beyond 1962.

As a result of the BTC refusal, the Pullman Car Co could not contemplate the construction of any more new Pullman cars, and the 1951 'Golden Arrow' cars were destined to be the last ordered by them completely to their own design and initiative. They were also the last straight-sided Pullmans to be built for British use. In 1954 the BTC bought out all the ordinary shares, giving it control of the Pullman Car Co activities. The company continued to operate separately, with its own management for the eight remaining years before the BTC contract expired. Before that, in 1958, Mr Stanley Adams retired from his interests in both Pullman and Wagons-Lits, and the chairmanship of the Pullman Car Co passed to Sir John Elliot,

152

153

152
The Pullman workshops at Preston Park, near Brighton were primarily concerned with overhauling, or remodelling existing Pullman cars, and their fittings. As a rule the company went to outside contractors for the construction of new cars. However, in 1952 three new Pullmans were built at Preston Park — first class kitchen car *Aries*, second class kitchen car No 303 and first class parlour car *Phoenix*. The underframes for *Aries* and Car No 303 were to LNER design with Gresley bogies, supplied by Metropolitan-Cammell, but *Phoenix* quite literally 'rose from the ashes' because the frames and bogies came from the burnt-out car *Rainbow*. Being of steel, these had survived the fire which destroyed the body of *Rainbow*, at Micheldever on 15 August 1936, and had been kept as 'spares' ever since. The bogies were of the Pullman prewar standard single-bolster type; the body of the car being to 1951 'Golden Arrow' specification. *Phoenix* became part of the 'Golden Arrow' formation, and was popular as a royal train car. Note the rectangular toilet window fitted.
Ian Allan Library

153
The interior of the 1952 car *Phoenix*. Although this closely followed the 1951 design, there were differences in the style of tablelamp, with simplified shade and stem, and the decorative panels, on which there were framed collages, by the artist Miss Mary Adhead, who was noted for collage work, using printed paper such as postage stamps, money, certificates, to build up pictorial effects.

who had already been Chairman of the BTC Railway Executive and of London Transport Executive, amongst other important posts. Sir John Elliot, although by training a journalist, was a true railwayman in his thinking and was sympathetic to the Pullman traditions. Under his chairmanship the Pullman services were to be extended to all BR Regions and record profits were made by the company — although of course it was the BTC that gained from his achievements, because all the profits went to them, after payments to the preference shareholders had been honoured.

The BTC pledged in 1954 to retain the Pullman Co intact for the duration of contract and not to operate any trains of its own under the name of 'Pullman', unless they were staffed and run by the Pullman Car Co, with a supplementary fare. They also pledged not to operate supplementary fare trains of their own, in competition. They avowed their wish to extend Pullman services to all BR Regions, whilst honouring existing contracts. From this stage onwards, the Wagons-Lits Co was no longer financially involved, but it retained a seat on the Pullman Board.

154
On Sunday 5 October 1952 the Railway Correspondence & Travel Society (RCTS) organised a special Pullman train known as the 'Centenary Pullman', to commemorate the 100 years-old Brighton Locomotive Works. The train ran from London to Brighton and back, hauled by former LBSCR 'Atlantic' No 32424 *Beachy Head*, in BR lined black livery. Here we see the return working near Patcham — livery apart, a fine reminder of the days of the old 'Southern Belle'! *J. H. Kent*

154

155

155
The Royal train conveying His Imperial Highness Haile Selassie, Emperor of Ethiopia, and HRH The Duke of Gloucester, from Portsmouth to London (Victoria) on the occasion of his Imperial Highness' visit to England. The gleaming locomotive is 'Battle of Britain' class 4-6-2 No 34088 *213 Squadron*, and the leading Pullman car is *Minerva*, followed by a 'Golden Arrow' kitchen car. This photograph was taken near Fratton on 14 October 1954. *Pursey C. Short*

156
Interior of a Mk 3b Pullman dining car with a steward pouring coffee at the table.

157
The 'Tees-Tyne Pullman' InterCity 125 set is seen here used to form the Sundays-only 09.50 Newcastle-King's Cross, leaves Washington, County Durham, on 13 October 1985. A large Pullman crest with legend 'Tees-Tyne Pullman' beneath is displayed on the front of the leading power car. This was specially applied for a high speed publicity run.
Peter J. Robinson

158

158
Preserved World War 2 American-built War Department 'USA Tank' No 22 *Maunsell* formerly used on the SR, heads the 'Wealden Pullman' near Cranbrook Road crossing on the Kent & East Sussex Railway on 28 April 1985.
H. Nightingale

159

159
The superb interior of the restored former narrow-bodied Hastings line Pullman car *Barbara*, now operating on the Kent & East Sussex Railway 'Wealden Pullman' service.
Jim Berryman (courtesy K&ESR)

160

160
BR Class 33 diesel-electric locomotive No 33008 *Eastleigh*, repainted in green livery, makes a fine sight arriving at Waterloo on 21 June 1986 with the return working of the 'Bournemouth Belle' formed of the VS-O-E Pullman set.
Chris Morrison

CHAPTER 16

A Chance to Modernise

Whereas the activities of the Pullman Car Co had progressed sufficiently well since the end of the war to enable it to pay a dividend to its ordinary shareholders each year since 1949, the fortunes of the nationalised rail network was by no means so healthy, and since 1948, government investment in the railways had rested only at a level that permitted them to continue to provide a service. The Railway Executive had proposed the continued use of steam locomotives, and a considerable fleet of new locomotives, to standard BR design began to emerge in 1951. It was not because the BTC, or the Executive wanted to remain firmly in the 'steam age', it was simply that the necessary funds for electrification (the ideal) or dieselisation (the compromise), were not forthcoming from Parliament.

At last, in 1955, the BTC was given government approval to embark on a massive Modernisation Plan, adding up to a capital expenditure of £1,240 million in 1955 money values. It was virtually at the eleventh hour, because by then the competition on the roads and in the air was building up to previously unknown proportions. Never had the public image of rail travel been at such a low ebb, and the morale of the staff so defeatist. Whatever the faults of the Modernisation Plan — and they were abundant, with hindsight — it gave the necessary adrenalin to the system, with its sweeping plans of replacing steam by diesel and electric traction, rebuilding stations, improving signalling, improving the tracks, and above all of establishing new levels of passenger comfort, able to match those of the alternative modes of transport which had posed such a threat.

The BTC looked to the Pullman Car Co

161

162

161
Apart from the electric 'Brighton Belle' and the occasional use of the pioneer LMSR (Ivatt) and SR (Bulleid) diesel-electric locomotives, Nos 10000/1, 10201-10203, on the 'Bournemouth Belle' and 'Golden Arrow' all the regular Pullman trains running in the early 1950s were steam-hauled, mostly by varied classes of locomotives of the 4-6-0 and 4-6-2 types. Moreover, the Pullman car fleet was almost entirely formed of prewar vehicles, in some cases even pre World War 1. Thanks to the splendid workmanship put into their original construction, and to the craftsmen at Preston Park who looked after their upkeep, the Pullmans were still giving satisfactory main line service, but no-one could present them as 'modern' in the public eye. On the contrary, their image was one of traditional railway travel — giving comfort and high-quality personal service, at a price. This lovely photograph shows the northbound 'Queen of Scots' all-Pullman train crossing the bridge over the River Ure at Ripon, headed by Gresley Class A3 4-6-2 No 60036 *Colombo* in 1955. The identical train, with the same locomotive and Pullmans, could have been photographed there 20 or more years earlier. Meanwhile civil aviation and road transport had undergone a vast development.
Charles Long collection/Leslie Overend

162
Maunsell 'King Arthur' class 4-6-0 No 30763 *Sir Bors de Ganis* is seen in charge of the 'Kentish Belle' all-Pullman train near Bickley on 1 September 1953, making an interesting comparison with picture No 93 which was taken a quarter of a century earlier. It said much for British steam locomotive builders that their products remained in top main line service year-in, year-out, likewise the splendid old Pullmans which they hauled. (The 'Kentish Belle' was previously named the 'Thanet Belle'.)
Charles Long collection

with a fresh eye, armed with the knowledge that there was ample finance to invest in an improvement of the services in which it held the controlling interest. The Pullman marque could be useful in establishing the new levels of comfort and amenity which the plan demanded. There was now money to spend on replacing the older cars, and there was scope to introduce a new concept of Pullman, intended to compete with the internal airline services in terms of speed and comfort. An important item in the BTC announcement of the 1955 Modernisation Plan was that new high speed luxury diesel multiple-units were to be built to operate between important cities and to contain all that was best to offer for the comfort and amenity of the passenger. They stated that, 'Public reaction to these would be watched carefully, before increasing the fleet'. Although it did not immediately specify that these were to be Pullmans the BTC implied as much, by the suggestion that they were to be luxury trains.

In 1955, when the Modernisation Plan was announced, the Pullman Car Co fleet was basically of prewar origin and even included several SR-based cars of pre-World War 1 origins, whilst all the East Coast cars were at least 26 years old, although a considerable amount of work had been done in remodelling or rebuilding cars. There were still some of the massive 12-wheelers (as the cars with six-wheel bogies were described), and the company was not in a hurry to rid itself of these older cars, which were useful standbys when peak demands occurred. The Pullman Car Co was always quick off the mark to provide special trains for special events, such as race meetings, and business parties. Pride of place, in this respect, was the constant use, already established before the war, by the Southern Railway, of Pullman cars and staff for members of the Royal Family, and visiting dignitaries of all nationalities. The Southern Railway did not possess carriages of its own suitable for such occasions, choosing instead to let the Pullman Car Co cater for such special traffic. A very interesting account of these special trains which the Southern Region perpetuated and of the staffing and menus involved, is given by Julian Morel in his book *Pullman*. The unique feature of Southern Railway Royal Specials was that the public often found themselves travelling very comfortably indeed in a Pullman car on an ordinary service, which had been specially groomed the day before for Royalty, and they were waited upon by the very same staff. Another important service which ran according to demand was the 'Ocean Liner Express' between Waterloo and Southampton, which required cars kept on standby. The old 12-wheeler *Grosvenor* ended her days on these services, and Behrend records seeing her in use as a buffet car on a Southampton boat train as late as June 1960; *Grosvenor* was then aged 52.

The BTC honoured its stated intentions to expand Pullman car services to other Regions, at first by introducing them, in June 1955, to the Western Region. For the very first time Pullman cars ran into Wales, on what was called the 'South Wales Pullman'. Some 25 years had elapsed since Pullman cars had operated out of Paddington station over what were until

163
The southbound 'Pennine Ltd' near Garsdale on 1 July 1986 consisting of the 'Pullman Rail' rake of former East Coast main line Metro-Cammell Pullmans and Mk 1 BR brake seconds, all painted in umber and cream livery. Also included is a standard British Rail restaurant buffet in blue and grey livery. The special train is hauled by Class 47/4 No 47449. *J. H. Cooper-Smith*

164

165

164
The Pullman kitchen car *Grosvenor*, built in November 1908 for Dalziel's new 'Southern Belle' train. A wooden-bodied 12-wheeler, remodelled in 1936 at Preston Park as a buffet car, with bar counter (with 11 stools and 8 seats at tables). This picture was taken at Newhaven on 3 August 1957, when it was still in main line service as the buffet car of the London Victoria-Newhaven boat train. The roofboards read: 'London-Paris. Newhaven-Dieppe'. This fine old car was still in occasional boat train service in 1960; then 52 years old!
W. M. J. Jackson

165
Standing in the Pullman Co's Preston Park workshop yard is the Pullman car *Maid of Kent*, on 18 April 1955, awaiting attention. The underframe and bogies of this car had been built for a World War 1 ambulance train carriage by the London & North Western Railway. In 1921 the carriage was purchased by the Pullman Car Co, converted to a wooden-bodied Pullman kitchen car by Claytons of Lincoln, and named *Formosa*. It was rebuilt at Preston Park in 1934 as a first/third composite car and again rebuilt at Preston Park in October 1948 when it became all first class, and was renamed as *Maid of Kent*. Many Pullmans had similar lengthy and varied careers to this one, because the company was adept at conversion and modernisation work to meet changing traffic patterns. Of added interest is the comparison of gangway ends and coupling seen here. *Maid of Kent* has bellows gangways and screw couplings. The car *Orpheus*, just visible on the right, has the Pullman standard gangway and drophead buckeye coupler (plus an emergency three-link coupling for the yard shunting engine to use). *R. C. J. Day*

1948 the tracks of the Great Western Railway.

The 'South Wales Pullman' was a success, because of the high standards of service given to passengers, and because the BTC, who had the controlling interest in its profitability wanted it to succeed. None of the Pullmans used on it was new, but a feature was the *Daffodil Bar*, staffed by an attractive lady attendant in full Pullman uniform. This bar was installed in the car *Diamond*, built in 1928, and which we have already encountered in the 1946 'Golden Arrow' running as the *Trianon Bar*, and then in the Ostend boat trains as the *One Hundred*, and then, the *New Century Bar*. The Pullman Car Co was certainly adept at changing the identities of their cars — and one wonders how many people using the *Daffodil Bar* between Paddington and South Wales in 1955 had the slightest idea of its true age and previous identities!

The year that saw the successful introduction of the 'South Wales Pullman', also witnessed the demise of the 'Devon Belle'. It was a popular train, but uneconomic to operate because of the need to keep two complete rakes of Pullmans (including the two special observation cars) specially for it, despite the fact it only ran over a long weekend period, and only during the summer season. Various imaginative uses had been found for the trains when not in 'Devon Belle' service, including Festival of Britain tours of the cathedral cities of the south, and later they became a static 'Pullman Club' for visitors to the British Industries Fair at Castle Bromwich, Birmingham where the ordinary cars were used as restaurants and the observation car (only one) as a lounge. There was a 'Pullman Club' there each year from 1953 until 1957, when the Birmingham Fair ceased. When the 'Devon Belle' was taken off, the two observation cars were sold to the London Midland Region and the Scottish Region and continued in use on tourist train routes; no longer carrying the name Pullman. At this time, the designation Third Class was changed to Second Class on all BR services.

The Modernisation Plan was to be responsible for both the promotion and the demotion of Pullman cars. The demotion took place on the Southern Region, where the plan extended the third rail electrification to the important Kent Coast services, placed diesel-electric multiple-unit trains on the restricted loading gauge route to Hastings, and later on replaced the prewar electric trains on the Brighton and South Coast routes. Only the 'Golden Arrow' (which was henceforth hauled by an electric locomotive) and the 'Brighton Belle' electric multiple-unit Pullman cars survived these changes and continued as before. All the SR's other Pullman cars, including those used on the 'Kentish Belle' and cross-Channel boat train services from Victoria (other than the 'Arrow') were replaced by buffet cars in ordinary train service. The Pullman Car Co was responsible for staffing the new buffet cars that replaced the Pullmans, but these vehicles were ROVs (Railway Owned Vehicles) and no Pullman supplement was charged for their use. In fact the Pullman Car Co had already been responsible for staffing existing ROVs on the SR Kent Coast in pre-electrification days, as well as operating its own cars. On the Hastings line the Pullman cars had been converted into non-supplement buffet cars — still company-owned. A new standard bogie design used under the ROV electric buffet cars unfortunately proved to be rough-riding in everyday service, and the cars did not prove very popular as replacements for the staid and stately Pullmans. The Waterloo-Bournemouth line was not an early candidate for electrification and the Pullman car services were not affected at this stage, although some of the earlier cars were replaced by more modern examples. At this time the policy of 'plating' wooden-bodied Pullmans with aluminium sheeting was being pursued, giving them a more up-to-date appearance. Not all the cars displaced on the SR were withdrawn from service immediately. The six narrow-bodied cars originally built for the Hastings route in 1926 and which had been converted into non-supplement buffet cars in 1947, were sold to the Southern Region in 1961 for use in the Waterloo boat trains in which they had run since 1958. They had previously (1958) lost their names and numbers, and they no longer carried Pullman livery or legend, being painted in a rather plain overall SR green livery (see list for details). It

was one of these cars which pioneered the use of bottled propane gas for cooking, a Pullman initiative in the 1950s.

The Pullman fleet still numbered 195 cars in 1959, plus 64 ROVs staffed by the company. From that date onwards the earlier types of car were swiftly withdrawn, leaving the more recent examples to be switched around from Region to Region to fill the gaps. Sir John Elliot evidently produced a convincing case for the BTC to finance construction of a batch of replacement cars, quite apart from the forthcoming diesel luxury trains. These 44 cars would replace those running on the East Coast route, whose existing cars would then take over the 'Bournemouth Belle' and other duties still including wooden-bodied Pullmans.

In September 1958 two further Pullman Car services using prewar cars had been added to the ER, called the 'Master Cutler' (morning and evenings) and the 'Sheffield Pullman' (midday runs), running between King's Cross and Sheffield (Victoria). With the introduction of main line diesel locomotives, with their potential for faster running, the 1924-29 cars were pushed to their limits, and no doubt this strengthened the case for replacements. An order for 44 new all-steel Pullmans for locomotive haulage was placed in June 1958 with the Metropolitan-Cammell Carriage & Wagon Co, with delivery to commence in 1960. The same company was awarded the contract to build the five prototype high speed de luxe diesel multiple-units — now referred to as the 'Diesel Pullmans' or 'Blue Pullmans' (these developments are described in the next chapter). To operate with the 44 new East Coast cars (for which no brake vehicles were designed) 12 existing 1928/29 parlour brake seconds (including three new conversions from kitchen cars) and other assorted 'spare' cars were retained for some years. The original 'Hadrian Bar', was — like the Hastings buffets — sold to BR after transfer to the SR in 1961. It became No S7879S.

As a result of the planned changes in the fleet, in 1960 a re-assessment of the fleet took place and 56 old cars were withdrawn from traffic and converted into camping coaches at Preston Park, and at the SR carriage works at Lancing. These were then sold to British Railways for about £900 each — 20 to the Southern; 11 to the Eastern; 11 to the Western; and 12 to the Scottish Region. These coaches (the only time a Pullman car has been officially described as a coach) were in a variety of liveries — blue, umber, umber and cream — according to Region, and some carried the name Pullman, whilst others were simply 'Holiday Coach'. They were fitted out like rather superior caravans or holiday chalets, and parked in sidings at suitable rail-served beauty spots around Britain, where they could be rented for the season. A stipulation was that the hirer had to travel by rail to the site. Although proving to be extremely popular, the holiday coaches were difficult to maintain, because of their remote locations, and in some cases the branch lines on which they were housed were subsequently closed to traffic, and the Pullmans were left stranded. The scheme was eventually abandoned, but it gave years of added life to some of these fine old cars, and some have survived (see list). We will meet some of them again, later in the story.

One traditional feature of Pullman operations already mentioned, which was left untouched by the BTC Modernisation Plan, was their use on Royal trains and other special workings over SR routes. Visiting Head of State, and other important personages, arriving at Gatwick Airport were comfortably transported to the reception committee and red carpet waiting at Victoria to greet them. The Royal Family still used Pullmans to visit the Derby Day Epsom race meeting, although for a period HM The Queen preferred to travel in one of the two vintage ER 12-wheel Royal saloons

166
Pullman cars returned to the lines of the former GWR in June 1955 and penetrated Wales for the first time — finally justifying the Pullman coat of arms! The 'South Wales Pullman' was to prove a very successful venture. The train is seen passing Ebbw Junction, Newport, on 26 August 1960, hauled by 'Castle' class 4-6-0 No 5006 *Tregenna Castle* and composed of the prewar cars: Car No 54, Car No 171, Car No 35, *Diamond* (the *Daffodil Bar*), *Cecilia*, *Zena*, *Chloria* and Car No 27. *G. England*

167
In the spring of 1958 the Eastern Region took delivery of 10 'pilot scheme' 2,000hp diesel-electric 1 Co-Co1 locomotives, built by English Electric, of which Nos D201/6-9 were allocated to Hornsey depot, London for services out of King's Cross. Amongst their duties was the 'Tees-Tyne Pullman' and a new Pullman Sheffield-London service, the 'Master Cutler' (with a midday 'Sheffield Pullman' link using the same stock) which commenced running in September 1958, using prewar Pullman cars of the all-steel 'K'-type. Even before the diesels commenced working the 'Master Cutler' service, the Pullman Car Co had realised that the acceleration and sustained high speeds of diesel power would show the need to design a new fleet of Pullman cars, with a bogie design and general structural insulation of noise and vibration to more sophisticated modern levels. No D207 is seen here at the head of the inaugural 'Master Cutler', with the 'K'-type cars in tow. *BR*

168
Scottish Region restaurant car No SC216M at St Rollox in December 1958. This was one of the former Pullman cars built in 1923 for LMSR Scottish services and sold to the LMSR in 1933, after which the car names were removed and they were repainted in crimson lake livery. *W. S. Sellar*

167

168

to Tattenham Corner, because these had wider vestibule doors than the Pullmans, and did not crease the dresses of the ladies, which was of 'full' fashion at the time. When the SR ceased to operate locomotive-hauled Pullman cars, the Pullmans were loaned from the ER for these duties, and hauled by diesel locomotives, or on occasions by SR electric locomotives. The 'Brighton Belle' cars were also used for Royal duties from time to time, and this found them at locations such as Portsmouth Harbour, where they were not normally seen. No single car could lay claim to being *the* Royal Pullman, but an early favourite was *Rosemary*, a parlour car built in 1923 and normally used in the 'Bournemouth Belle'. This was replaced by *Phoenix*, but *Aries*, *Aquila*, *Hercules*, *Malaga*, *Orion* and *Zena* were other locomotive-hauled cars used by Royalty, and the electric cars of 5-BEL and 6-PUL sets, *Ethel*, *Naomi and Peggy* were also used. According to Charles Long, in postwar days the normal procedure was simply to take one that was recently overhauled out of service for cleaning, and if necessary retouching, beforehand.

As Britain prepared to launch itself into the 'Swinging Sixties' the affairs of the Pullman Car Co, with good profits, were in fine shape, under the dynamic control of Sir John Elliot CBE, FCIT, and the Joint Managing Director Col F. D. M. Harding OBE, FCIT. But the BTC contract had only two years to run, and the BTC was itself in trouble with a mounting deficit, due to a number of unforeseen factors in the over-ambitious 1955 Modernisation Plan. The only glimmer of light on the horizon was the forthcoming 'new blood' for the Pullman pedigree — the diesel-electric de luxe Pullman trains, and the 44 replacement East Coast route cars. These would bring Pullman affairs into the limelight just as the contract ran out of wind.

169

169
A further use was found in 1960 for 56 vintage Pullman cars, mostly displaced by SR electrification, and by the arrival of some new cars on the East Coast main line (see next chapter) which allowed a transfer of the ER prewar steel-bodied cars to the SR, in turn displacing older wooden-bodied Pullman cars. These 56 cars were converted into holiday 'Camping Coaches'. (This was the only time that the Pullman Car Co used the word *coach* instead of *car*.) This picture shows 'Pullman Camping Coach No CC161' parked at Heacham station in Norfolk, in an attractive blue and gold decorated livery (later examples were mostly umber and cream) No CC161 was the old *Maid of Kent* (II) originally *Formosa*. It survives to this day in preservation. *BR*

170
The Royal train from Gatwick Airport station conveying Their Majesties The King and Queen of Siam, approaching East Croydon en route for London (Victoria), worked by rebuilt 'Battle of Britain' Pacific No 34088 *213 Squadron*. The leading Pullman car is *Niobe* — the second unidentifiable. The third vehicle is one of the two vintage 12-wheel ER Royal saloons that HM The Queen requested be used because of their wide doorways with double doors, which did not crease the full-skirted dresses which were fashionable during the early 1960s. The fourth vehicle is the 1952 car *Phoenix*; the fifth is unidentified. *Derek Cross*

170

CHAPTER 17

New Cars and New Owners

The famous 'Blue Pullmans' were a BTC initiative, whereas the 44 East Coast route replacement cars stemmed from a Pullman Car Co request, subsequently authorised by the BTC. This alone explains why two completely different types of new Pullman car were built at the same time, and by the same builder for the same customer. It did not really make much sense, except that the 'Blue Pullmans' were regarded as prototypes requiring evaluation before any further orders were placed, whereas the East Coast cars were basically like-for-like replacements, suitably up-dated for the 100mph running made possible by the delivery of 22 'Deltic' class Co-Co diesel-electric locomotives. The difference was emphasised by the use of a brand-new Pullman livery of Nanking blue and white for the diesel cars, whereas the East Coast cars were entirely traditional in umber and cream, although with simplified lining-out. The Pullman coat of arms was redrawn at the request of Sir John Elliot, and featured on both new car designs.

Both the car designs came under the scrutiny of the BTC Design Panel, who were requested to ensure that the highest possible standards of design, in terms of amenity and appearance, were achieved. To effect this they appointed well known freelance industrial design consultants for both orders. Jack Howe RDI, MSIA was given the diesel Pullman project, and John Carter MSIA that for the East Coast cars, both working in conjunction with the Pullman Car Co's officers, and with Metropolitan-Cammell, the builders. To liaise between the designers and the BTC was the responsibility of George Williams, the BTC Design Officer. It was an exciting project, particularly in the case of the diesel Pullmans, for which the BTC brief was quite simply 'make them the best'. Their sights were set upon beating the challenge of aircraft, and the long-distance coach, the latter being aided by an enormous government-sponsored programme of new motorway construction. In my book *British Rail 1948-1983 A Journey by Design*, I had this to say about the 'Blue Pullmans':

'The luxury Pullman diesel-electric trains, intended as the prestige symbol of the BTC Modernisation Plan, and as the British counterpart to the highly successful European TEE (Trans Europe Express) trains, were the first of the Design Panel's real success stories. Right from the start of planning the consultant designer, Jack Howe, had collaborated closely with the builders, Metro-Cammell, to produce an integrated aesthetic solution to the engineers' requirements, and one, moreover, which was a considerable advance upon anything which had previously run on British rails. Two six-car all-first class sets were produced for the London Midland Region, and three eight car first and second class sets for the Western. The LMR sets were delivered first, and began trial running in October 1959, before going into service between St Pancras and Manchester.

'The traditional Pullman livery was replaced by a striking Nanking blue and white scheme, with a specially redesigned Pullman coat of arms, and shaded Egyptian slab-serif lettering. The nose ends of the power cars were rounded off in a semi-streamlined form, with very neat detailing (a four-character headcode panel was at one stage included in the front end shape, but this was replaced by headlights with marker discs), and the bodysides were smooth and clean in detail. Less

171

171
The new coat of arms of the Pullman Co, designed by Mr Milner Gray of Design Research Unit, for the 1960 debut of the 'Blue Pullmans' and East Coast replacement cars.
Charles Long collection

172

172
The six-car diesel-electric 'Midland Pullman' was the first of the 'Blue Pullmans' (as they were quickly nicknamed) to run trials on BR in October 1959. Virtually everything about these trains broke away from the Pullman tradition, in a deliberate attempt to create a new image of luxury train travel, but the name Pullman was so respected that it was decided by the British Transport Commission (BTC) to market the new air-conditioned trains as Pullmans, and to charge a supplement for their use. In a brilliant Nanking blue and white livery with the new coat of arms prominently placed on the front ends, and with a new style of lettering featured on the car sides (Egyptian slab-serif) in white, shaded black, it was unfortunatly not decided to name each car individually. The driving power cars (one at each end of each set) housed a North British/MAN diesel engine of 1,000hp and a GEC main generator. Two GEC traction motors were on the rear bogie of the power cars, and two on the adjacent bogie of the nearest trailer cars.
Sam Lambert

satisfactory were the gaps between carriages, where the gangways and various jumper cables were visible; a sheeting of indiarubber, as on Gresley's prewar streamliners would have helped to produce a cleaner profile.

'Internally, Pullman traditions were virtually swept aside. Gone was the last vestige of stolid mock-Edwardian comfort and in its place were interior schemes of an extremely high standard, for both first and second class, with full air-conditioning and comfortable aircraft-type seats. For the first class saloons, two schemes were employed; one had grey leathercloth walls, blue and navy-blue striped upholstery, a red and black carpet and polished ebony partitions, while the other had grey leathercloth walls, red and navy-blue upholstery, a blue and black carpet and polished rosewood partitions. Inlaid murals by William Mitchell were a feature of the partitions, these were a slight echo of the traditional Pullman marquetry, but used a variety of modern materials instead of wood. The second class cars (on the WR trains) had fixed aircraft-type seating, whereas all the first class seats were adjustable for reclining, and the walls and partitions were finished in grey plastics; carpets and seat upholstery were the same as in the firsts.

'Wide windows had adjustable Venetian blinds between the panes and the lighting in the saloons was by diffused fluorescent tubes for the main ceiling, with incandescent lights below the luggage racks, and individual table lamps, controlled by the passenger, below the windows. These small table lamps were not strictly necessary, but the Pullman Car Co expressed a wish to retain a semblance of their traditional style, against the Design Panel's advice. In fact, they were not strictly speaking *table* lamps, because they were cantilevered from the bodyside, below the windows. A public address system was installed, with loudspeakers above the luggage racks, down one side of the ceiling.

'Perhaps the greatest single improvement was the internal design of the gangway between carriages. This feature had long been one of the least attractive areas of British coaching stock design, being usually dark, draughty and dirty. In its place was a clean wide passage, mounted upon pivots at each carriage end, which formed a semi-floating rubber-sealed platform, without loss of air-conditioning. One scarcely noticed the transition from one saloon to the next.

'The Panel also advised on all the small items for the diesel Pullmans, such as carriage litter bins, magazine trays, instruction leaflets for the adjustable seating, antimacassars, silver plate and glassware, signs and notices, toilet fittings, etc. A full scale mock-up of part of a carriage was constructed by the manufacturer, and all aspects of the design were displayed to the General Managers of the two Regions and the Pullman Car Co. The toilets, which had a mosaic wall finish, were exceptionally well designed, and had passenger controlled water temperature, delivered to a capacious wash basin by a single spray nozzle.

'Needless to say, trains of such a high standard were immediately acclaimed by the public, and it was interesting to note that, despite the fact that a supplementary fare was charged, as well as the business travellers who flocked to the comfort of the firsts, the second class cars also attracted a full complement more often than not. The usual Pullman standards of service were maintained, with a push-button and adjacent indicator light to summon the attendants to individual tables.

'These fine trains had one disappointing feature; the quality of their riding at speed. This was particularly sad because great efforts had been made to secure an improved ride, by fitting the Swiss-type Metro-Schlieren bogie widely used on the Continent, where it gave an excellent ride. The design just did not suit British track conditions, and the powered bogies in particular were lively over crossings and pointwork. At this time BR were searching for a new bogie for all their main line corridor stock, and as an interim measure they had started to fit the heavy cast steel Commonwealth bogie (very expensive to build) whilst research went on for a cheaper and lighter alternative. Had the Swiss bogie come up to expectations it might well have been adopted as a BR standard.'

I shall return to the 'Blue Pullmans' at the end of this chapter. Meanwhile, the East Coast replacement cars must be examined, because they began to emerge from Metropolitan-Cammell alongside the so very different 'Blue Pullmans' in 1960.

The 44 East Coast route replacement cars were of all-steel construction, with the standard curved body profile of the 1951 BR standard carriage design, although thicker gauge steel was used for their

173

174

bodysides, and Pullman-type inward-opening doors were used at each end of the car, instead of the normal outward-opening BR swing doors. The cant rail was extended over the door to the car end.

Commonwealth-type bogies were fitted to ensure good riding at speeds up to 100mph (they were the first 100mph Pullman cars in Britain) and these proved to be excellent indeed. Double-glazed windows and Stone's dual heating (electric and steam) and ventilating system produced very well warmed and insulated car interiors, with excellent sound proofing into the bargain (except when the sliding toplights in the windows were opened wide in hot weather). The interior colour schemes varied between cars, but a disappointment was the lack of the traditional wood veener and marquetry, which was considered to be too expensive. For the first class, with 1+1 a side seating accommodation, the standard type of loose Pullman chair was used. In the majority of cases these were old chairs from withdrawn cars, renovated and re-upholstered. Two new features were the use of capacious aircraft-style luggage racks, and glass saloon partitions, giving a more 'open' saloon layout than hitherto found in Pullmans; although for the clientele who sought privacy, some coupés were retained. For the second class cars, a new design of 2+1 seating was produced, and it is significant that such an arrangement (albeit with reclining backs) was now already considered suitable for the *first* class in the

173
A corner of a first class parlour car of the 'Midland Pullman', showing the adjustable reclining seats, aircraft-type luggage rack with reading light, double-glazed windows with the venetian blinds set between the two panes of glass, diffused central fluorescent lighting inset into the ceiling, small lamp above table, and decorative inlaid motif on the partition wall. The cars were fully air-conditioned. A change of policy was the seating density for first class of 2+1 instead of 1+1. *BR*

174
Every detail was considered for the 'Blue Pullmans', and even the designs of the menu, crockery, table linen and wine glasses were new. This view shows the single seat side of the aisle of a first class 'Midland Pullman' parlour car. The small handle on the wall was to adjust the venetian blinds. Note that the table lamp was in fact fixed to the wall, and not moveable. *BR*

'Blue Pullmans'. The economics of any longer providing the traditional low density payload of one seat each side of the aisle were becoming suspect in BTC eyes, and the Metropolitan-Cammell East Coast replacement cars were to be the last Pullmans built new with this degree of space per first class passenger. The kitchen equipment on both types of new Pullman functioned on bottled propane gas, and the layout and decor of the kitchen were both functional and attractive, giving the most up-to-date appearance, with stainless steel and formica finishes.

In many respects these two designs for Pullman travel represented the parting of the ways, because the affairs of both the BTC and the Pullman Car Co were to undergo very considerable change during the first two or three years that the new Pullmans were running. The East Coast replacement cars were in theory the last bastions of Pullman tradition, whereas the 'Blue Pullmans' were the heralds of a new age. It did not work out quite that way in reality, as the ensuing chapters will reveal.

As its name implied, the 'Midland Pullman' (the first 'Blue Pullman' introduced) was intended for the Midland line services of the London Midland Region. It was the first Pullman service to run on the Midland lines out of St Pancras since George Mortimer Pullman's contract had expired in 1888. Alas, even before it ran, it provoked hostility amongst the Region's own restaurant car staff, who resented the intrusion of the profitable Pullman Car Co into this territory. At this time the BTC still held only the ordinary shares, and profits had to be shared with the preference shareholders, to which the union-led railway restaurant car staff objected. An argument between the British Rail Catering Services and Pullman had already blown up when the Pullman 'Master Cutler' service was being schemed. BRCS staff who had formerly worked the withdrawn Great Central route train of the same name were being made redundant at the same time as plans were being made to transfer Pullman employees to Sheffield.

175
The three eight-car 'Blue Pullman' sets allocated to the Western Region incorporated both first and second class accommodation. Two went into service on 12 September 1960, forming the 'Birmingham Pullman' and 'Bristol Pullman' workings. The third set was initially a spare, but in September 1961 it took over the 'South Wales Pullman' service, now based at Swansea, while some of the locomotive-hauled Pullmans previously used (plus — at first — two BR coaches) were formed into a general standby train. Pictured here is a 'Blue Pullman' press demonstration special, leaving Paddington station on 5 September 1960, prior to entering public service. *BR*

176
First class parlour car No M60740 of the 'Midland Pullman' showing clearly the Swiss-type Metro-Schlieren bogies fitted, which were to prove very disappointing in service. The large jumper connections at the car ends were a curiously ugly oversight by the designers of what was, otherwise, an elegant new train. *Sam Lambert*

175

(Incidentally, at this time, Pullman rates of pay were *lower* than their BRCS counterparts received — although the gratuities were no doubt higher.) This argument was sorted out by an agreement to assimilate grades and pay rates, while BRCS staff could henceforth apply for vacancies in the Pullman services, and vice versa. To staff the new 'Blue Pullmans', extra men had to be recruited, and it was agreed beforehand with the National Union of Railwaymen that these should be obtained from existing BRCS restaurant car staff, suitably trained in Pullman procedure by Pullman staff sent from other Regions. Although this was agreed, some other NUR men still resented the extension of Pullman services to the Midland lines, seeing them as a threat of drain away the customers from their own best trains, including of course the best-tipping clientele.

The first 'Blue Pullman' unit was ready in the autumn of 1959, and a suitable press demonstration was planned, before the public service commenced. This 'launch' was prevented by the new BRCS staff refusing to work it, and the new Pullmans were not able to go into service until June 1960, after prolonged negotiations had taken place with the NUR.

Trouble flared up again on the 'Midland Pullman' when a midday service between St Pancras and Nottingham instead of Loughborough was introduced in 1961, in an attempt to make better daytime use of these expensive new trains. The NUR-led staff immediately reacted, protesting that it represented an unauthorised extension of Pullman services; which of course it was! The union had the previous year, as its annual Blackpool conference passed a resolution for the NUR Executive to work for the abolition of Pullman services and for their assimilation into the BTC hotel and catering services. The result was that for a few weeks the catering staff withdrew into the train kitchens between Loughborough and Nottingham — and reappeared at Loughborough on the return run. During this period of dispute, a rota of Pullman clerical staff travelled on the train to Nottingham to open and close the doors and to pick up the Nottingham seat reservation chart. It was the first time in British Pullman Car history that the staff had refused to work a train. In due course the problems were ironed out, but a lot of publicity of a damaging nature had surrounded the name Pullman because of the NUR's opposition. In fact, the midday service to and from Nottingham was very poorly loaded (often as few as 12

177

177
Quite unlike the 'Blue Pullmans' in virtually every respect, the 44 East Coast route replacement cars were nonetheless built by the same builder (Metropolitan-Cammell) for the same customer, and at the same time! Finished in Pullman umber and cream livery (with simplified lining-out) and with aluminium/silver roofs, the East Coast cars perpetuated Pullman traditions, but used the standard BR Mk 1 underframe and body construction, all in steel. The bogies (which proved excellent) were of the Commonwealth type, widely used by the world's railways because of their rugged qualities. Standard Pullman-type gangways and buckeye couplers were fitted. The new coat of arms and redesigned table lamps were two features in common with the 'Blue Pullmans'. Pictured here is the handsome *Hadrian Bar* built in 1961, a second class bar car No 354. *Julian Morel collection*

176

passengers!) and the whole business was something of a storm in a teacup. The morning and evening services between Manchester Central and London were, by contrast, very popular indeed.

At about this time (1960) an interesting proposal was made to convert the Western Region 'Golden Hind' service to full Pullman status. The intention was to use the 'Golden Arrow' cars — the 'Arrow' itself being either withdrawn altogether, or else replaced by electric multiple-units, formed of Kent Coast 4-CEP motor coaches with composite Pullman trailer cars from the 1933 Brighton 6-PUL units. The proposal came from Col F. D. M. Harding, Pullman Joint Managing Director. The following objections were raised, which killed the idea:

(1) The arguments with the NUR on the spread of Pullman operations were at their height — and this proposal would have been seen as a further provocation.

(2) Several of the 'Golden Arrow' cars would have to be converted to second class to provide the right mix of accommodation for WR needs.

(3) The Pullman composite cars were approaching the end of their useful lives and rode badly. It would not be worth investing in refurbishing or up-dating them.

178

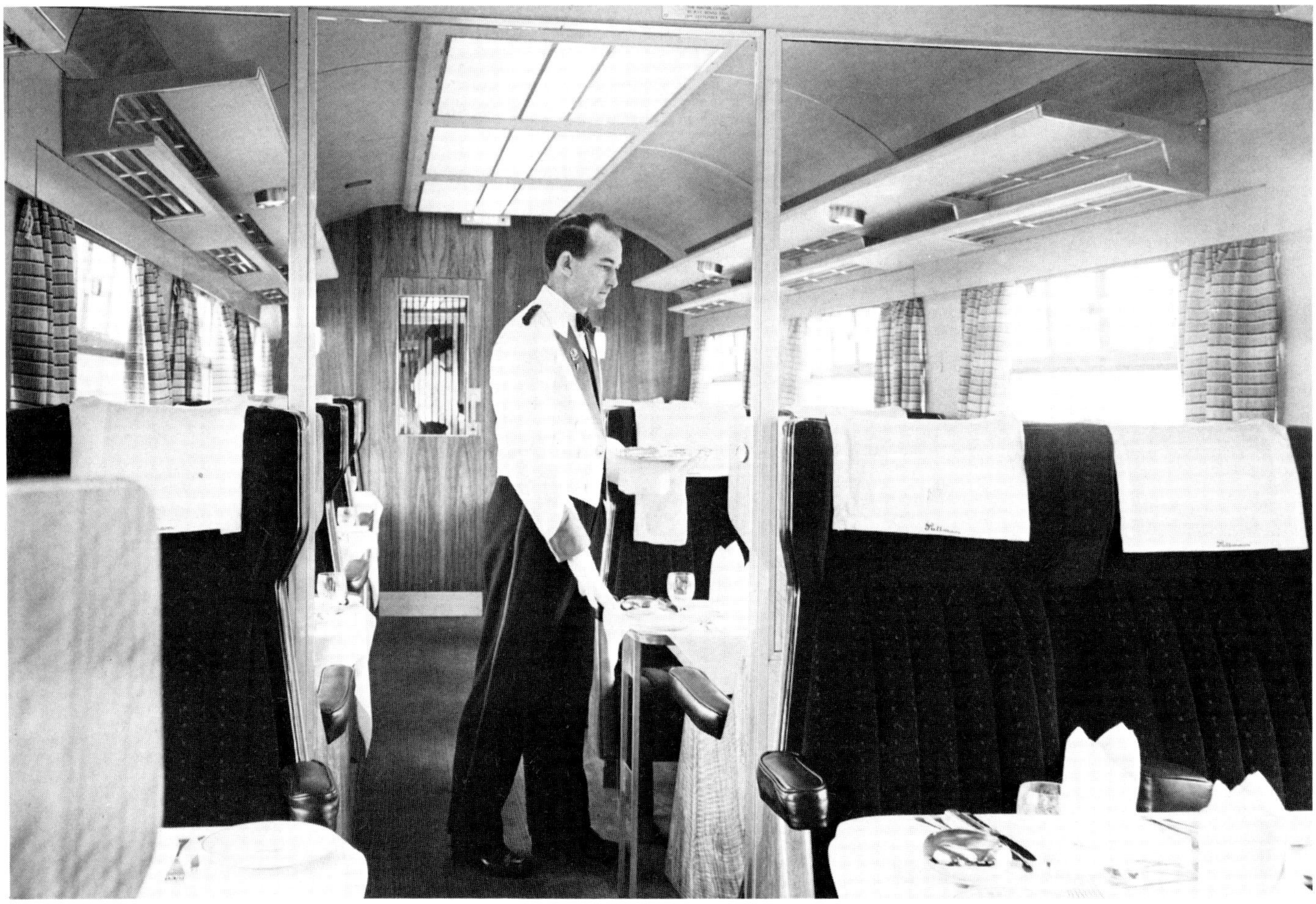

179

178
The interior of a first class kitchen car of the East Coast route replacement Pullmans of 1960. These featured 1+1 seating in traditional loose armchairs, but innovations in the lighting and in the design of the luggage racks. A spacious and pleasing layout, in true Pullman style, but one which the BTC was beginning to think of as uneconomical. The armchairs were from older cars, refurbished by Preston Park workshops for the new vehicles. *BR*

179
The 2+1 fixed seating layout of the second class accommodation of the 1960 East Coast route Pullmans was in fact remarkably well-appointed and spacious, very much the equivalent of a BR *first* class saloon carriage of the period. The glazed partitions gave an added feeling of space, plus very good lighting. No decorative panels were used in these cars, although the wall finishes were all of a very high standard. Note the two-way mirror in the end door, permitting a view into the saloon for attendants carrying trays, before opening the door. *BR*

(4) Although the 'Golden Arrow' was scarcely profitable from Pullman's point of view (its timings precluded much in the way of meal service) it was still regarded as an important prestige service by British Railways. The idea of a Pullman service to Plymouth was again considered in 1966 — as recounted on page 144 — but once again failed to get off the ground.

Unfortunately, even as the public demonstrated their enthusiasm for the new Pullmans, in particular the 'Blue Pullmans' the government of the day had a far less appreciative view of British Railways, and the BTC was being censured for an increasing deficit. The promises of the Modernisation Plan were not being fulfilled and something had to be done. On 10 March 1960 the Prime Minister, the Rt Hon Harold Macmillan had informed the House of Commons that:

> 'The railway system must be remodelled to meet current needs and the modernisation plan must be adapted to this new shape. This will involve certain sacrifices of convenience, for example in the reduction of uneconomic services. . . The Commission (BTC) must accept a radical alteration of its structure, so as to secure a more effective distribution of functions and a better use of all its assets.'

The action taken was to appoint an advisory group of industrialists, to recommend in a report, a strategy for the railways in the years ahead. One of the industrialists was the technical director of Imperial Chemical Industries (ICI), Dr Richard Beeching. His observations in the report impressed Transport Minister Ernest Marples so much that when it was decided to abolish the BTC and to create a new British Railways Board (BRB), the chairmanship was offered to Dr Beeching, who accepted. The BRB was created by the Conservative Government's 1962 Transport Act and came into being on Monday, 1 January 1963. The very same day the Pullman Car Co — whose contract with the BTC had just expired — was integrated as a division of British Transport Hotels (BTH). It was now wholly nationalised, and the company name, Pullman, passed into the hands of the newly-created British Railways Board. It was 93 years since George Mortimer Pullman had brought his ideas to Britain.

Only three months later, on 25 March 1963, the Pullman Division, as it was now called, was allied to the Chief of BTCS refreshment rooms and restaurant cars of BTH. It was to prove to be a bitter pill to swallow for many faithful Pullman employees.

Although the assimilation of Pullman had in effect now taken place, much as the NUR had wished, there were still some activists amongst members of BTH and its subsidiaries who disliked the very name Pullman, and who had already created the problems with the operation of the 'Blue Pullman' trains. The Pullman marque was resented because it represented the fruits of private enterprise and capitalism. The Pullman coat of arms — proud and symbolic — had already been dubbed 'the badge of shame' by some of the activists. But for the moment Dr Beeching was content to retain the Pullman identity because he was busy formulating a new overall corporate identity for British Railways, into which the Pullman services would be fitted, and identified as an integral part of the whole. The Corporate Identity programme was revealed to the press by Dr Beeching in January 1965. With a new name style 'British Rail' and a new double-arrow symbol, or logotype, a series of house colours: blue, grey and red, replaced the various Regional liveries and the two Pullman car liveries. The name Pullman was retained, but as cars were in due course repainted blue and grey, they lost their individual names and simply carried a number. The oldest cars, in 'Bournemouth Belle' service were never included in the new livery scheme, and ran out their days in umber and cream, as we shall see later.

The same month that Dr Beeching revealed the Corporate Identity also witnessed a very sad event, in which Pullman cars played a prominent role: a special train to carry the body and the mourners of the late Sir Winston Churchill to Bladon near Blenheim, where he had asked to be buried. Sir Winston died on Sunday 24 January 1965, and a state funeral took place the following Saturday, 30 January. The special train, all in umber

180

180
The chef of a 'Queen of Scots' Pullman at work. The kitchens of the 1960 cars were models of compact efficiency, and one always marvelled at the quantity and variety of foods prepared in such modest areas often whilst on the move, and at speeds of up to 90mph. Cooking was by bottled gas on both the East Coast design of kitchen and on the 'Blue Pullman' design of kitchen, despite the fact that the diesel-electrics generated their own power supply, in the case of the 'Blue Pullmans'. *BR*

and cream livery, including a specially repainted SR bogie parcels van (PMV) to carry the coffin*, ran from London, Waterloo station to Handborough on the Western Region, the nearest station to Bladon. As the immaculate rake of Pullmans and hearse progressed on their solemn journey, the lineside was thronged with spectators, and the nation watched on television. History has a way of repeating itself, because all the people who witnessed the funeral train cannot have failed to recognise the use of Pullmans, rather than ordinary BR rolling stock. Just as in the case of the funeral cortege of the late President Abraham Lincoln, one hundred years earlier, in 1865, only the stately Pullmans possessed the air of dignity appropriate to the occasion. During his lifetime Sir Winston had been a Pullman passenger on numerous occasions, and the company had valued his desire to use their services, even in the darkest years of World War 2.

*The PMV used as a hearse, or catafalque van, for the train, had in fact been repainted for this melancholy task at least two years before, when Sir Winston had suffered a serious fall whilst on holiday in France. It was kept hidden away at Stewarts Lane shed, governed by strict instructions that it was not to be used under any circumstances whatsoever — until authorised — its intended use being undisclosed to BR or Pullman staff.

A 1966 scheme to convert the 'Midland Pullman' sets to gas-turbine power was proposed by Rolls-Royce Ltd. This would also have entailed new light-weight Budd-type bogies being fitted. It is perhaps a pity that this idea was not deemed worthy of some practical experiment.

The Beeching reshaping was a period of great change in Pullman affairs. While the existing SR ROV services were passed to BRCS control, the Pullman Division of British Transport Hotels acquired responsibility for the prime restaurant car services on the other regions. There was talk of replacing (or supplementing) first-class accommodation on many InterCity services with supplementary fare vehicles

181

182

181
The special funeral train conveying the body of Sir Winston Churchill, and family mourners from Waterloo to Handborough on 30 January 1965; seen shortly after passing Virginia Water. Onlookers trespass on the lineside as 'Battle of Britain' class Pacific No 34051 *Sir Winston Churchill* heads the train in fine style, its three headcode discs arranged to form the famous 'V for Victory' sign so beloved by Sir Winston. The train was composed as follows: Pullman Car No 208 next to the locomotive, hearse van No S2464 (with paint date 7/62), then Pullman cars *Carina*, *Lydia*, *Perseus* and *Isle of Thanet*. *Brian Stephenson*

182
Visiting African dignitaries are greeted by the Prime Minister, The Rt Hon Harold Macmillan (with Lord Home walking forward on the right to greet them). Pullman attendants, with white gloves, flank the car doorway, which is exactly opposite the red carpet on the platform. The Pullman car is one of the 'Golden Arrow' cars (note the rectangular toilet window), exact identity not known, possibly *Phoenix*. *Graham Warburton collection*

— and, in fact, the first class section of one of the East Coast main line 'Talisman' sets was replaced by Pullman cars in 1964, only to have them taken off the following April. There was a constant background of trade union opposition to these developments, and in no time at all threats of industrial action raised their ugly head once again. The Pullman 'badge of shame' was realised to be rubbing salt into the wound when worn by NUR men, and carried on BR-owned vehicles. Accordingly Dr Beeching decided to replace it with the new BR double arrow logo, for all applications, from liveries to uniforms and menus. For the moment the name Pullman was retained on the sides of the cars.

A whiff of private enterprise entered the political arena in March 1964 when Lionel Burleigh commenced charter runs, using SR Pullman cars, under the title of the 'Regency Belle'. A lavish publicity exercise launched the idea and the hand-bill is worthy of reprinting in full because, although the scheme soon failed — after a promising start — it was, in reality, ahead of its time. If such a train could run again today, using 'Brighton Belle' cars, it would be a winner! The Pullmans were decorated with floral displays, and added feminine charms:

'The Regency Belle

'If you would like to get away from it all for a few hours and enjoy a really carefree night out in good company, reserve a seat on "The Regency Belle" Leave your ignition key at home. Go to Victoria Station on Saturday or Sunday night at 7.00 for 7.15 departure. Pass under the gay awning, along the red carpet to the "Regency Belle", the most luxurious train in the world staffed by efficient stewards and six of Britain's most beautiful girls in immaculate Regency costume, who will escort you to you reserved seat in an atmosphere of soft lighting and very quiet Gershwin music. There will be an orchid for every lady, and on the journey down you sip your vintage champagne and relax. Our charming hostesses will be glad to talk and give advice or help to visitors, on where to dine, dance or play in this Country, and other point of interest.

'They are conversant with the Ballet, the Theatre and the Arts, and have been specially chosen for their knowledge and ability to put people completely at ease. In fact, if this is your first visit to these shores, you will get a wonderful insight into our way of life, or at least one facet of it. Why not make this your initial foray!

'Upon arrival you will be given a "V.I.P." welcome by one of Brighton's celebrated personalities of whom there are many. A fleet of luxurious cars will then drive you to a first class hotel where you will partake of a recherché supper in delightful surroundings. After this, dancing; a midnight swim? or a mild flutter at the gaming tables and then—idle chatter until your carriage arrives at 1.45 to transport you back to your softly heated luxury train.

'Then the perfect English breakfast will be served as only this Country knows how to cook it. Finnon Haddock, crisp Wiltshire bacon, Sussex eggs, buttered toast and chunky marmalade with tea or coffee. All this and the morning papers too!

'A night to remember and the cost £7 7s. 0d.

'Interested?

'Tickets may be obtained at the General Enquiry Office at Victoria Station, the British Railways Travel Centre, Lower Regent Street, London, W.1, or you may write to the Line Manager, B.R., Essex House, College Road, Croydon, Surrey.

Direction of Lionel Burleigh'

Although the intention was always to use the 'Brighton Belle' electric multiple-unit Pullmans, weekend engineering works meant diverting the train on occasions on to non-electrified lines, in which case a locomotive-hauled Pullman set (based at

183

184

183
In 1964 Pullman cars formed the first class portion of one of the sets for the ER 'Talisman' London-Scotland workings. The down afternoon 'Talisman' is seen here at Hadley Wood, hauled by 'Deltic' class Co-Co No D9015 *Tulyar*. The second of the four Pullmans in the formation (third vehicle in train) is of particular interest because it is a 1951 SR 'Golden Arrow' kitchen car, on Gresley bogies, presumably on loan due to a temporary shortage of 1960 Metropolitan-Cammell Pullman kitchen cars at the time.
E. Treacy

184
'Deltic' No D9013 *The Black Watch* stands at Darlington with the up 'Tees-Tyne Pullman' on 12 February 1964. The train is formed principally of 1960 Metropolitan-Cammell cars, but next to the locomotive is one of the 1928/29-vintage parlour brake seconds retained for use with the new East Coast cars, there being no brake vehicles in the 1960 batch of Pullman cars.
B. J. Ashworth

185

185
The 'Wells Fargo' was the unofficial name given to the 'Blue Pullman' standby trains, used when the rostered diesel-electric Pullman units were out of service for some reason. Portrayed here is the 13.00 'Birmingham Pullman' from Birmingham (Snow Hill) to London (Paddington) arriving at High Wycombe on 26 August 1964, hauled by Brush Type 4 Co-Co No 1690. The formation of the standby train on this date was as follows: Car Nos 54, 106; 340; 352; 344; 105 and 55. The three second class East Coast route replacement cars (340, 352 and 344) were acting as *first* class accommodation in this instance; matching the higher seating density of the 'Blue Pullman' firsts, with their 2+1 layout.
H. K. Harman

186
The introduction of 100mph 25kV electrified working between London (Euston) and Manchester robbed the St Pancras line 'Midland Pullmans' of their previous speed advantage. Nonetheless the Pullman services were popular, so the LMR decided to retain them, using new locomotive-hauled Pullman cars, but running from both Manchester and Liverpool to Euston and back on weekdays. The new Pullmans were designed around the basic BR Mk 2 all-steel integral carriage, then coming into the vanguard of BR Intercity services. The livery was grey and blue with a semi-matt finish and the cars were not named. Class 86 Bo-Bo electric locomotive No E3141 heads the 'Manchester Pullman' in this picture, taken in 1966. *BR*

Stewarts Lane) was substituted. This comprised: Car No 64, *Evadne*, *Octavia*, *Lydia*, *Phyllis*, Car No 334 and Car No 84, and was usually hauled by Bulleid 'Battle of Britain' class Pacific No 34088 *213 Squadron*.

'Blue Pullmans' and East Coast cars apart, the Pullman fleet was now showing its age, especially when compared to the vastly improved ordinary passenger carriages of the BR Mk 2 design. Many 'Blue Pullman' features were taken by BR and modified in the light of experience, and it could be said that the Mk 2 open saloon with adjustable seats, and (in later batches) full air-conditioning, was a direct development of the 'Blue Pullmans'. The enhanced comfort of these carriages, with excellent bogie design, made a considerable commercial impact, and Inter-city services began to witness a market upsurge; in particular where 95 or 100mph electric and diesel-electric, or diesel-hydraulic locomotives were used with the new carriages. The railway's passenger managers began to view the Pullman trains as unnecessarily costly to run because of their extra staffing and array of kitchen equipment spread down the length of the train. They decided that they could offer equivalent comfort in a new generation of main line open saloon carriages, and technical development of the Mk 3 carriage was commenced in 1966. A project for a luxury lounge car, with spaciously arranged groups of 10 seats in coupé-style compartments, housed within a first class Mk 2 carriage (for which a supplementary fare was envisaged) never got beyond the prototype stage, because the policy was now to improve the ordinary accommodation, rather than offering special *de luxe* features for a minority only.

The operating cost per passenger seat of a Pullman car plus the staff wages, was now the weak point in any argument for their retention, or extension, in the BRB's view and by the mid-1960s it was already looking to ways to improve or else cut back Pullman services, and in some cases to abandon them completely. The BRB recognised that certain Pullman trains were popular with the travelling public, and it was decided to retain the more profitable examples for as long as the rolling stock lasted in good condition and the demand continued. To some extent the use of buffet cars was considered a suitable replacement, to reduce operating costs, although it was realised that these could not offer the same standards of comfort or service.

Two other factors influencing the retention, or otherwise, of BR Pullman services were the changing pattern of business travel demands, and the general acceleration of train schedules on Inter city routes made possible by modern diesel and electric locomotives, and improved tracks and signalling. Business people, in particular, increasingly wanted an earlier start and finish to their day — and, for example, the once heavy demand for lunch (southbound) and dinner (northbound) on the 'Tees-Tyne Pullman' fell off considerably. A spin-off from this changing pattern was the increased demand for breakfasts of a very substantial nature — known today on BR as 'The Great British Breakfast'.

The biggest setback to Pullman affairs at this crucial stage was the disappointing performance of the 'Blue Pullmans' in everyday use; in particular their bad riding. The 'Blue Pullmans' were a victim of their

186

187
In concept the Mk 2 Pullmans were an amalgamation of the two 1960 Pullman designs. The integral bodyshell and B4 and B5 type bogies were in fact due to BR developments since the Mk 1 era, while the gangways and couplings, remained to BR/Pullman standard. From the 'Blue Pullman' concept came the full air-conditioning, wide double-glazed windows and the interior design. The passenger doors opened inwards in true Pullman fashion, whereas BR Mk 2 stock had outward-opening doors. Brake parlour first No M581 was photographed at Manchester on 28 January 1966, when new. *C. R. Whitfield*

188
The interior design of the LMR electrically-hauled Mk 2 Pullman cars was closely based upon that of the 1960 diesel-electric 'Blue Pullmans' except for the seats, which were to a new and more comfortable design, developed by Wilkes & Ashmore for the BTC in the meantime. The seat density was 2+1 for the first class, as seen here. One small feature missing was the traditional table lamp — deemed unnecessary by BR's designers. *BR*

187

188

own specialised unit-formed concept, and in retrospect it would have been better if they had been built as locomotive-hauled carriages. (Who knows, they might have ridden better without the 'push and pull' effect of the power bogies at each end of the set?) The real trouble was utilisation. On the 'Midland Pullman' service one whole set was forced to stand idle as a 'reserve' in case of the failure of the other one, and on the WR one of their three sets was often similarly idle, although in time the WR did assemble a motley collection of old locomotive-hauled Pullman cars to act as a substitute, which gained the nickname of the 'Wells Fargo' because of its obvious antiquity by comparison with the 'Blue Pullmans'. When electrification linked Euston with Birmingham and Manchester in 1966 it robbed the 'Blue Pullmans' of the speed advantage, and the new generation of Mk 2 carriages entering service offered quite comparable standards of comfort (and a better ride) without a supplementary fare, but without the personal, at the table, service of course. The LMR also opted for some new locomotive-hauled Pullman cars for its electrically-powered London-Manchester/Liverpool routes and the interior of these was virtually identical to the diesel sets, but housed in the new Mk 2 bodyshell, and the new B4 and B5 type bogies were used. These were built at Derby, and entered service in 1966. The first class accommodation was now 2+1 each side of the aisle as in the 'Blue Pullmans'.

The two LMR six-car 'Midland Pullman' diesel sets were offered around to the other Regions and finally found a home on the WR. At this time there was again some thought about using Pullmans to Plymouth on the 'Golden Hind', but possible lack of power and adhesion on the notorious South Devon banks ruled this out for the 'Blue Pullmans' — wisely! Instead, the two ex-LMR sets were modified at Swindon works, with multiple-unit jumper cables added to the power car nose-ends, and with some conversion to second class accommodation. In this guise they formed a 12-car train, used on the busy morning and evening 'Bristol Pullman' service. In between these duties, the two six-car sets were divided; one working to Bristol at midday and back, and the other one to Oxford and back (until 1969 in the latter case). The ex-'Birmingham Pullman' set was switched over to the 'South Wales Pullman' service, which was augmented, now having two sets at work; the third eight-car set again becoming spare.

The fact that these luxury trains were non-standard units did not help their cause, and the steady progress being made with passenger accommodation in the standard carriage designs meant that there was little justification for charging the supplementary fare. The final years saw them painted in a new grey and blue livery (sadly finished in a semi-matt, which lacked the traditional Pullman 'sparkle') and it was in the spring of 1973 that all the sets were withdrawn in one operation. None was preserved, but their contribution to today's comfortable rail travel should never be forgotten. They were the true precursors of the High Speed Trains on BR, and in their time they were classic.

For the Pullman services that were to be retained, the BRB commenced some refurbishment, and application of the new livery; the Pullman coat of arms, as we have seen, being quietly dropped at the same time, as a gesture to the NUR. In September 1967 the periodical *Modern Railways* carried this report:

'**East Coast Pullmans modified**
'Metro-Cammell Pullmans built in 1960 for the East Coast Route services are being modified. The vehicles involved are the 13 first-class kitchen cars and eight first-class parlour cars, which are having their original one-and-one armchair seating replaced by a two-and-one layout employing seating of the type in the modern BR-built Pullmans on the LMR. At the same time there has been some internal re-styling of the décor—but not removal of the old Pullman hallmark, table lights—and the exterior of the cars has been repainted in BR blue and grey; in the latter operation the cars' bird and jewel names have been erased. The seating reorganisation has raised the capacity of the kitchen cars from 20 to 26 and that of the parlour cars from 24 to 29. By the end of July three cars had been treated. To maintain commercial services the modification programme has had to be spread through the winter to a rigid schedule, under which the conversions are to be completed at the end of March next year. The five WR "Blue Pullman" DMU sets are also to be repainted in the new livery.'

The BRB was evidently anxious to improve the costs per seat ratio of these cars, in order to justify their continued use as Pullmans.

The ever-popular 'Brighton Belle', still making its runs day-in, day-out, next came under the scrutiny of the BRB, and it was decided to refurbish the interiors of the cars. All the other electric Pullman cars had been replaced by buffet cars in the new 4-BIG electric multiple-units, introduced in 1964, but the 'Belle' cars still had some useful years of life in them (due to wartime storage) and the SR took all three sets into the workshops for refurbishing, commencing with No 3052 which was outshopped in December 1968. When it emerged it was almost unrecognisable as a Pullman externally because now even the *name* Pullman did not appear on the sides! It was painted in the standard BR blue and

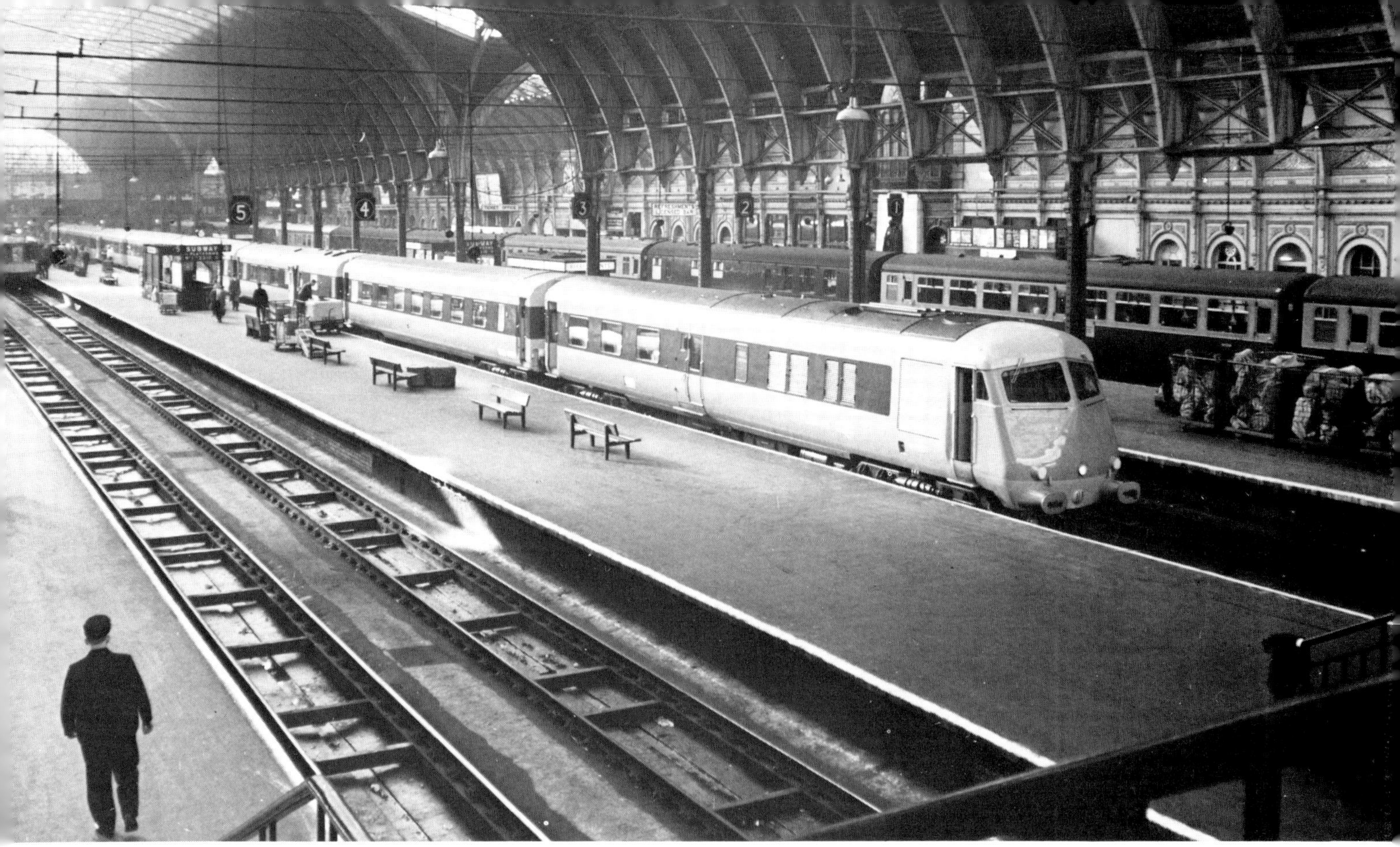

189

190

191

189
This September 1967 photograph shows one of the eight-car WR 'Blue Pullmans' in the revised livery of grey with a blue window band, and yellow nose ends. No Pullman coat of arms was featured, simply the name 'Pullman' on the power cars only, together with the double-arrow BR logotype. The scene is Paddington, with the train being prepared for departure. *BR*

190
Also in 1967, the Metro-Cammell 1960 East Coast replacement cars began to receive the new livery and to lose the Pullman coat of arms and their individual car names. This was linked to a programme of interior refurbishment for the first class cars, which gave them 2+1 seating of identical design to that used in the LMR's Mk 2 Pullmans. Comparison with Picture No 151 shows the extent of the change, which included removal of the glass partitions within the saloon, but retained the table lamps. *BR*

191
The Pullman electric composite cars of the SR Central Section services had given stalwart service since their introduction in 1933. There were 23 of them, each one marshalled into a unit of five railway-owned vehicles, oddly contrasting in style and livery, and known as 6-PUL, or (originally and for three units only) 6-CIT. The composite cars seated 12 first class (four in a coupé) and 16 third class (later, of course seconds) and included a kitchen, pantry and lavatory — a minor miracle of compact layout. Cooking was by electricity as on the 'Brighton Belle' cars, with the supply taken off the third rail and metered, to be charged to the Pullman Car Co's account by the SR. Ideas of extending their lives were dropped largely because of the bad riding of their later years, and they were phased out by modern buffet cars (ROVs) over a period from September 1965 until June 1966. Car *Daisy* is seen here (not in use) at Eastbourne on 13 June 1963. *Alec Swain*

grey, complete with double-arrow logotype, and each car carried the legend 'Brighton Belle' on the sides in large white lettering. No Pullman insignia of any description was carried on the front ends, just the train name in black on the yellow warning colour which had by then been adopted as standard for the front ends of BR trains. The names were removed from the first class cars, and all the cars were identified solely by numbers, as follows:

Unit No 3051: 289 (formerly 89), 286 (86), 282 (*Doris*), 279 (*Hazel*), 288 (88).
Unit No 3052: 291 (91), 287 (87), 284 (*Vera*), 280 (*Audrey*), 290 (90).
Unit No 3053: 293 (93), 285 (85), 283 (*Mona*), 281 (*Gwen*), 292 (92).

Inside the cars the seats were re-upholstered, the first class in charcoal and grey check, the second (ex-third) class in blue and green check with navy blue arm and head rests. The carpets were mustard coloured, and the curtains, like those in the latest Southern Region express stock, were light orange.

The 'Golden Arrow' suffered a similar demotional treatment at about the same time, with the removal of the name Pullman and the car names (although these could still be found *inside* the cars). At the same time the uniforms of the staff on both trains were changed, and even the famous Pullman job title of 'attendant' was changed to 'steward'. Prior to this at least one car had been repainted in the reverse grey and blue livery, with the word Pullman retained, but the final livery was the same as carried by the 'Brighton Belle'. It did not suit the cars at all, and they were not identified as Pullmans.

Since June 1961 the 'Golden Arrow' had been hauled by SR electric locomotives, and the number of Pullman cars had been steadily reduced, as demand for them dwindled. Ordinary carriages outnumbered the Pullmans by June 1965, when the second class Pullman cars were withdrawn, and replaced by ordinary stock. Earlier, the car *Pegasus*, with the *Trianon Bar*, had been removed from the train and sent (together with the 1961 Metro-Cammell *Hadrian Bar*) to the London Midland Region, both for use as *Nightcap Bars* on the London-Glasgow overnight sleeping car services. To begin with they ran in full Pullman livery, but this was later changed to the standard BR blue and grey. At the end of the 1960s the 'Golden Arrow' was but a shadow of its former glory, due to the huge new Channel car-ferry developments, and the competition of the jet aeroplane.

On 9 July 1967 the 'Bournemouth Belle' all-Pullman car train had run for the last time. The cars remained in full Pullman umber and cream livery until the very last day, and never suffered the demotion of the 'Brighton Belle' or 'Golden Arrow'. New electric train services then took over the Bournemouth line, and the refreshment facilities were of a buffet/restaurant type, rather unfortunately placed in carriages with motor bogies beneath them. The new trains were fast, but they lacked the qualities of the stately 'Bournemouth Belle'. The age of the 'Belle' cars had alone justified their withdrawal, because the cost of providing new ones was out of the question. The

192

193

192
The gauche new livery of the 'Brighton Belle' units, applied in 1968 by BR, without a trace of Pullman identity in the form of either the name, or the coat of arms, and with each car an anonymous blue and grey vehicle. Only the name of the train itself was retained, with the BR logo used on the motor cars. The picture was taken on 19 December 1968. Note the twin air horns mounted on the driving cab roof, instead of the original whistle. *BR*

193
From June 1961 the 'Golden Arrow' boat trains worked by Bo-Bo electric locomotives, of Class 71 (late classification) and a real effort was made by Stewarts Lane depot to keep these in spotless condition; complete with headboard, British and French flags, and bodyside arrows. No E5015 is seen entering Dover Marine in June 1961 on the first day of electric haulage. *BR*

194
Demand for Pullman facilities on the 'Golden Arrow' declined as more people owned their own vehicles. By 1965 the ordinary BR carriages outnumbered the supplementary-fare Pullmans on the train. The *Trianon Bar* car *Pegasus* had already been removed and sent to work as the 'Nightcap Bar' on Euston-Glasgow sleeping car services. It is seen here at Rugby on 27 February 1965. *John Edgington*

Bournemouth line electrification was only justifiable by a narrow economic margin, which was achieved by making the 'new' electric trains from rebuilt 1951-design BR standard Mk 1 carriages. The added cost of providing Pullman-type accommodation was prohibitive, even if there was a demand for it — which there was.

The same year that witnessed the withdrawal of the classic 'Bournemouth Belle' also witnessed the announcement of the futuristic Advanced Passenger Train (APT) project. It began to look very much as if Pullmans on BR would soon be a thing of the past.

194

195

195
Time was running out — not only for the SR Pullman trains but also for steam power — when this picture was taken of the empty stock of the 'Bournemouth Belle' being hauled out of Waterloo station on 16 May 1967. The following July, the 'Bournemouth Belle' ran for the last time, and new electric multiple-unit trains with buffet car facilities took-over. Ivatt Class 2MT 2-6-2T No 41298 struggles to get the heavy load underway, bound for the carriage sidings, where the train will be stabled and cleaned, ready for the next working. *L. A. Nixon*

CHAPTER 18

Kippers and Final Bows

196

The popularity of the electric 'Brighton Belle' Pullman really achieved new heights when, in what were to prove to be its final years, it was based in Brighton instead of London, as had in fact been advocated by Col Harding for some years. This allowed no less than eight trips per day to be run between the two termini — London and Brighton — and it produced an interesting off-shoot, in true Pullman tradition. Because it was possible to take a late full breakfast on the morning up service to London, and a full supper on the late-night down run it became established as a sort of travelling theatrical club, due to so many eminent actors and writers choosing to live on the South Coast, but having to be at the theatres each night in London, or even earlier for rehearsals.

One morning in 1970, the famous British actor Lord Olivier, a regular passenger on the 'Brighton Belle', took his seat and ordered as usual his pair of 'marvellous, juicy and succulent' grilled kippers for breakfast. He was told by the steward that they were no longer available on the menu, because of 'economies' ordered by British Rail Catering (later Travellers Fare) who had taken over control of the Pullman management from the Pullman Division in 1967. There had already been rumours that the 'Brighton Belle' might be withdrawn, but for the moment it ran as usual. However to find no kippers was the final insult, and Lord Olivier organised a petition for their restoration to the menu. It quite naturally attracted the attention of the national press

196
The interior of the 'Brighton Belle' first class parlour car *Audrey* as refurbished by British Rail workshops in December 1968, using standard BR upholstery, carpets and curtains. By some oversight the Pullman coat of arms and the car name were left in place on the circular panel above the partition doorway, although the exterior of the car had lost such individuality, and simply carried a number and the train name 'Brighton Belle'. *Audrey* survives today in main line service, as one of the Venice Simplon-Orient-Express British Pullmans. Indeed, although the 'Belle' itself was withdrawn by BR in April 1972, all the cars have been preserved, as is related in Part 3 of this book. *BR*

and television. The 'Belle' received a good deal of publicity and BR Catering a good deal of criticism, until forced by the hue and cry together with the petition, the kippers were solemnly reinstated. This action, by the great man of the theatre, would surely have appealed to the late George Mortimer Pullman! It highlighted the dilemma which confronted what, for convenience, I will henceforth refer to by its present name of Travellers Fare — the BR Catering Services into whose hands the Pullmans had been placed.

Unlike the former, profitable Pullman Car Co (apart, that is from Pullman-*staffed* catering services on the SR's Central and South Eastern Divisions, which had not been profitable for some years — the ROVs — and which had been financially supported by the Region since 1960), the BR restaurant and buffet car services regularly showed an annual deficit, and it was the station catering and other activities of Travellers Fare which subsidised the costs of providing meals on trains, and which showed a profit. On the SR London-South Coast services, wherever Pullmans had formerly run (except for the 'Brighton Belle') the refreshments were now of a light snack nature only, served in a buffet car, often staffed by only one man. The first class seating accommodation was for some reason located well away from the buffet, necessitating a longish walk through the train. By contrast the 'Belle' still had a full complement of staff, giving service to each seat, and chefs to prepare meals. The operating overheads were enormous compared to the buffets, but the 'Belle' was popular and well patronised. Travellers Fare no doubt thought that it could 'streamline' the menu to reduce the losses; the passengers thought otherwise. It was a stalemate, and the only possible move which would satisfy Travellers Fare was the complete withdrawal of the 'Brighton Belle' as being uneconomical to run. They realised this was likely to arouse a furore, and it did!

The background to the decision to abolish the 'Belle' was, however, more concerned with technical problems than with the catering, and various options were examined before the final decision was made. The real problem was that the train was becoming quite alarmingly rough-riding when running at speed, and at certain times on traversing junctions. This had developed to the stage where it could not be allowed to continue, not so much on safety grounds, but because the old cars were unable to maintain the schedules in reasonable comfort for the passengers. Whereas the new electric trains with their buffet cars, could and did reach 85-90mph and still ride well, the 'Belle' cars did not take to being 'flogged', and could not adhere to the 55-minute timing from London to Brighton possible with the new trains. All the staff, and the regular passengers knew that the 'Belle' was showing its age at the time of the 'kipper revolt', and the bogies were giving everyone a fairly rough time, at any sort of speed. The 'regulars' (the Author included) knew when to hang on to glass or cup as the 'Belle' took the curves at Clapham Junction, Balham, and Earlswood, but as in earlier years the uninitiated were taken by surprise. A notice had been printed upon each menu card on the 'Brighton Belle' prior to 1955, since its postwar restoration to service. This notice (which in fact appeared on *all* menu cards for SR diesel and electric catering services) read quite simply:

> 'Our staff take every possible care and precaution in the service of refreshments, and the Company cannot be held responsible for minor accidents of spillage etc, which may occur on account of excessive movement of the train.'

The 'excessive movement of the train' on the 'Brighton Belle' was caused by a deterioration of the original suspension system of the bogies and in 1955 the Pullman Car Co had invested some £70,000 in renewing the bogies. Unfortunately the new type soon proved to be little better than the originals, and by 1970 they were giving a ride far inferior to that provided by the latest SR electric trains of the 4-CIG/BIG/VEP types. In BR's eyes these bogies were spelling out the death sentence for the 'Brighton Belle'. Consideration was given to providing new bogies, or even replacing the complete trains with suitably adapted 'Blue Pullman'

197

198

197
The institutionalised BR 'Golden Arrow' of 1969, shorn of Pullman dignity to satisfy trade union pressures. By consulting the lists we are able to identify this car, No S301, as *Perseus*. What one is unable to identify is the mentality that thought it better to remove the name and Pullman splendour in favour of such a cheapjack image! The livery was standard BR blue and grey, with thick white lining and lettering; photograph taken on 31 March 1969 at Dover. *Perseus* is another of the cars rescued by the VS-O-E enterprise, and is now running again, restored to pristine condition in umber, cream and gold. *BR*

198
Apparent dislike of the Pullman car image did not prevent those responsible on BR from recognising their excellent qualities as rolling stock, at least, with regard to the latter-day examples. Hence, when it was decided to cease offering second class Pullman supplement seating to passengers, BR quietly up-graded the second class East Coast 1960 cars to offer them as first class ordinary accommodation; simply by means of a coat of standard blue and grey paintwork and some white numerals to denote the class on doors and windows!
Ian Allan Library

199

199
The prototype of the High Speed Train (HST) which was intended as a stop-gap InterCity express diesel-electric multiple-unit, pending development and production of the futuristic tilting Advanced Passenger Train (APT). The HST was capable of boosting rail speeds to 125mph and shortening journey times accordingly. In a number of respects it was a direct line of development from the 'Blue Pullmans', but no supplement was charged to use it. The prototype set No 252001, in a livery practically identical to the BR Pullman livery, is seen here climbing swiftly towards Box Tunnel, en route from Bristol to Paddington on 5 May 1975, the first day in passenger service. *G. F. Heiron*

200

200
The last Pullman service remaining on BR in 1981 was the 'Manchester Pullman', an all first-class train. It is seen here at Camden Bank summit on an earlier date, 6 June 1973. The Mk 2 Pullmans were no longer the front rank passenger accommodation on the LMR main line and it was widely rumoured that when they were replaced by new Mk 3 carriages the Pullman marque would disappear completely from BR InterCity services, and the title of 'Executive' would be adopted instead. *Brian Morrison*

cars (but one doubts whether they would have provide a better ride). Finally the sad announcement was made that the 'Brighton Belle' would be withdrawn from service on Sunday 30 April 1972. No replacement was to be provided. Although the 'Belle' died, the cars did not (as we shall see presently). Moreover, it died in a blaze of publicity which only served to remind people what they were missing. To this day, the 'Brighton Belle' is recalled with affection by all who used it, even if the 'spillages' left their mark on smart evening dress, or summer frock.

Just five months after the 'Brighton Belle' made its final bow, the 'Golden Arrow' ran for the last time; on Saturday 30 September 1972. No more regular Pullman car services ran on the lines of the Southern Region, which in the past had been the true bastion of the Pullman Car Co. Now, on the SR only the still-majestic dark blue and gold sleeping cars of the Wagons-Lits Co, which were still running on the 'Night Ferry' service, gave a reminder of the once important links between Pullman, Wagons-Lits and the Southern Railway system.

I will be describing subsequent BR developments, as an essential background to the Pullman story, in Part Three (Chapter 20). However to end this part of the book, a brief résumé of the remaining Pullman services, during this bleak period, is presented.

As mentioned in the previous chapter all the 'Blue Pullman' trains were withdrawn from service on the Western Region in 1973. This left only the London Midland and the Eastern Region still operating Pullman cars, all smooth-riding, 100mph capability locomotive-hauled cars of comparatively recent design. Travellers Fare evidently found the returns to be more satisfactory than on the 'Brighton Belle', and Pullman service continued, but on the Eastern Region the 'Queen of Scots' had reverted to a London-based King's Cross-West Riding service in 1964. Known as the 'White Rose Pullman', it survived only to 1967. The 'Master Cutler', too, had gone in October 1968. The 'Tees-Tyne Pullman' and the 'Yorkshire Pullman' remained, however, the latter running to Harrogate. The former Hull portion of this was still operating, but renamed as the 'Hull Pullman'. The trains also included normal second class non-supplementary fare carriages of modern design, and had lost their all-Pullman status; only the first class accommodation being in Pullmans, with the increased 2+1 seating layout. The Pullmans continued to run until Friday 5 May 1978 when the last journeys were made to Harrogate and to Hull and Newcastle. They were replaced by new High Speed Trains without Pullman accommodation, but fully air-conditioned, and in many respects direct descendants of the 'Blue Pullman' prototypes.

Although not strictly speaking part of our story, the 'Night Ferry' service, resumed in 1947 as already mentioned, lasted until 31 October 1980. Various schemes to provide rebuilt BR Mk 1 sleeping cars to keep the service running came to nothing, and the old sleeping cars ran the service till the end. The 'hot-and-cold' Channel Tunnel schemes had always affected the future of the train, and it seemed to hang on by the skin of its teeth for years, offering Britain a tantalising glimpse of the great Wagons-Lits continental organisation, with all its associated romance and nostalgia, so richly suggested by the elaborate lettering and crest carried on their sides. Since 1976, when the Wagons-Lits contract had expired the staffing, equipping and victualling of the cars had been done by BR Southern Region. At the time of withdrawal some of the cars were 44 years old, and one car, No 3792, is now preserved at the National Railway Museum, at York.

At the commencement of 1981, only one Pullman car train was still running on British Railways. This was the 'Manchester Pullman' between London (Euston) and Manchester Piccadilly. There were persistent rumours that it would be withdrawn when the cars fell due for replacement and then the link between Pullman and BR would be severed. But the name Pullman was not destined to be so easily forgotten, thanks to the work of the private preservationists, who refused to let it die.

TABLE TWO:

British Pullman Car List 1906-1962

British Pullman Car List 1906-1962 (Including all Preserved Examples and Fleet Changes up to 1966)

In Table 1 I have acknowledged my debt to the intrepid researches of Messrs J. T. Howard-Turner, George Behrend, Julian Morel, and others, who have previously published lists of the British Pullman cars. The most recent of these *complete* lists appeared in George Behrend's classic book *History of Trains de Luxe*, first published in 1982 (see Bibliography) and this list follows the very legible style he has created. Quite considerable changes have taken place since then, in particular with the preserved cars, and I have endeavoured with the help of various owners of cars and with Julian Morel's book *Pullman*, published in 1983, to bring Behrend's list up-to-date. Quite remarkable are the numbers of modern Metro-Cammell 1960 East Coast cars that have been rescued, and which in some cases are once again running on BR tracks in full Pullman finery. At the time of writing it was not clear how many more Pullmans would be restored to BR running, although some preserved examples are possibles in this respect, if thoroughly overhauled, and inspected and approved by BR beforehand.

The Pullman Car Co seems to have taken a delight in changing the roles, the names and the numbers of the fleet, to suit its needs, and it is a tribute to the various intrepid researchers that most changes are recorded here. There are some question marks around the names and numbers of cars in the late 1930s/early 1940s period, when the war broke out, and contemporary (and historical) records were destroyed in the blitz.

When Mr Davison Dalziel's private Pullman Co Ltd gave way to the Pullman Car Co Ltd in 1915, he handed over 74 cars. In 1918 Dalziel revalued the company's assests at postwar prices, and every car on the company's books received a schedule number, quite distinct from the running number or name by which cars had been identified until then. No schedule numbers were allocated to the 40 Pullmans withdrawn before the company was formed, and these are usually referred to by historians as the 'pre-schedule' cars. Under Dalziel's scheme, each car had the schedule number mounted on a plate, or painted on the car end. The list that follows uses the schedule numbers as identification, but not in strict numerical order, because the cars are grouped in the order of the names chosen and batches built or cars rebuilt, which did not happen so precisely as a strict numerical list would suggest was the case.

Only first class (and non-supplement dining cars in Scotland) bore names. Third-class cars carried 'Car No 00 Third Class' in the name panel. No third class ran in Britain after 1956, and the terms 'Third Class' or 'Second Class' disappeared and the term 'Car No 00' sufficed. Before 1956 second class existed only on 'Continental Expresses' from London to the Channel Ports. Pullmans on these trains bore the words 'Car No 00 Second Class' (also on the 'Golden Arrow' when second-class Pullmans ran in it). The cars were identical to third class. The numbers bore no relation to the schedule number in many cases, and identification became even more problematic with the change of some schedule numbers in 1960 (here ignored). Cars frequently changed status, but not all changes are recorded for reasons of space. Names often recurred, but no two cars with the same name ran at the same time during the period covered by this list.

A favourite trick of the Pullman Car Co was to try to persuade the public that a rebuilt car was *new*, by giving it a new name or running number; it should be noted, however, that the schedule number was not changed on the company records for these cars. The abbreviations used, are those that were used by the Pullman Car Co itself, as follows;

P= Pullman Parlour Car, usually two saloons and two end coupés.*
K= Pullman Parlour Car with kitchen, two saloons and one coupé.*
GP= Parlour Car with hand-brake wheel in one end vestibule, tip-up seat for the guard and 'Guard' painted on the vestibule doors, used on SR.
BP= Parlour Car with luggage van, hand brake and seat of guard inside the van (known as Brake Parlour).
MBP= Brake Parlour with driving compartment at one end instead of vestibule, used at each end of 5-car 'Brighton Belle' electric sets.
OBS= Parlour Car with small buffet, large rear window and no vestibule at one end;
D= Non-supplement dining car in Scotland, with kitchen, dining chairs in the saloons, not Pullman armchairs.
RC= Some non-supplement buffet cars in SR service in England bore the title 'Refreshment Car' or 'Buffet Car' in the name panel, others, painted green, with no name and lettered 'Buffet Car', or 'Buffet'.
CC= Withdrawn Pullman Cars transferred to BR ownership, with BR Regional Number, converted to 'Camp Coach' ('Holiday Coach' on SR), stationary, at country stations; numbers are BR Regional Nos.

*Second class Pullmans rarely had coupés, except those converted from firsts.

Note: Three cars were delivered from America, and assembled at Brighton for the LBSCR during the protracted negotiations between Mr Davison Dalziel and the American company, for the purchase of the British company. These were the last to feature clerestory roofs, and the last built for Britain in the USA.

Schedule Nos	*Original names or numbers*	*Type of car*	*Year into service*	*Builder*	*Notes*
21-23	*Duchess of Norfolk*	K (6)	1/1906	Pullman USA	Withdrawn in 1932; became bungalow at Lancing, Sussex.
	Princess Ena	K (6)	1/1906	Pullman USA	Scrapped in 1932.
	Princess Patricia	P (6)	1/1906	Pullman USA	Rebuilt to K by Pullman Longhedge in 1925; scrapped in 1934.

Notes: (a) Henceforth all new cars were British-built.
(b) The figure (6) alongside the type of car, ie K (6) indicates six-wheeled bogies fitted.

24, 29 25, 27 26, 28	*Verona, Alberta,* *Belgravia, Cleopatra,* *Bessborough, Princess Helen*	BP (6) P (6) P (6)	11/1908	Metropolitan Amalgamated Carriage & Wagon Co	All except *Princess Helen* rebuilt to K by Pullman Longhedge in 1914. *Verona* withdrawn in 1930; became wood and marquetry store, Preston Park. Rest withdrawn in 1935.

Schedule Nos	Original names or numbers	Type of car	Year into service	Builder	Notes
30	*Grosvenor*	K (6)	11/1908	Metropolitan Amalgamated Carriage & Wagon Co.	Rebuilt to bar car in 1936 by Pullman Preston Park; withdrawn in 1960; BR (E) camping coach (CC) No CC 169.
38, 35 37	*Sorrento, Corunna, Savona*	P	3/1910	Birmingham RC&W Co	*Sorrento* renamed *Baroness Nairn* 1919-21 in service on CR. *Sorrento* and *Corunna* rebuilt to K in 1924; withdrawn in 1962; CC BR(S) Nos P63 and P62. *Savona* became K in 1924; mobile stores supply car in 1933; rebuilt to First class in 1951; withdrawn in 1960; CC BR (Sc) No SC44.
39, 36	*Valencia, Florence*	K	3/1910	Birmingham RC&W Co	Withdrawn in 1962; CC BR(S) Nos P54 and P53.
40	*Clementina*	K	3/1910	Birmingham RC&W Co	Withdrawn in 1960; CC BR(E) No CC166
33, 34	*Galatea, Mayflower*	K	6/1910	Birmingham RC&W Co	Built without gangways for Metropolitan Railway in 1910; London Transport in 1933-39, withdrawn in 1939; became wood merchant's store/office at Hinchley Wood, Surrey, scrapped in 1948.
32	*Emerald* (I)	K	6/1910	Birmingham RC&W Co	Burnt out in 1955; became Instruction Car No 101; preserved as *Emerald* (I) at Conwy Valley Museum, Betws-y-Coed, Wales.
42	*Regina*	K	10/1910	Birmingham RC&W Co	Withdrawn in 1962; CC BR(S) No P55.
43, 44	*Sapphire, Palermo*	P	10/1910	Birmingham RC&W Co	Became Army (NAAFI) Canteens Nos 37 and 21 in 1942-48; withdrawn in 1962; CC BR(S) Nos P51 and P64. *Sapphire* preserved by the Lavender line, Isfield, Sussex and restored to SECR crimson lake livery, for sale mid-1987.
78	*Shamrock*	D	3/1911	Birmingham RC&W Co	Became *Duchess of Gordon* in 1918 for Caledonian Railway (see main text) then sold to LMSR in 1933 (No 210).
31	*Vivienne*	P (6)	7/1911	Cravens Limited, Sheffield	Scrapped in 1935.
46	*Myrtle*	K (6)	7/1911	Cravens Limited, Sheffield	Rebuilt in 1936 to bar car by Pullman Preston Park; withdrawn in 1962; CC BR(Sc) No SC47.
48	*Cosmo Bonsor*	K	5/1912	Cravens Limited, Sheffield	Named after SECR chairman; renamed *Rainbow* (III) in 1948; withdrawn in 1962; CC BR(Sc) No SC46.
41	*Leghorn*	P	12/1912	Cravens Limited, Sheffield	Withdrawn in 1962; CC BR(S) No P61.
45, 47	*Seville, Alicante*	K	12/1912	Cravens Limited, Sheffield	*Seville* became Army (NAAFI) Canteen No 3 in 1942-48; withdrawn in 1962; CC BR(S) No P47. *Alicante* withdrawn in 1962; CC BR(W) No W9874. *Alicante* intact at Marazion, Cornwall.
49, 54	*Glencoe, Scotia* (II)	P (6)	1/1914	Cravens Limited, Sheffield	In War Office Special Train, London-Dover, in 1914-18. *Scotia* used by King George V, Lloyd George and other VIPs. *Glencoe* scrapped at Shoreham in 1955. *Scotia* withdrawn in 1960; CC BR(E) No CC167.
51, 53	*Hibernia, Orpheus*	K (6)	1/1914	Cravens Limited, Sheffield	*Hibernia* and *Orpheus* withdrawn in 1960; became CC BR(S) No P46 and BR(E) No CC164.
56, 58, 59, 50	*Ruby* (I), *Daphne, Hawthorn, Mimosa*	K	3/1914	Birmingham RC&W Co	*Ruby* (I) was last composite first and second class car in Le Havre boat train, London (Waterloo)-Southampton Docks until rebuilt to first class in 1956. All withdrawn in 1962; CC BR(S) Nos P45, P60 and P58. *Mimosa* BR(W) No W9869 and still intact at Marazion, Cornwall.
60	*Topaz* (I)	P	3/1914	Birmingham RC&W Co	Withdrawn in 1960; preserved at NRM York (armchairs removed for 1960 cars); ran in Travellers Fare Centenary Express in 1979 and Rainhill Rocket 150 Cavalcade in 1980. Now restored to SECR crimson lake livery.
55, 57, 64, 67, 63, 66, 61, 62, 52	*Mary Hamilton, Mary Beaton, Mary Seaton, Mary Carmichael, Annie Laurie, Helen MacGregor, Flora MacDonald, Fair Maid of Perth, Lass o' Gowrie*	D (6)	5/1914-8/1914	Cravens Limited, Sheffield	For Scotland, Caledonian Railway. All sold to LMSR, 1933 (Nos 200-208), who removed names and repainted them in crimson lake livery, lettered as 'Dining Car'. All scrapped.
65	*Maid of Morven*	K, OBS	7/1914	Cravens Limited, Sheffield	Specially designed observation car for Glasgow-Oban; stored 1915-9; sold to LMSR in 1933 (No 209).
72-75, 17, 19	Car Nos 1-4 Third Class, Car Nos 17 and 19 Third Class	K	–/1915 (rebuilds)	Pullman USA (1875-1895)	Converted to third class by Pullman, at Longhedge in 1915 from old American cars. No '1' was formerly *Jupiter*, '2' was *Adonis*, then *Victoria*, '3' was *Alexandra* (II) (1877), '4' was *Albert Edward*, '17' was *Duchess of York* and '19' was *Princess of Wales*. All scrapped in 1932 except '4', which became store at Preston Park, then scrapped in 1950.
68-71	Car Nos 5-8 Third Class	K (6)	–/1917	Pullman, Longhedge	'5' became *Trianon Bar* and then *New Century Bar* in April 1946, and then once again *Trianon Bar* from July 1946 to 1951; renamed 'Buffet Car No 5' and withdrawn 1960. '6' and '8' withdrawn in 1962; CC BR(S) Nos P57 and P56. '7' scrapped (Shoreham) in 1955.
79	*Tulip*	P	3/1919 (rebuild	Jackson Sharp USA (1897)	Ex-SER first class 'Folkestone Vestibuled Limited' car No 171 rebuilt by Pullman Car Co; All scrapped in 1930.
80, 87, 84, 86, 83	*Dorothy, Stella, Dora, Mabel, Venus* (II)	P	–/1919 (rebuilds)	Metropolitan C&W Co (1897)	Ex-SER 'Folkestone Vestibuled Limited' cars, built in 1897 Nos 205/4/2/3/6 rebuilt by Pullman Car Co at Longhedge. All scrapped in 1930 except *Venus*, which was scrapped in 1929.
82, 85, 81	*Hilda, Albatross, Thistle*	K	–/1919 (rebuilds)	Metropolitan C&W Co (1897)	Ex-SER 'Folkestone Vestibuled Limited' cars built in 1897; Nos 201/8/7 rebuilt at Longhedge. *Albatross* and *Thistle* became mobile stores supply cars in 1931-38, withdrawn in

Schedule Nos	Original names or numbers	Type of car	Year into service	Builder	Notes
					1938; became office at Lancing SR Carriage Works in 1940. *Hilda* scrapped in 1930.
117, 106	Car Nos 45 and 47 Third Class	K (6)	2/1920	Clayton Wagons, Lincoln	Nos '45' and '47' rebuilt with 'left-handed' kitchens as D by Pullman Longhedge in 1923 (opposite side corridor and kitchens to other standard Pullmans). Later third class K. Withdrawn in 1962; CC BR(Sc) No SC45 and BR(S) No P50.
173, 107	Car Nos 56 and 57 Second Class	K (6)	2/1920	Clayton Wagons, Lincoln	For 'Hook Continental' London-Harwich GER, in 1920-23; rebuilt to third class D and third class BP by Pullman Longhedge in 1923; withdrawn c1938. '56' sold to LNER.
108, 109	*Ansonia, Arcadia*	K (6)	2/1920	Clayton Wagons, Lincoln	Remodelled to BP first class in 1924 by Pullman Longhedge; became third class BP Nos '94' and '95', at Pullman Preston Park in 1934. '94' scrapped at Shoreham in 1955. '95' withdrawn in 1962; CC BR(Sc) No SC50.
88, 89 90, 105, 103, 104	*Carmen, Constance,* *Figaro, Diana, Dolphin, Falcon* (I)	K P	1920 (rebuilds)	Gilbert Car Co USA (1891)	Rebuilt at Longhedge in 1920 as Pullmans, from ex-SER, 'Hastings Car Train' cars. *Carmen* destroyed in derailment at Sevenoaks in 1927. Rest withdrawn in 1928-30.
99	*Padua*	P (6)	10/1920	Birmingham RC&W Co	Became 'No 99 Second Class' GP in 1946, GP '99' in 1947, withdrawn in 1962; CC BR(S) No P59, preserved by Flying Scotsman Enterprises, Carnforth, as *Padua*.
101	*Portia*	K (6)	10/1920	Birmingham RC&W Co	Withdrawn in 1960; CC BR(E) No CC163.
76	Car No 9 Third Class	K	11/1920 (rebuild)	Pullman USA (1874)	Formerly *The Queen*; scrapped in 1932.
119, 120, 126	*Cambria, Catania, Corsair*	K (6)	11/1920	Clayton Wagons, Lincoln	For 'Hook Continental' London-Harwich GER. *Cambria* rebuilt to BP in 1924 by Pullman Longhedge. All withdrawn in 1938. *Cambria* became BR departmental coach No DE 960820 and has now been preserved at Tenterden, K&ESR for future restoration.
97, 98	*Calais, Milan*	P (6)	1/1921	Birmingham RC&W Co	For SECR London-Dover boat trains and the Kent coast services. Became 'Nos 97 and 98 Third Class' in 1947; withdrawn 1962; CC BR(W) No W9870 and BR(S) No P52. *Calais* intact at Marazion, Cornwall.
100, 102	*Palmyra, Rosalind*	K (6)	1/1921	Birmingham RC&W Co	*Palmyra* withdrawn in 1960; CC BR(E) No CC168. *Rosalind* became Army (NAAFI) Canteen No 18 in 1942-48; withdrawn in 1960; CC BR(S) No P47. Preserved at Steamtown, Carnforth, by Flying Scotsman Enterprises.
110	Car No 46 Third Class	K (6)	2/1921	Clayton Wagons, Lincoln	For London-Harwich boat trains GER; rebuilt to BP in 1924 by Pullman at Longhedge; scrapped in 1932.
91, 96, 95	*Cadiz, Sylvia, Sunbeam*	P (6)	1921	Pullman, Longhedge	For SR 'Continental Express'. *Cadiz* and *Sylvia* became 'Nos 294 and 96 Second Class' P in 1947. *Cadiz* and *Sunbeam* withdrawn in 1960; CC BR(E) No CC170 and BR(Sc) No SC41. *Sylvia* withdrawn in 1962; CBR(S) No P48.
92, 93, 94	*Malaga, Monaco, Neptune*	K (6)	1921	Pullman, Longhedge	*Monaco* and *Neptune* withdrawn in 1960; CC BR(Sc) No SC43 and BR(E) No CC162. *Malaga* rescued by Ian Allan; preserved at Shepperton, Surrey, (boardroom), in 1962.
121-125	Car Nos 40-44 Third Class	K (6)	8/1921	Birmingham RC&W Co	Nos '40' and '41' rebuilt as BP by Birmingham RC&W in 1924. '40' destroyed by air raid on Preston Park in 1943. '41' withdrawn in 1962; CC BR(Sc) No SC42. For LNER services. All withdrawn in 1939.
15	Car No 3 Second Class	K	9/1921 (rebuild)	Pullman USA (1893)	Ex-*Prince Regent*; rebuilt at Longhedge; became No '18 Third Class' in 1925; scrapped in 1932.
118, 138, 174, 199, 201	*Albion, Alexandra* (IV), *Nevada, Atlanta, Columba*	K (6)	9/1921	Birmingham RC&W Co	*Albion* and *Alexandra* (IV). Scrapped c. 1936. Rest rebuilt to D in 1927 by Pullman Longhedge for LMSR Scottish services. *Nevada* renamed *Jenny Geddes; Atlanta* renamed *Diana Vernon; Columbia* renamed *Jeannie Deans*. Sold LMSR 1933 (Nos 212, 213, 211).
132, 133, 134 135, 136, 137	*Anaconda, Erminie, Coral, Elmira, Formosa, Maid of Kent* (I)	K	11/1921 (rebuilds)	Clayton Wagons, Lincoln	New bodies on ex-LNWR ambulance train underframes. Rebuilt as composite cars at Pullman Preston Park in 1934. *Anaconda, Erminie, Elmira* and *Maid of Kent* (I) became third class 'Nos 132, 133, 135 and 137' in 1948. *Formosa* became *Maid of Kent* (II) first class in 1948 for Thanet Belle. All withdrawn in 1960; CC BR(LM) No 022 260; BR(Sc) No SC 40; BR(S) No P43; BR(LM) No 022261; BR (E) No CC161; BR(LM) No 022262. *Elmira* and *Maid of Kent* (I) preserved as 'Nos 135 and 137' (static) by Ravenglass & Eskdale Railway, Ravenglass, Cumbria. *Maid of Kent* (II) preserved by the Great Central Coach Group at Banbury, Oxon; moved to Lavender Line, Isfield, Sussex in 1986.
111-116, 113, 114	Car Nos 11-16 Third Class 13, 14	K OBS	1921 1947 (rebuilds)	Clayton Wagons, Lincoln and Pullman, Longhedge (Nos 10, 13, 14) Pullman, Preston Park	New bodies on ex-LNWR ambulance train underframes. All except '12' became bar cars in 1937 at Pullman Preston Park; '12' scrapped in 1943 after damage in aid raid. '11, 15 and 16' rebuilt in 1949 to BP by Pullman Preston Park; withdrawn in 1960; CC BR(S) Nos P40, P42, and P41. '13 and 14' rebuilt to OBS with 'Pullman Observation Car' in cream below windows. Sold to BR in 1958; Nos SC281 and M280. '13' preserved in San Francisco USA by Victoria Station restaurant chain. '14' preserved on Dart Valley Light Railway, originally as No 280, later repainted in GWR livery and named *Devon Belle*.

Schedule Nos	Original names or numbers	Type of car	Year into service	Builder	Notes
127-129	Car Nos 18, 20 and 21 Third Class	K	2/1922 (rebuilds)	Midland RC&W Co	New bodies on ex-GWR ambulance train underframes. '18' became 'Car No 3 Second Class' in 1925 and 'No 30 Third Class' in 1927; withdrawn in 1960; CC BR(S) No P44, '20 and 21' scrapped in 1943 after damage in air raid at Preston Park.
131, 130, 139, 140, 141	Car No 23 Third Class Car Nos 22, 24 Third Class Car Nos 25,26 Third Class	K P B	1922 (rebuilds)	Pullman, Longhedge	New bodies on ex-LYR ambulance train underframes. '25 and 26' scrapped in 1935; rest scrapped in 1938.
142, 143	*Lady Nairn, Bonnie Jean*	D	6/1922 (rebuilds)	Midland RC&W Co	New bodies on ex-GWR ambulance train underframes. *Lady Nairn* delivered as *Lady Nairne*; name corrected before entering service. For CR services. Sold to LMSR in 1933 (Nos 214/5).
77	Car No 10 Third Class	K	12/1922	Pullman USA (1895)	Ex-*Her Majesty*; scrapped in 1932.
153-156	*Aurora* (II), *Flora, Juno* (II), *Montana*	GP	3/1923	Birmingham RC&W Co	End cars of 1924 'White Pullman'; 'Golden Arrow' from 1929. *Flora* became 'No 154' in 1946-48. *Aurora* (II) and *Juno* (II) became 'Nos 503 and 502' in 1950-52. All third class. GP for 'Devon Belle', Nos screwed on board over names. *Montana* withdrawn in 1960; CC BR(E) No CC165. Rest withdrawn in 1962; CC BR(W) Nos W9873, W9871 and W9872. *Montana* in use as a private dwelling; Barnwell Junction, near Cambridge. *Aurora, Flora* and *Juno* (II) intact at Marazion, Cornwall.
148, 149, 151, 152	Car Nos 48, 49, 52 and 53 Third Class	K (6)	6/1923	Clayton Wagons, Lincoln	For LNER 'Harrogate Pullman', later rebuilt to D. Sold to LNER in 1939.
147, 150	Car Nos 50 and 51 Third Class	K (6)	6/1923	Clayton Wagons, Lincoln	Rebuilt by Pullman Longhedge to 'Nos 2 and 1' Second Class 1923; renumbered '17 and 19' Second Class in 1927. '17' withdrawn in 1962; CC BR(Sc) No SC48. '19' scrapped at Shoreham in 1955.
Note: The Clayton Wagons, Lincoln cars delivered in June 1923 were the last batch of new Pullmans to feature six-wheeled bogies (12-wheelers). All cars listed henceforth ran with four-wheel bogies, although of several different designs over the years.					
157, 158	Car Nos 54 and 55 Third Class	K	7/1923	Clayton Wagons, Lincoln	'54' rebuilt to BP by Pullman Preston Park in 1937; became LNER No 490 1942-6; withdrawn in 1963; preserved at Birmingham Museum Tyseley; sold to VS-O-E in 1981; not used and moved to Bluebell Railway in 1986. '55' became D first class (no name) in 1925; rebuilt to BP third class by Pullman Preston Park in 1937; became LNER No 491 1942-6; withdrawn in 1963. First of Standard 'K'-type.
161, 162, 163, 164	*Fortuna, Irene, Iolanthe,* *Rosemary*	K P	6/1923	Midland RC&W Co	*Fortuna* and *Irene* rebuilt by Pullman Preston Park to BP 'Nos 161 and 162' in 1946 as third class for 'Devon Belle'. No '161' rebuilt again in 1962 to first BP, named *Fortuna* again, for 'Midland Pullman' standby. All withdrawn by 1966.
159, 144, 145 146	*Meg Dods, Lass o' Ballochmyle* *Mauchline Belle* Car No 80 Third Class (I)	K K K	1923 1923 1923	Clayton Wagons, Lincoln Clayton Wagons, Lincoln	For LMSR Scottish services. Sold to LMSR in 1933 (Nos 217, 216, 218) who removed names and painted them crimson lake. Became No '27' in 1928; rebuilt as BP before 1939; became LNER No 489 1942-6; scrapped about 1966.
165, 166, 167, 168, 169, 170, 171	*Argus, Geraldine,* *Marjorie,* *Sappho, Viking, Medusa,* *Pauline*	K K K	1924 11/1924 11/1924	Midland RC&W Co	For 'White Pullman' ('Golden Arrow' from 1929). *Geraldine, Marjorie, Viking* and *Pauline* remodelled to third class. 'Nos 166, 167, 169 and 171' for 'Devon Belle' in 1947-48. All withdrawn about 1964-66.
160, 172, 175, 179, —, —, —, —, —, —, —, —, —, —	*Rosamund, Aurelia,* *Fingall,* *Cynthia, Adrian, Ibis,* *Hermione,* *Lydia, Rainbow* (I), *Leona* (I), *Minerva* (II), *Niobe* (I), *Octavia* (I), *Plato* (I)	K K K P P	10/1924 7/1925	Birmingham RC&W Co	For 'White Pullman' ('Golden Arrow' from 1929). *Cynthia* built in 1925 after those in batch without schedule numbers, which were all sold to CIWL for service Milan-Nice. Cars sold to CIWL; returned 1928 (below), except *Hermione* and *Rainbow* (I). Withdrawn in 1962-63. *Fingall* preserved by Bluebell Railway, Sheffield Park, Sussex. *Ibis* preserved by Sea Containers for VS-O-E British train.
186,	Car No 58 Third Class	K	7/1925	Midland RC&W Co	Car No '58' remodelled to second class D in 1928-36; then third class K; became LNER No 492 1942-6. Withdrawn in 1962; CC BR (Sc) No SC49.
187, 188	Car No 65 Third Class Car No 66 Third Class	D P			'65' rebuilt to third class BP in 1937 by Pullman Preston Park; became LNER No 496 1942-6. No '66' became LNER No 497 1942-6, and *Avon* for WR 'Blue Pullman' standby train, 1962. Both withdrawn c1964.
180, 181, 182, 183, 184, 185	*Camilla, Latona,* *Madeline, Pomona,* *Theodora, Barbara*	K K K	1/1926	Metropolitan Carriage, Wagon & Finance Co	For London-Tonbridge-Hastings line. Narrow bodies. *Madeline, Pomona* and *Barbara* renamed *Refreshment Car* in 1946. All repainted green, no names in 1958 and labelled 'Buffet Car'. Sold to BR(S) in 1960; Nos S 7872/3/5/6/4/7S for ocean liner boat trains Waterloo-Southampton Docks. Withdrawn by BR(S) in 1964. *Theodora* and *Barbara* preserved by K&ESR; *Barbara* restored and used in 'Wealden Pullman' wine and dine special, Tenterden, Kent.
189-191	Car Nos 31-33 Third Class	K	8/1926	Birmingham RC&W Co	Withdrawn about 1963.
195-198	Car Nos 100-103 Third Class	K	6/1926	Birmingham RC&W Co	For Ireland. Bogies supplied by LMSR (NCC) for Irish Pullmans Ltd. Used only in Southern Ireland; sold to GSR in 1937; Nos 100-103. Chassis of 100 used in 1981 of Inchicore works as mobile rail/timber store, CIE (Irish Railways.).
192-194	Car Nos 34-36 Third Class	P	7/1926	Birmingham RC&W Co	'35-36' became second class in 1946. All second class in 1951, with square lavatory windows, remodelled for 'Golden Arrow'. All withdrawn in 1962-63. '35' became, car *Wye* in the 'Blue

Schedule Nos	Original names or numbers	Type of car	Year into service	Builder	Notes
					Pullman' standby train, and was afterwards preserved at Haven Street, Isle of Wight; now scrapped. '36' preserved by H. P. Bulmer Ltd, named *Morella*; sold to VS-O-E in 1986.
200 176-178	*Cassandra,* *Rainbow* (II), *Plato* (II), *Octavia* (II)	K K	11/1926 1/1927	Metropolitan Carriage, Wagon & Finance Co	*Rainbow* (II), burnt out at Micheldever, Hampshire, in 1936. Chassis used for *Phoenix* in 1952. Others withdrawn about 1963.
202 203 204	*Queen Margaret,* *Kate Dalrymple,* *Helen of Mar*	D D D	1927	Metropolitan Carriage, Wagon & Finance Co	For LMSR Scottish services with first and third class dining saloons. Sold to LMSR in 1933 (Nos 219-221), who removed names and painted them crimson lake.
205, 206 207 208, 209 213	*Cecilia, Chloria,* *Zenobia,* *Leona* (II), *Niobe* (II) *Minerva* (III)	K K P K	11/1927 11/1927 1927 12/1927	Midland RC&W Co	For LNER Pullman trains. *Leona* rebuilt at Pullman Preston Park as third class GP 'No 208' in 1947; withdrawn in 1967; preserved at Cressing, Essex. *Niobe* rebuilt at Pullman Preston Park as third class BP 'No 209' in 1948; renamed *Niobe* BP first class in 1959. All withdrawn in 1967. *Minerva* rebuilt by Pullman Preston Park in 1951 as GP first class with square lavatory windows for 'Golden Arrow'; withdrawn in 1965; preserved at Lytham St Annes; bought by Sea Containers in 1981 for VS-O-E.
210-212	*Marcelle, Sybil, Kathleen*	K	1927	Metropolitan Carriage, Wagon & Finance Co	For LNER Pullman services. Became third class Nos 105, 106 and 107 for 'Yorkshire Pullman' in 1946. In 1962 'No 105' became *Hebe*; '106' *Thalia* and '107' *Thetis* for LMR 'Blue Pullman' standby train. Withdrawn in 1966 but *Hebe* preserved as the 'Pullman Car Restaurant' 26 Kensington High Street, London.
214-216, 218 217, 219	Car Nos 59, 60, 61 and 63 Third Class Car Nos 62 and 64 Second Class	K D	1928	Metropolitan Carriage, Wagon & Finance Co	No 59 rebuilt to D c1930; became LNER No 493 1942-6; rebuilt in 1948 by Preston Park to become 'Car No 59' *The Hadrian Bar*, for the 'Tees-Tyne Pullman'. Sold to BR(S) in 1961 repainted green and branded 'Buffet' as No S7879S. '63' became LNER No 495 1942-6; rebuilt to BP by Preston Park, 1950. '62' and '64' used for London-Harwich boat trains; rebuilt to P third by Preston Park in 1937. '62' became LNER No 494 1942-6; rebuilt to BP by Preston Park, c1950. Nos '60' and '61' became *Severn* and *Thames* for the 1962 BR(W) 'Blue Pullman' standby train. All withdrawn by 1966. '64' preserved at Hereford by H. P. Bulmer Ltd, named *Christine*; sold to VS-O-E in 1986.
220-225 226-229 230-233 234, 235, 237 236, 239 238, 240, 241 242, 243	Car Nos 67-72 Third Class Car Nos 73-76 Third Class Car Nos 77-80 (II) Third Class *Nilar, Belinda, Thelma* *Sheila, Agatha,* *Phyllis, Penelope,* *Philomel* *Ursula, Lucille*	K P BP K P K P	1928	Metropolitan Carriage, Wagon & Finance Co	All-steel standard Pullman design first introduced for the 'Queen of Scots' service. *Agatha, Sheila* and 'Nos 75-80' used by LNER 1942-48 repainted brown, with LNER Nos 468/9/81/2/5-8; returned to Pullman Car Co 1946. Car Nos 67-72 all converted to BP by 1961 and retained on East Coast route to work with 1960 Metro-Cammell cars (as were Nos 77-80). Car Nos 73 and 74 became *Ceteia* and *Melandra* for 'Blue Pullman' standby train. *Phyllis* and *Lucille* in 'Bournemouth Belle' 1963-66; then preserved at SE Steam Centre, Ashford, Kent when withdrawn 1966. *Lucille* now owned by VS-O-E for use in 'British Pullman train'. *Ursula* and '75' preserved at the 'Spot Gate' pub at Hilderstone, by Ind Coope (Allied Breweries). '76' preserved by H. P. Bulmer Ltd, named *Eve*; sold to VS-O-E in 1986. '79' preserved by North Yorkshire Moors Railway, Grosmont, for use on their 'Yorkshire Pullman'. *Agatha* purchased by Sea Containers for VS-O-E.
244, 245, 246 247 248 249, 250 251	*Adrian, Ibis, Lydia* *Princess Elizabeth* *Lady Dalziel* *Pearl* (I), *Diamond* *Onyx*	K K GP K P	(1928) (1928) (1928)	Midland RC&W Co (Rebuild) Birmingham RC&W Co (Rebuild) Midland RC&W Co (Rebuild)	All built by Birmingham in 1925 (see above); bought back from CIWL in 1928. Only *Adrian, Ibis* and *Lydia* bore their old names. *Princess Elizabeth* was ex-*Leona* (I); *Lady Dalziel*, ex-*Minerva* (II); *Pearl*, ex-*Niobe* (I) and *Diamond* was ex-*Octavia* (I). In 'Golden Arrow' in 1946. *Princess Elizabeth* rebuilt to GP *Isle of Thanet* by Pullman Preston Park, 1950(?), for 'Thanet Belle'. In inaugural electric-hauled 'Golden Arrow' in 1961, and with *Lydia* in Sir Winston Churchill's funeral train in 1965. Both withdrawn in 1966; now preserved at National RR Museum, Green Bay, Wisc, USA. *Lady Dalziel* became third class GP 'No 248' for 'Devon Belle' in 1948. *Pearl* (I) became third class P 'No 249' in 1947 for 'Devon Belle'. *Diamond* became *Trianon Bar* for 'Golden Arrow' then *One Hundred Bar*, then *New Century Bar* — all three renamings in 1946. Renamed *Diamond, Daffodil Bar* in 1955. 'No 248' to first BP in 1962 (with proper 1+1 seating) for 'Midland Pullman' standby train, renamed *Athene*. All withdrawn in 1966-67. *Ibis* preserved at Birmingham Railway Museum, Tyseley; and then bought by Sea Containers; restored in 1981 for VS-O-E.
252, 253, 254 255, 256, 257 258	*Eunice, Juana, Zena* *Ione, Joan, Loraine,* *Evadne*	P K	1928	Metropolitan Carriage, Wagon & Finance Co	For GWR London-Plymouth ocean liner boat trains and 'Torquay Pullman' (1929). Later in 'Cunarder' boat train, London (Waterloo)-Southampton Docks, and in LNER Pullman trains. All-steel cars. *Joan* used by Gen Eisenhower during World War 2. All withdrawn by 1967. *Zena* preserved by Steamtown, Carnforth, and *Ione* by Birmingham Railway Museum, Tyseley, in 1967-80 and then both bought by Sea Containers for VS-O-E.

Schedule Nos	*Original names or numbers*	*Type of car*	*Year into service*	*Builder*	*Notes*
259, 260 261, 262	Car Nos 81 and 82 Third Class Car Nos 83 and 84 Third Class	K P	1930/1	Birmingham RC&W Co	For ocean liner boat trains and LNER Pullman trains. 'Nos 81 and 82' rebuilt by Pullman Preston Park to second class BP for East Coast route in 1959-60, possibly later used for 'Bournemouth Belle'. 'Nos 83 and 84' in LNER colours in 1942-46 as LNER Nos 483 and 484. All withdrawn c1967. '83' preserved by H. P. Bulmer Ltd, named *Prinea*: sold to VS-O-E in 1986. '84' preserved by Keighley & Worth Valley Railway, Yorkshire and named *Lorna*, and then *Mary*.
263-278	*Ida, Ruth, Rose, Violet, May, Peggy, Clara, Ethel, Alice, Gwladys, Olive, Daisy, Anne, Naomi, Lorna, Bertha*	K	1932	Metropolitan-Cammell Carriage, Wagon & Finance Co	Composite first and third class all-steel 'electric' cars for SR '6-Pul' and '6-Cit' rakes of five SR coaches and one Pullman per rake. For London-Brighton, Eastbourne, Hastings, Worthing and Littlehampton services. *Ruth* preserved at Hereford, by 6000 Locomotive Association. *Bertha* preserved on Bluebell Railway, Sheffield Park, Sussex, by Mr R. Martin. All withdrawn in 1965-68.
279-293			1932	Metropolitan-Cammell Carriage, Wagon & Finance Co	For 'Southern Belle' of 1933; renamed 'Brighton Belle' in 1934. SR '5-Bel' sets: '5-Bel' Set 2051 (later 3051) '88-*Hazel-Doris*-86-89'; '5-Bel' Set 2052 (later 3052) '90-*Audrey-Vera*-87-91'; '5-Bel' Set 2053 (later 3053) '92-*Gwen-Mona*-85-93'. All withdrawn in 1972.

Note: All 15 of the 'Brighton Belle' Pullman cars have survived for preservation in one form or another. Herewith their individual locations:

279	*Hazel*	K	Preserved as Black Bull Inn (pub), Moulton, North Yorkshire.
280	*Audrey*	K	Preserved by Mr D. Lowther. Ran in Rainhill 1980 'Rocket 150' Cavalcade with 'Brighton Belle' roofboards. Bought by Sea Containers in 1981 and altered for loco-haulage with dynamo, etc, for London-Venice train, UK portion (London-Folkestone Harbour), in 1982.
281	*Gwen*	K	Preserved on Colne Valley Railway, Castle Hedingham, Essex.
282	*Doris*	K	Preserved at Finsbury Park BR (London) by City Industrial Ltd.
283	*Mona*	K	Preserved at Brighton Belle Inn (pub), Winsford, Cheshire.
284	*Vera*	K	Preserved at Westleton, Suffolk, by M Amadée Turner, Member of European Parliament; bought by VS-O-E in 1985 for restoration as loco-hauled car.
285	Car No 85 Third Class	P	Preserved at Nags Head (pub), Mickleover, Derbyshire.
286	Car No 86 Third Class	P	Bought by Sea Containers in 1980 for VS-O-E. To be altered for loco-haulage, first class, for London-Folkestone train, not yet undertaken.
287	Car No 87 Third Class	P	Preserved at North Norfolk Railway, Sheringham, by Allied Breweries.
288	Car No 88 Third Class	MBP	Preserved at Swanage, Dorset, on loan from Stour Valley Railway.
289	Car No 89 Third Class	MBP	Preserved as 'Derbyshire Belle', Little Mill Inn (pub), Rowarth, Derbyshire.
290	Car No 90 Third Class	MBP	Preserved by Travellers Fare on Nene Valley Railway, Peterborough.
291	Car No 91 Third Class	MBP	Preserved at North Norfolk Railway, Sheringham, by Allied Breweries.
292	Car No 92 Third Class	MBP	Preserved at Preston Park, former Pullman works, by Brighton Belle Society
293	Car No 93 Third Class	MBP	Preserved at Preston Park, former Pullman works, by Mr M. Allen.

294-300	*Brenda, Elinor, Enid, Grace, Iris, Joyce, Rita*	K	1932	Metropolitan-Cammell Carriage, Wagon & Finance Co	Composite first and third class 'electric' cars for SR '6-Pul' and '6-Cit' sets, like schedule Nos 263-278. All withdrawn in 1963-68.

Note: No new Pullmans were constructed after delivery of the cars for the 1933 electrified SR services, up until the outbreak of war in 1939. However a new fleet of cars for the LNER was planned, and some work had been done for it, such as the marquetry panels, when the outbreak of the war stopped it completely. The project was revived, using the stored prewar marquetry, and the same basic car design, to be built in time for the 1951 'Festival of Britain'. These steel-panelled, timber-framed cars were notable for having square lavatory windows in place of the traditional oval shape, and for having LNER Gresley-type steel underframes and double bolster bogies (except *Phoenix*). Their schedule numbers were 301-310, and they were the last cars ordered by the Pullman Car Company before the BTC purchased a controlling interest. Although associated primarily with the 'Golden Arrow' and Royal Special workings, at times of pressure, spare vehicles were used in Waterloo-Southampton Docks 'Ocean Line' trains and in the 'Bournemouth Belle'. In 1964 at least one car was on the East Coast route in use in 'The Talisman'. The last ones to run on BR were withdrawn in 1972. The individual case histories are as follows:

Schedule Nos	*Names or numbers*	*Type of car*	*Year built*	*Builder*	*Notes*
301	*Perseus*	P	1951	Birmingham RC&W Co	Bought by Sea Containers in 1978 for VS-O-E British train, and Oval lavatory windows fitted during restoration.
302	*Phoenix*	P	1952	Pullman, Preston Park	Built upon the underframe of *Rainbow* (II) of 1927; see text for details. Used for royal trains. Preserved at Hôtel Mercure, Dardilly, Lyon (France), in 1972-80. Bought by Sea Containers in 1980 and restored for VS-O-E British Pullman train.
303	Car No 303 Third Class	K	1952	Pullman, Preston Park	In 'Golden Arrow', 1952-65; used also in special trains; withdrawn in 1967.
304	*Aries*	K	1952	Pullman, Preston Park	Withdrawn c1967; preserved as 'Yew Tree Inn' (pub), Rochdale, Lancashire.
305	*Aquila*	K	1951	Birmingham RC&W Co	Preserved by H. P. Bulmer Ltd, sold to VS-O-E in 1986.
306	*Orion*	K	1951	Birmingham RC&W Co	Preserved by Peco Ltd 'Modelrama', at Beer, Seaton, Devon.
307	*Carina*	K	1951	Birmingham RC&W Co	Preserved at Hôtel Mercure, Dardilly, Lyon, France; bought by Sea Containers in 1980 for VS-O-E British Pullman train.

Schedule Nos	Names or numbers	Type of car	Year built	Builder	Notes
308	*Cygnus*	P	1951	Birmingham RC&W Co	Bought by Sea Containers in 1979 for VS-O-E British Pullman train. Restored, with oval lavatory windows in 1981.
309	*Hercules*	P	1951	Birmingham RC&W Co	Fate uncertain but thought to have become a BR Research Department vehicle for a while.
310	*Pegasus*	Bar	1951	Birmingham RC&W Co	Incorporated 'Trianon Bar' in 'Golden Arrow' to 1965, then became 'Nightcap Bar' on LMR until sold and preserved at Birmingham Railway Museum, Tyseley.

Note: Known as the East Coast route replacement cars, these 44 Metro-Cammell-built all-steel Pullmans were the last to feature the classic interior layout for the first class, complete with single armchairs (recovered, and refurbished from old Pullmans) and the traditional umber and cream livery, with simplified lining-out, was applied. However, their underframe and bodywork was based upon — although in the body panelling somewhat better than — the 1951 BR standard Mark 1 carriage, with its curved body profile; and Commonwealth cast steel bogies were fitted. Because they were introduced during the changeover period from steam traction to diesel traction, they were dual-heated, for both steam and electricity, as required. Suitable for 100mph running, and fitted with vacuum brakes (some later received air brakes as well). The first class cars had schedule Nos 311-331, and the second class cars Nos 332-354. The individual case histories follow. (All first class cars were withdrawn from BR by 1978, when the last ones were running without their names and with BR blue and grey livery, and with their seating altered to 2+1 layout.)

311	*Eagle*	K	1960	Metropolitan-Cammell C&W Co	Preserved at National Railway Museum, York, with *Emerald* (II) (below). In Travellers Fare Centenary Exp, 1979, and Rainhill 'Rocket' Cavalcade, 1980.
312	*Falcon* (II)	K	1960	Metropolitan-Cammell C&W Co	To King, Snailwell, scrap; bogies removed in 1980.
313	*Finch*	K	1960	Metropolitan-Cammell C&W Co	Privately owned by Mr M. Bailiss, and located on Severn Valley Railway; under restoration.
314	*Hawk*	K	1960	Metropolitan-Cammell C&W Co	BR, CM&EE Instruction coach No ADB975876, Longsight.
315	*Heron*	K	1960	Metropolitan-Cammell C&W Co	Preserved by Great Central Railway, Loughborough.
316	*Magpie*	K	1960	Metropolitan-Cammell C&W Co	Preserved as restaurant at Muswell Hill, London.
317	*Raven*	K	1960	Metropolitan-Cammell C&W Co	Privately owned by Mr M. Bailiss, and located on Severn Valley Railway; restored to service June 1986.
318	*Robin*	K	1960	Metropolitan-Cammell C&W Co	Privately preserved in 1980 on North Yorks Moors Railway, Grosmont. For NYMR 'Yorkshire Pullman' (Grosmont-Pickering) in 1981 with '79', *Opal* and *Garnet*.
319	*Snipe*	K	1960	Metropolitan-Cammell C&W Co	Privately owned by Mr M. Bailiss, and located on Severn Valley Railway; under restoration.
320	*Stork*	K	1960	Metropolitan-Cammell C&W Co	Preserved as restaurant at Muswell Hill, London.
321	*Swift*	K	1960	Metropolitan-Cammell C&W Co	Preserved on the Northampton Steam Railway.
322	*Thrush*	K	1960	Metropolitan-Cammell C&W Co	To King, Snailwell, scrap; bogies removed in 1980.
323	*Wren*	K	1960	Metropolitan-Cammell C&W Co	Used as BR Laboratory coach No RDB975427 *Wren*, at Derby.
324	*Amber*	P	1960	Metropolitan-Cammell C&W Co	Privately owned by Mr M. Bailiss, and located on Severn Valley Railway; under restoration.
325	*Amethyst*	P	1960	Metropolitan-Cammell C&W Co	Preserved by Strathspey Railway, Aviemore, Scotland, 1978.
326	*Emerald* (II)	P	1960	Metropolitan-Cammell C&W Co	Preserved at NRM, York. In Travellers Fare Centenary Exp, 1979.
327	*Garnet*	P	1960	Metropolitan-Cammell C&W Co	Preserved by North Yorks Moors Railway, Grosmont, for 'Yorkshire Pullman' NYMR.
328	*Opal*	P	1960	Metropolitan-Cammell C&W Co	Preserved by North Yorks Moors Railway (as above).
329	*Pearl* (II)	P	1960	Metropolitan-Cammell C&W Co	Privately owned by Mr M. Bailiss, and located on Severn Valley Railway; under restoration.
330	*Ruby* (II)	P	1960	Metropolitan-Cammell C&W Co	To King, Snailwell, scrap; bogies removed 1980.
331	*Topaz* (II)	P	1960	Metropolitan-Cammell C&W Co	Privately owned by Mr M. Bailiss, and located on Severn Valley Railway; under restoration.

Note: The second class cars featured 2+1 seating, and when no longer used in East Coast Pullman services (after the BR decision to cease offering Pullman amenities to second class ER travellers) they saw some use in a variety of roles. Some were included in the 'Regency Belle' set; some were used as Lounge cars in 'Motorail' trains, and at least three ran in WR 'Blue Pullman' standby trains. BR then decided to upgrade them as open firsts; until withdrawal.

332	Car No 332	K	1960	Metropolitan-Cammell C&W Co	Platform 1 Public House, Pannal, Yorks; formerly on NYMR. Now named *Mae*
333	Car No 333	K	1960	Metropolitan-Cammell C&W Co	Preserved at the Casterton Taverner Motor Inn (Motel), Stamford, Lincolnshire.
334	Car No 334	K	1960	Metropolitan-Cammell C&W Co	To King, Snailwell, scrap; bogies removed in 1980.
335	Car No 335	K	1960	Metropolitan-Cammell C&W Co	Preserved by Strathspey Railway as mess van 1979; bought by SLOA in 1981 for SLOA Pullman train. Kitchen converted to boutique for BR running. Now owned by Pullman Rail.
336	Car No 336	K	1960	Metropolitan-Cammell C&W Co	To King, Snailwell, scrap; bogies removed in 1980.
337	Car No 337	K	1960	Metropolitan-Cammell C&W Co	Preserved privately at Wansford, Nene Valley Railway.
338, 339	Car Nos 338, 339	K	1960	Metropolitan-Cammell C&W Co	To King, Snailwell, scrap; bogies removed in 1980.

Schedule Nos	Original names or numbers	Type of car	Year into service	Builder	Notes
340	Car No 340	K	1960	Metropolitan-Cammell C&W Co	Owned by Mr and Mrs Palmer and used at Avon Causeway Hotel, Hurn, near Christchurch, as pub extension.
341-345	Car Nos 341-345	K	1960	Metropolitan-Cammell C&W Co	To King, Snailwell, scrap; bogies removed in 1980.
346	Car No 346	K	1960	Metropolitan-Cammell C&W Co	Preserved at the Casterton Taverner Motor Inn (Motel), Stamford, Lincolnshire.
347-353	Car Nos 347-353	P	1960	Metropolitan-Cammell C&W Co	In service with BR on football specials, etc, until 1980. Bought by Steam Locomotive Operators' Association (SLOA) for their steam-hauled specials on BR. Now owned by Pullman Rail. The train can be hired with steam or with BR traction, when not running SLOA of other specials. It is based at Carlisle Upperby, maintained by BR for Pullman Rail.
354	*The Hadrian Bar* (II)	Bar	1960	Metropolitan-Cammell C&W Co	For 'Tees-Tyne Pullman' replaced No 59; name removed later; renamed *The Nightcap Bar* and used in 'Night Limited' all-sleeper BR London-Glasgow train, until 1979; withdrawn in 1980; bought by SLOA. Now owned by Pullman Rail and converted to 'Disco Car'.

Note: The design of the 'Blue Pullmans', as they became popularly known, was totally different to anything the Pullman Car Company had ever proposed, and it was intended to be the very last word in modern luxury rail travel. The British Transport Commission (BTC) sponsored the design of these trains, as spearheads of the modernisation plan, and use of the name Pullman was to give emphasis to their de luxe qualities. Apart from the 'Brighton Belle' electric Pullmans, these trains were the only multiple-unit Pullmans in Britain, using diesel-electric power cars at each end of the set formations, and with some of the passenger seating located above powered bogies (not a good idea, as events proved). The diesel cars carried no names, although the power cars of the sets carried the name 'Midland Pullman' on the LMR, and simply 'Pullman' on the WR. Within each train, the running numbers of the individual cars within each *half*-unit all ended in the same digit, and they were, for convenience, referred to as 'Half Sets 0-9'. These half-units were changed-over, as necessary, for maintenance purposes. A new livery of Nanking blue and white was specially devised for these trains. For reasons explained on page 144 the 'Blue Pullmans' only ran in BR service for some 12 years. No cars have been preserved.

Schedule Nos	BR running numbers	Type of car	Formation of cars within set	Make-up of sets
355-358	M60090-60093	Power car, with first class saloon	1st and 6th	Two 6-car sets for London Midland Region, 'Midland Pullman'
359-362	M60730-60733	First class saloon with kitchen	2nd and 5th	
363-366	M60740-60743	First class saloon	3rd and 4th	
367-372	W60734-60739	First class saloon with kitchen	3rd and 6th	Three 8-car sets for Western Region.
373-378	W60744-60749	First class saloon	4th and 5th	
379-384	W 60094-60099	Power car, with second class saloon	1st and 8th	
385-390	W60044-60049	Second class saloon	2nd and 7th	

All the 'Blue Pullman' cars were built by Metropolitan-Cammell C&W Co in 1960. All were withdrawn in 1973.

Notes

(1) Schedule Numbers

All the schedule numbers shown above are those carried by the cars up to and including 1960, based upon lists commenced in 1915 by Mr Dalziel. On 1 April 1960 a new schedule was prepared, in which some unfortunate discrepancies arose. This 1960 list is therefore not used in this book. Some renamings and renumberings took place after 1960, but by good luck these retained their original schedule numbers!

(2) 1962-1967 'Blue Pullman' Standby Trains

It proved necessary by 1962 to provide a standby train for the 'Blue Pullmans', in the event of their failure. I am indebted to Mr Charles Long for the following details:

'When the third, originally spare, "Blue Pullman" train entered service as the "South Wales Pullman", the locomotive-hauled cars that had formed the service previously — with their spares — plus two BR standard open firsts (which had been specially painted in umber and cream — *not* WR chocolate and cream) continued to be held at Old Oak Common for use in emergency on any of the three WR workings. A seven-car set was kept made up ready for immediate use. This was formed as follows: Pullman brake second; Pullman kitchen second; BR first open; Pullman kitchen first (1+1 seating); BR first open; Pullman kitchen second; Pullman brake second. The seats in this set were individually renumbered so that they corresponded as closely as possible with the seat numbers in an eight-car "Blue Pullman" unit for reservation purposes — eg the London-end brake second, which had (I believe) 30 seats compared with 18 in the motor brake second of the diesel unit, incorporated reservations A1-9 (face and back) and B1-6 (face and back). Besides this set, a number of additional loose Pullmans were also initially retained at Old Oak Common for incorporation in a second scratch train (which would likewise have included BR stock) had the need arisen. I have no complete record of all the second class cars involved, but they included Pullman brakes Nos 27 and 54 and kitchen No 171. The first class cars were: *Aurelia*, *Cecilia*, *Chloria*, *Medusa* and *Rosamund* (all kitchen cars).

'In September 1962, the "instant standby" set was reformed: the BR FOs were released to normal traffic and all the first class cars listed above returned to the SR. These vehicles were replaced by the "River" series of named cars, "upgraded" seconds all retaining 2+1 seating, namely: parlours *Wye* (No 35) and *Avon* (No 66), and kitchens *Severn* (No 60) and *Thames* (No 61). The standby set itself was then formed as follows: brake second; kitchen second; kitchen first; parlour first; kitchen first; kitchen second and brake second (all Pullmans). Again the seats were individually numbered to match the "Blue Pullmans" (I know this for a fact — I produced the chart). The second first class parlour car was held as a general spare vehicle, together with the existing spare second class cars retained from the former "South Wales Pullman".

'Following the BTC buyout of the remaining Pullman preference shares towards the end of 1962, it was planned to introduce a London-based "Midland Pullman" working early in 1963, using the spare six-car diesel set. It was in order to cover a failure or routine maintenance of either "Midland Pullman" set that a second standby locomotive hauled train was prepared. Stabled at Derby, this was of five cars, all nominally first class — although with a mix of 1+1 and 2+1 (upgraded seconds) seat vehicles. The formation was as follows: brake *Athene*; kitchen *Thetis* (No 107); parlour *Ceteia* (No 73); kitchen *Thalia* (No 106); brake *Fortuna* — and once again seats were individually numbered to match those in a diesel unit. This set train was joined by three additional "loose" spare cars: parlours *Juana* and *Melandra* (No 74) and kitchen *Hebe* (No 105).

In the event, the second diesel Pullman working was not introduced and, so far as I am aware, the locomotive-hauled Midland standby train was never used for its intended purpose. However, in January 1963, all eight cars (set train+spares) were pressed into service on the "Birmingham Pullman" working when the WR standby train was already in use on the "Bristol Pullman". In these circumstances the carefully prepared equivalent seating plan was useless, and passengers were seated on a first-come, first-served basis.

'At some time after I left Pullman employment, the first class cars in the WR standby train were replaced by redundant 1960-61 East Coast second class cars.'

The standby sets gained the unofficial name of the 'Wells Fargo' and the 'Santa Fé' (the latter being the scratch set which included BR Mk 1 stock) and they were distinctly unpopular with regular first class travellers, who had grown accustomed to the air-conditioned, reclining seat comfort of the 'Blue Pullmans'. The 'Wells Fargo' ceased to run when the LMR 'Midland Pullmans' were transferred to the WR in 1966/67.

Note: New Pullman cars built under BR auspices after 1962 feature in Table 4, on pages 187/188.

Part Three:

The Pullman Revival

CHAPTER 19

Retirements, Rescues and Restorations

To tell the story of the Pullman revival, it is necessary to retrace our steps a little, and reconsider some of the events of the years since 1960, seen from the viewpoint of the railway enthusiast.

Throughout its long history up until 1960, the British Pullman Car Co did not interest itself in the preservation of old cars, once they were no longer required to run. Perhaps it was a deliberate policy, because many cars in everyday use were in fact rolling period pieces, and it would have been difficult to draw a line between what was the past, and what was the present! As a result no *intact* British Pullman car of the pre-1910 era exists today; although there are countless steam locomotives and other relics in existence, dating back far beyond the Pullman car era in Britain. What has survived from the pre-1910 years has survived by hazard, and will require substantial restoration work to make it truly representative of its time. These cars were not laid aside with history in mind, but as useful covered accommodation of one sort or another.

It was because they were so well built, and well cared for that some of the old wooden-bodied Pullmans, once retired from service, have lived on for a great many more years in use as bungalows, studios, mess rooms and stores. Most became 'grounded bodies', having lost their wheels and bogies, and now resting upon the

201

201
The wooden body of *Midland*, the very first Pullman sleeping car in Britain, assembled at Derby in January 1874, survives to this day. After withdrawal from service, the Midland Railway used it as a 'grounded body' at Skipton, as a well-appointed local Carriage & Wagon Office, mess room and stores. It remained there until June 1970 when attempts to preserve it were started by the Midland Railway project and Derby Museums, which finally resulted in its removal from the site in March 1974. Stored until 1981, it is now at the Midland Railway Centre at Butterley, Derbyshire, where it is seen being placed in a safe position, awaiting the lengthy job of restoration. What a tribute to George Mortimer Pullman's design, and the original Detroit USA and Derby, England workmanship, that it has survived so long! *Frank Ashley*

202

203

202
Mrs Thornhill, in the garden of her home — the Pullman car *Montana*, at Barnwell Junction, near Cambridge. Quite a lot of early Pullmans had ended their days as holiday bungalows, or even artists' and writers' studios. However, time and the weather, or wartime neglect, have written them off and the author believes that the Thornhill's home is the only example still in use as such in Britain today, although the two short-underframed 'inferior' type sleeping cars *Balmoral* and *Dunrobin* existed *inside* a bungalow at Seaford, Sussex until 1987, but not on bogies, unlike *Montana*. (See text for further details.) *R. R. Dursley*

203
Ian Allan Ltd hostess Mrs Anne Peacock shows the executive dining saloon of *Malaga* used by Ian Allan directors on suitable occasions (and in which the idea of this book was first conceived). The car *Malaga*, still on its six-wheel bogies, is now joined to and plumbed into, the Shepperton office block of the Ian Allan Group, publishers of this book. *Ian Allan*

ground. So long as the roof was kept watertight, the body would last. Three old Midland Railway Pullman cars have thus survived until this day, and are now rescued for preservation. One of them is the pioneer sleeping car *Midland*, and it is hoped that one day this will be restored to all its original glory, and run again, as a tribute to George Mortimer Pullman's enterprise way back in 1874. The other two are former 'day cars', never used as true Pullmans, and for the moment their identities are not established.

The Pullman Car Co itself had used two old Brighton line Pullman cars at its Preston Park workshops: the car *Devonshire*, built 1900 and withdrawn in 1931 was used as a staff canteen, and the much older car *Albert Edward*, built in 1877 and withdrawn in 1932, served as the canteen kitchen and as a store for window panes; sadly both were scrapped. At Lancing SR Carriage Works could be found the old cars *Albatross* and *Thistle*, curiously joined together after being condemned in 1940, and still on their bogies and in use as offices in the early 1960s. The two Metropolitan Railway Pullmans, *Galatea* and *Mayflower*, withdrawn when war broke out in 1939, later became the offices of a wood merchant at Hinchley Wood in Surrey, and survived until 1948. The lists published from time to time of these grounded bodies, and their original identities are probably incomplete. The old Pullmans that had gone to ground as seaside holiday bungalows in prewar days were, alas, dismantled by the military forces in the course of beach defence work, early in the World War 2. One wonders whether any relics or photographs of them have survived?

The first official preservation of a Pullman car was in fact largely motivated and financed by a London businessman, Mr Henry Maxwell, who donated a large sum of money in September 1961 towards the cost of renovating and preserving the 47-year-old car *Topaz* (I) at the Museum of British Transport, Clapham, London. *Topaz* (I) was built for the SECR services and it was restored to 1920s condition, complete with the original wooden-framed bogies. The livery was the 1920s umber and cream (of whitish shade). Since transfer to York the National Railway Museum have had it fully refurbished and changed to the original SECR crimson lake livery.

The use of old Pullmans as 'Holiday Camping Coaches' has already been described in Chapter 16. Some of these still survive, and in Marazion, Cornwall, there are no less than six of them, still on wheels. These survived because, when no longer used for public holiday letting purposes, they were taken over by the British Railways Western Region Staff Association, who continued to maintain them in good condition. These are now to be incorporated into a new holiday camp development on the site, and it looks as though their future is secure; provided that the new owners care for them, and keep the rain out. The roof is always the weakest point of such vehicles.

One former camping coach which commenced life as the 1923 brake parlour car *Montana*, is still in use as a private residence at Barnwell Junction, near Cambridge. It is, I think, the only free-standing Pullman car still in use as a home, although some have been used as camping dormitories for volunteer staff working on railway preservation schemes. The car *Montana*, dating from 1923 and first of all used on the SR London-Dover boat trains, then in postwar days on the 'Thanet Belle' and 'Kentish Belle', had been withdrawn and converted to a camping coach in 1960/61, and placed at Heacham, Norfolk, together with the former Pullman car *Maid of Kent*, where they were let to holidaymakers until the end of the 1965 season. They were then redundant and moved to Kings Lynn and put up for disposal, in 1967. *Montana* was purchased by Mr Thornhill, then assistant BR area manager at Cambridge, and moved to an orchard site close by the closed Barnwell Junction station, where it has been lived in by his parents ever since in splendid comfort; complete with original panelling.

The idea of preserving Pullmans as static exhibits was the next stage after the original deployment of them as stores, or useful accommodation. The publishers of this book have the car *Malaga* (as related on page 9) located adjacent to Shepperton station, on the SR. Ian Allan, the Chairman of Ian Allan Ltd, in a letter to the Author in June 1984, had this to say about the car:

'*Malaga* was constructed in 1922, is mounted on six-wheel bogies and was for many years on the 'Golden Arrow' service. In 1949 it was completely refurbished for use by HM King George VI for his journeys on the Southern Region where, as you know, Pullmans were always used for royal journeys.

'The car was a standard kitchen unit with a large kitchen, comprising a parlour section of 12 seats and a coupé of 4 — 16 first class seats in all. The interior panelling is unique being walnut veneer on aluminium and there are hand engraved copper plaques in the parlour saloon and coupé depicting the various goodies available in Pullman cars; meat,

204
The since disbanded static 'Bournemouth Belle' exhibit at Lord Montagu of Beaulieu's museum in Hampshire. The Maunsell 'Schools' class 4-4-0 No 30928 *Stowe* is now preserved on the Bluebell Railway and the nearest two Pullman cars are also preserved (*Agatha* and *Fingall*, see Table 2, page 151), but the one furthest from the camera, Car No 35, was subsequently destroyed. *Montagu Motor Museum*

205
Former 'Brighton Belle' motor brake second car No S289S, of set No 3051, went to the 'Little Mill Inn' at Rowarth, Stockport. A problem with placing old railway vehicles in such surroundings is that they will soon deteriorate if left fully exposed to all weathers and all seasons, unless constant painting and varnishing is undertaken — or better still, an overall protective roofing is built. *Keith Smith*

206
The famous Gresley Class A3 Pacific No 4472 *Flying Scotsman*, complete with cowcatcher, bell, headlamp and American-style front coupling, heads Alan Pegler's 'American Flying Scotsman Tour Train'. It is pictured between Springfield (Mass) and Hartford (Connecticut) on 12 October 1969. *Robin Russell*

204

205

fish, wheat, grapes, etc. The luggage racks and window surrounds are uniquely of anodised copper.

'When we built the offices in Shepperton in 1963 I told Frank Harding (General Manager of the Pullman Car Co) that we wished to obtain a Pullman car and he was very delighted to make *Malaga* available to us as he felt it was an historical vehicle which should be preserved. Accordingly it came to Shepperton in the early Autumn of 1963, hauled by a steam locomotive accompanied by a steam crane, and it was carefully transferred to its own private piece of railway line where it has been ever since. It is one of the few vehicles where the lavatories can be used when the train is standing in the station, and to mark the occasion Frank Harding gave me a free pass to travel on all other Pullman cars which was rapidly clawed back when the Pullman was absorbed by BR.

'PS. *After* the Pullman was in situ, we could build the offices!'

At about the same time Lord Montagu of Beaulieu, placed three Pullman cars *Agatha*, *Fingall* and Car No 35 on static outdoor display at the Montagu Motor Museum, Hants, attached to preserved SR 'Schools' class 4-4-0 No 30928 *Stowe*, calling the exhibit the 'Bournemouth Belle' Unfortunately the upkeep of this attractive exhibit proved to be too expensive, and the collection was subsequently dispersed; two cars have survived (see list) as well as the locomotive, which is now on loan to the Bluebell Railway; which also has the car *Fingall*.

The Bluebell Railway deserves special mention at this point because it was the first standard gauge preserved line in Britain. It showed what could be done to prevent the rail heritage from disappearing under the reshaping plan of Dr Beeching, in a moment in history when much was being destroyed on the grounds of economy. As other preserved lines were created, the list of preserved rolling stock began to grow to amazing proportions, and it was quite obvious that more Pullman cars would be considered very worthy candidates for rescue from the scrapdealers' torch.

For enthusiasts, Pullman cars started to become very desirable acquisitions. Each withdrawal from BR service was a possible candidate for preservation, although to begin with attention was more upon the older cars which were lying around the countryside since they had ended their days as camping coaches, or were still in use as BR departmental vehicles. A lot did go for scrap, of course, because funds could not be raised in time. The price of a redundant Pullman car was around £1,500 in the early 1960s (steadily increasing as the demand grew) which was almost twice the price BR then wanted for an ordinary carriage. The price included moving the car to the nearest railhead for collection by the purchaser. At this stage the cars were only attracting the attention of private individuals, and railway preservation societies.

Then came the famous 'kipper revolt' of Lord Olivier, and Pullmans were big news; all the nation now knew that Pullmans were in danger of extinction on BR, and the attention of many interested parties was focused upon the idea of saving cars for very commercial purposes. Suddenly the purchasing power of the railway enthusiast (unless fortunate enough to be very wealthy) was challenged by big concerns such as brewery companies, who had their eyes on the possible further use of the cars as static restaurants and steak houses; amongst other things. As a direct result of the 'Brighton Belle' furore, all 15 cars were subsequently purchased by a variety of private individuals and societies and companies. Nothing quite like it had happened on BR before. The 'Golden Arrow' cars followed suit and BR realised it had a very saleable product (although one which was curiously unwanted in its own jaded eyes!). The list in Table Two, on pages 151-158, gives a complete breakdown of the preserved Pullmans, and space does not allow me here to do more than single out some exceptional examples of what almost became 'Pullman Mania'.

Until quite recently — the autumn of 1985, to be precise — in a meadow under impressive East Anglian skies at Westleton, Suffolk, stood the former 'Brighton Belle' car *Vera*, complete with bogies. In front of it was a large lake; sometimes with visiting herons fishing in it, momentarily disturbing the car's reflection in the water. In the foreground were large pampas grasses, and pieces of sculpture, setting off this almost theatrical backdrop. *Vera* was owned by M Amédée Turner and his wife who used the car for entertaining, and on the occasion of fêtes in their garden, and Red Cross open days, to serve teas, etc. It also featured in Conservative Party events held at their home, because M Amédée Turner QC, MEP, is the Member of the European Parliament for Suffolk and Harwich.

Happily retired to green pastures, who, in their wildest flights of fancy would imagine that *Vera's* wheels would ever turn again? Memories of its former life on the Brighton line became more remote with each passing year, and *Vera* assumed a new static personality; rather like a huge piece of sculpture. And then, in the autumn of 1985, when crops of wheat and barley had been harvested from the surrounding fields, a 60-ton telescopic crane arrived to lift the 29-ton Pullman car on to an articulated lorry, for transportation back to Stewarts Lane, London! For *Vera* life begins again at the age of 54, because she has been purchased for a new career in the prestigious Venice Simplon-Orient-Express. Once restored to all her former glory, at a cost of around £250,000 *Vera* will join the Pullman train that forms the British section of the VS-O-E. I shall return to this train later in our story.

Another 'Brighton Belle' car serves as a VIP dining room rather like *Malaga* at Shepperton. This is *Doris*, located in the private sidings of City Industrial Ltd at Finsbury Park, London. Her owner, Mr Sam Morris, Chairman of the CIL Group, wrote to the Author in October 1983 as follows:

'I own *Doris* and she is situated on my company's private railway sidings here at Finsbury Park station. I am afraid she is no longer in running order — I believe "static display" would be the correct term.

206

207

207
Another view of the 'American Flying Scotsman' train showing the three Pullman cars which made up the rear portion, the final vehicle being the former 'Devon Belle' observation car, still carrying its British Railways Scottish Region numbering as SC281. *Robin Russell*

'However, apart from the fact that she is exposed to the weather, her condition is good as she was refurbished to her original condition and colours by the British Rail Workshops at Derby. The interior is exactly as it was in the 1930s and the exterior is in the original colours of cream and brown.

'*Doris* is used as a VIP dining room and my guests find it a unique experience especially as once aboard it is not possible to be reached by telephone, or by any other disturbance. The atmosphere, therefore, is most conducive to business negotiations.'

Allied Breweries, Trumans and Whitbreads have all purchased former 'Brighton Belle' cars and these are located at various public houses in England in use as restaurants (see list), although two are leased to the North Norfolk Railway, by Allied Breweries and one, bought by Trumans is now at Swanage on loan from the Stour Valley Railway. This is former motor coach No 288 of unit No 3501. Before sale, BR removed all the electrical gear, and traction motors from all the motor coaches, so that today they can only be hauled by a locomotive, if required to run, and certain modifications have to be made to the brake system before this can be done. One 'Brighton Belle' motor coach, No 290, sees quite frequent use in passenger trains hauled by preserved steam locomotives on the Nene Valley Railway, near Peterborough. It arrived at the Nene Valley in 1979, after it had been in store at Wolverton Carriage Works for preservation by British Transport Hotels (BTH) in 1972. Nowadays it is often used in evening charter trains, which feature an evening meal, and a 25p supplementary fare is charged, in true Pullman style. A feature of much interest is the driver's cab, which still contains all the controls, and passengers often ask to be allowed to visit the cab during their journey. Another 'Belle' car *Gwen*, which served as a restaurant at the 'Horseless Carriage' public house at Chingford Hatch in Essex after sale by BR, is now located on the Colne Valley Railway at Castle Hedingham station, on loan from the owners, Beefeater Steakhouses. This former first class parlour kitchen car now serves afternoon English teas and luncheons and can cater for private functions by prior arrangement. Most of the 'Brighton Belle' cars have been restored to umber and cream livery, but with varying degrees of authenticity and accuracy, by their owners.

The 'Golden Arrow' cars had similar new careers, once sold off by BR. *Carina* and *Phoenix* were shipped across to France to become roadside restaurants at the Hôtel Mercure near Lyon (more about these later) and *Orion* was purchased by Peco, the Pritchard Patent Product Co Ltd, of Beer, Seaton, Devon, about which Mr S. C. Pritchard, the Chairman of Peco had this to

208
Beautifully restored by BR at their Wolverton carriage works, the former 'Golden Arrow' car *Orion* was then lifted from its bogies and taken on special road wheels to Beer in Devon, in January 1978. It is seen here entering Mare Lane from the Causeway, Beer, on a gradient of 1 in 5. *Orion* was soon to be in its new home at the Beer Victoria station of Peco Modelrama, where it is now a high class licensed restaurant, once again upon its bogies but as a static display.
Peco Modelrama

208

209

210

209
Former 'Devon Belle' Pullman observation car No 14 is on the left-hand side of this picture; forming the outer wall of the building, with an enclosed roof and patio connecting the entrance to the car, which is used as an executive office by Western Athletic Clubs Inc at 150 Chestnut Street, San Francisco, California, USA. Car No 14 went to America as part of Alan Pegler's 'Flying Scotsman' tour train and was left in San Francisco when the tour had to be abandoned for lack of funds. *Western Athletic Clubs Inc*

210
The other 'Devon Belle' observation car, No 13, is now preserved in Britain by the Dart Valley Railway, and has been refurbished and named *Devon Belle*. Sliding toplights have been added to both of the large rear windows, but the car is otherwise much as it appeared in BR days, apart from livery details. Internally, the seat layout has been changed, but the bar area is retained. The Dart Valley charge a supplement to travel in the car, which is licensed. *Dart Valley Railway*

say in a letter to the Author in November 1983:

'*Orion* is from the "Golden Arrow" service and was built in 1951 by the Birmingham Railway Carriage & Wagon Co and although of postwar origin, all the fittings and equipment were from prewar stocks of the Pullman Car Co. The bogies are of Gresley design.

'The location is in Platform One at Beer Victoria station, part of the Beer Modelrama complex. The station is of course mock but is built out of original materials and is used as a Buffet. It is called Beer Victoria to associate it with the train which left Victoria station, London. It is beautifully restored in its original colour scheme of umber and cream, and carries the destination boards "Golden Arrow" and "Flèche d'Or".

'Minor alterations have been made in so much that the kitchen was pulled out and completely re-fitted. The little pantry or stillroom remains intact but was converted to a dispensary, ie bar. Reverting again to the kitchen, the most awful improvement was that the window which is against the wall of the setting and not seen by the public, was taken out and a bay window attached to contain the sink unit in order to make more room in the kitchen — but nobody sees this.

'The Coupé at the far end has had its passage wall removed so as to accommodate a table with six chairs. The toilet has not been altered in any way but is not used for its original purpose, being fitted with removable shelves and used as a linen cupboard. The original loudspeakers for destination announcements are still in use for background music which includes many train noises.

'Only four original seats were included with the car and these now reside inside our office building so it has been completely furnished with new seats — miniaturised versions of the real thing — and new tables seating four one side and two the other. These tables however can be rearranged to make a long table down the centre for special functions or a buffet table on one side.

'*Orion* is owned by the Pritchard Patent Product Co and is used as a restaurant, but the catering is done by an independent company. It is of course static but nevertheless in runnable condition.'

One 'Golden Arrow' car, *Aquila*, was purchased in 1968 by H. P. Bulmer Ltd, the Hereford cider makers, together with four older cars from the 'Bournemouth Belle' to make up a five-car 'Bulmer Cider Train' and although a Pullman-style livery was applied, the colours were green instead of umber below the waist, white instead of cream for the window areas, and red instead of umber for the panel above the windows to the roof, which was grey. The lettering was in Pullman tradition, but read 'Bulmers Cider' on the red panel and the car names, in Pullman style on the green, were those of the wives of Directors of the company, except for *Aquila*, which kept its name. Each car carried the arms of the City of Hereford and the royal coat of arms, held by Bulmers since the granting of the royal warrant in 1911, instead of the Pullman coat of arms. I shall return to this train in the next chapter, because it played an important part in what I choose to call the 'Pullman revival'. Meanwhile, a sixth car at Hereford must be mentioned: the former SR 6-PUL composite car *Ruth*, which was purchased in 1973, and subsequently sold to the 6000 Locomotive Association. This has been repainted in umber and cream, and is used by the Association as its members' clubroom and headquarters office, located at the Bulmer Railway Centre at Hereford.

In 1969, British railway enthusiast and businessman, Alan Pegler created his 'American Flying Scotsman Tour Train', to travel around North America promoting Great Britain and British products. A complete exhibition train was formed, and included were three Pullman cars: *Lydia*, *Isle of Thanet* and the former 'Devon Belle' observation car No 14 (purchased from British Rail, Scottish Region, who had numbered it No Sc281 after it was sold by the Pullman Car Co). The complete train was painted in Pullman umber and cream, with huge British and American flags on the sides of all the carriages, except the Pullmans. The famous LNER Pacific locomotive No 4472 *Flying Scotsman*, designed by Nigel Gresley in 1922, which Alan Pegler had purchased from British Rail in 1963 was specially prepared with headlight, bell and cowcatchers, to run in steam on the American railroads. For the first time, working Pullman cars were shipped *from* Britain *to* America; whereas a considerable number had made the voyage in the opposite direction since 1874.

Unfortunately, Pegler's enterprise ran into financial problems (as I have related in the book *Flying Scotsman*, published by Ian Allan Ltd) and the complete train was marooned in San Francisco, after extensive travels in Canada and the United States. The locomotive was rescued, at very great expense, by The Hon William McAlpine, and safely shipped back to England, but the Pullmans and the other carriages remained in San Francisco. I do not know the subsequent fate of the exhibition carriages (which had all been converted from LNER bogie vans of Gresley design) but the Pullmans have survived; in one case in a most remarkable form. The cars *Lydia* and *Isle of Thanet* have found a home at the National Railroad Museum at Green Bay,

Wisconsin, which is believed also to possess the car *Hercules*. Pullmans *Lydia* and *Isle of Thanet* had both been used in Sir Winston Churchill's funeral train, which no doubt gave them added historical interest for Americans, and perhaps aided their salvation.

The most remarkable fate befell the former 'Devon Belle' Pullman observation car No 14, which became part of a building. I am grateful to Sonja Scott, of the Western Athletic Clubs Inc for the details which follow. The car is actually owned by Victoria Station, a restaurant company, whose restaurants are in old American railroad cars. Victoria Station has leased the building (which is also owned by them) and the car, to Western Athletic Clubs Inc. Sonja Scott wrote to me in January 1983, as follows:

'The car is located at 150 Chestnut Street, San Francisco, California, USA 94111, which is at the foot of the famous Telegraph Hill. The car is stationary, set in front of our office building, and an enclosed patio and roof connect the building to the entrance of the train car.

'The outside of the train car is maroon, the inside is used as executive office space. The carpet is a print of brown/tan blue, the furniture is of tan fabric and the curtains which are on the windows the length of the car are of tan velvet. The ceiling is painted white.

'Since I do not know how the car looked while in service it is difficult to say what changes have been made. The inside has a restroom (water closet), electrical plugs and a bar. The bar has sinks, and the mirrored wall behind the bar has shelves for glasses and an advertisement for Watneys beer on the wall. There is a plaque stating that the "Pub Corner Designed by Ayala Designs Ltd. England." The windows of the train car have had a solar film applied to reduce heat from the sun in the summer.

'Outside, at the end of the car, is a typical red British telephone booth, but there is no phone inside.'

Both the 'Devon Belle' observation cars have survived, the other one, No 13, now being owned by the Dart Valley Light Railway PLC, and named, very suitably, *Devon Belle*. It is in GWR 'chocolate and cream' livery and a supplementary fare is charged to travel in it. The small bar serves draught beer, and there is a new seating layout which helps to give passengers a superb view of the Devon countryside. The original 'Devon Belle' route map is still on the wall, in excellent condition.

As the preservation societies developed, so did the idea of actually running Pullman cars on these private lines, and a lot of work had to be put into various cars, which had been purchased as obsolete, in order to make them fit to run again. At this stage BR had imposed a total ban on steam traction on its routes, and it did not view the use of privately-owned rolling stock with much enthusiasm either. Thus it was only on certain preserved lines that the sight of a steam-hauled train, perhaps with a Pullman car included, could be a possibility. This state of affairs continued until 1971, when after years of protracted negotiations — in particular by H. P. Bulmer Ltd — the ban on steam traction was lifted for a special 'Return to Steam' tour, using the magnificently restored GWR 'King' class 4-6-0 No 6000 *King George V* and the Bulmers Cider Train Pullman cars. The tour was a complete success, and as a result steam-hauled tours have been run over selected BR routes ever since, and many

211

211
The National Railway Museum's preserved Great Northern Railway Ivatt small Atlantic No 990 *Henry Oakley*, in steam on the Keighley & Worth Valley Railway on 1 June 1977, seen passing over Mytholmes Viaduct with a special train. The leading vehicle is a preserved third class Pullman car, now named *Mary*. Next to it is a standard BR Mk 1 carriage repainted in umber and cream to match the livery of the Pullman, and named *Jubilee Bar*. *Rodney Wildsmith*

have featured Pullman cars, as we shall see in the next chapter.

To end this chapter, one more preserved Pullman car deserves to be featured, because it is so superbly restored, and because it now carries the name *Mary*, after the wife of the late Bishop of Wakefield, Eric Treacy, so renowned for his wonderful railway photography. Bishop Treacy was President of the Keighley & Worth Valley Railway Preservation Society (KWVRPS) at the time of his death. The car *Mary* is of additional interest because it is believed to be the most complete and authentic example of a third class Pullman car in the UK today. It was built in 1930, according to the maker's plates, not 1931 as given in some histories, as Car No 84, by the Birmingham Railway Carriage & Wagon Co Ltd and first ran on the 1931 Southern Railway 'Bournemouth Belle' service, later being moved to the LNER. During the war it was painted brown, and became LNER No 484, as a first class carriage. When restored to Pullman ownership it was retained on the LNER, and then British Railways Eastern Region, until 1960 when it was once again sent to the 'Bournemouth Belle' service, and it ran on the Southern Region until withdrawal in 1966. Car No 84 was then purchased (from York) by Mr R. Ainsworth of Leeds, who had it moved to Haworth on the Worth Valley line. In 1968 he named it *Lorna*, after his wife. When Mr Ainsworth died the KWVRPS purchased the car, and in March 1977 it was renamed *Mary*. The interior of the car is in immaculate condition, and the car is now used about eight times a year for railway functions or private hire. With two other vehicles it forms a three car rake, and refreshments are served from the adjacent vehicles. When not running, *Mary* is kept in the Oxenhope Exhibition shed, and is certainly well worth a visit, because in every detail it is authentic — only the name has been changed.

So far in this chapter, we have seen the early work of the preservation movement, chiefly concerned with the rescue and restoration of what might be called vintage Pullmans; the most recent being the 1951/52 'Golden Arrow' cars. On BR there were still the modern Pullmans in service and these included until 1973, the 'Blue Pullmans', which were then withdrawn; leaving only the East Coast replacement cars, and the most recent of all, the LMR Manchester service cars. In Chapter 18 I have already shown how the BR Pullman services went into decline in the 1970s. For the moment there was a curious state of affairs, whereby BR seemed anxious to cast off the Pullman mantle altogether, whilst the preservationists were painstakingly restoring it!

212

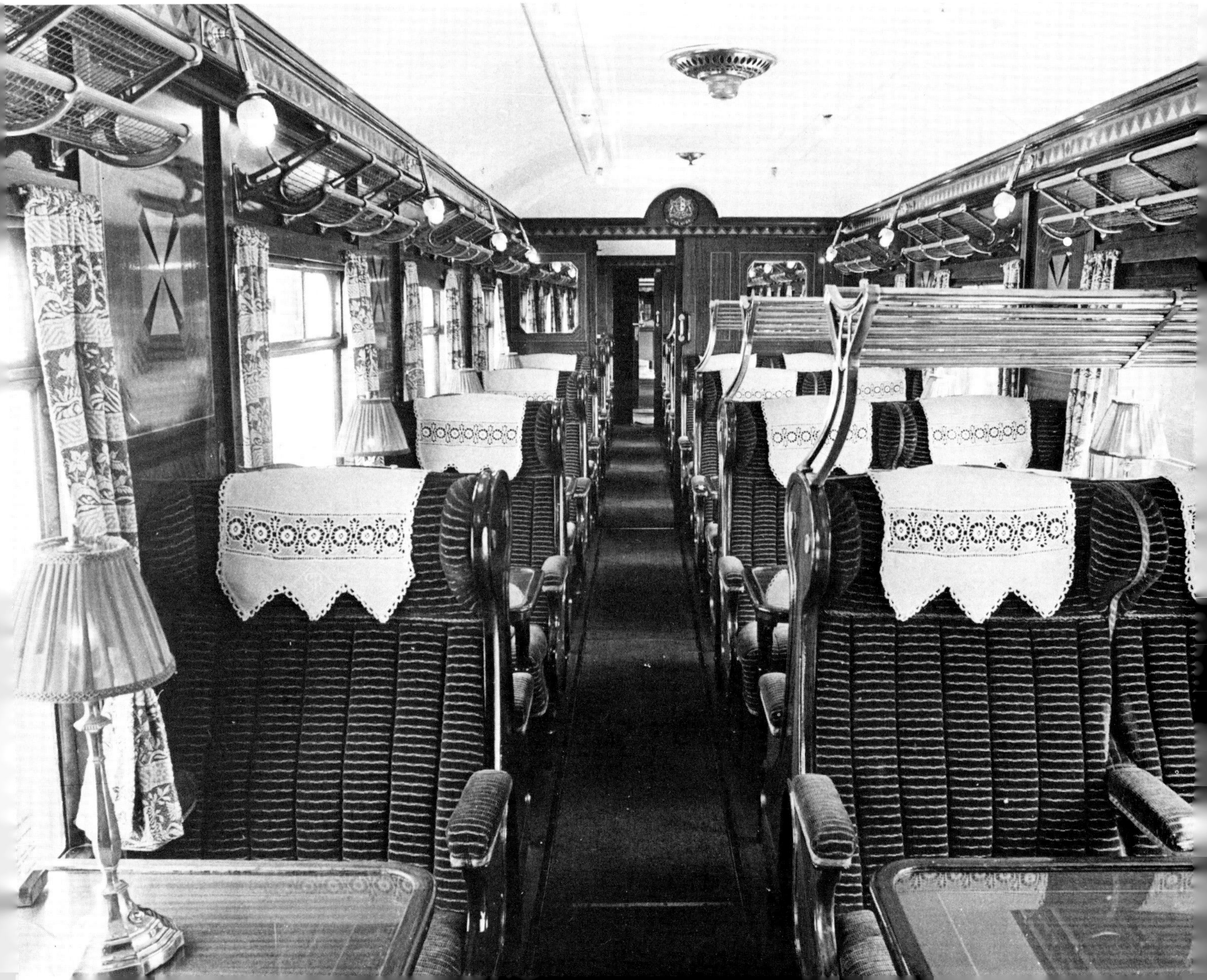

212
The magnificently restored interior of the Keighley & Worth Valley Railway Pullman car *Mary*, a 1930 third class car at first used on the 'Bournemouth Belle' and later transferred to the LNER. The additional luggage racks above the double seats are an interesting feature. The Pullman coat of arms features on the plaque above the doorway, together with the inscription 'Car No 84'. This space was originally occupied by a clock, but the Pullman Co removed the clocks from their cars in postwar days due to the expense of keeping them in good repair. The conductor was responsible for winding each clock. Prominent in this picture (left-hand foreground) is a typical Pullman table lamp, with pink silk shade. After the war, plastic shades replaced the silk, in green, pink or white. Each table lamp had the name or the number of the car to which it belonged stamped on the base. Each car had its own set of lamps. An added use was the weight of the lamps, to hold table linen in place! *K. Roberts*

CHAPTER 20

The Pullman Revival Begins

The return to main line steam running on BR, with the use of privately preserved or National Railway Museum owned locomotives, also encompassed the use of privately-owned carriages, provided these were passed as fit to run on BR tracks by their Inspectors. Chief amongst these in the early 1970s was the Bulmers Cider Train, with its Pullman cars, in Bulmers green, red and white house colours. Three of the cars had been converted internally for exhibition purposes as a museum, a cinema, and a bar, and four of them had been named after Director's wives. *Morella*, originally Pullman Car No 36 was the cinema, and was named after the wife of Mr Esmond Bulmer; *Eve*, originally Car No 76 was the bar product display coach and named after Mrs Brian Nelson; *Prinia*, originally Car No 83 was named after the wife of Mr Peter Prior and was the museum car displaying the development of H. P. Bulmer Ltd; and *Christine*, originally Car No 64 had been retained in its original form as a dining car, and was named after Mrs Bertram Bulmer. The fifth car *Aquila*, with its original name retained was in virtually original condition except for the fitting of a new kitchen range and new carpeting. Coupled to the immaculate 'King' class locomotive No 6000 *King George V* the train made a splendid spectacle wherever it ran.

The withdrawal of the 'Blue Pullmans' in 1973 did not go unnoticed by the preservationists, but the idea of preserving such a train intact was out of the question in terms of cost, and in any case the MAN diesel engines in the power cars were notoriously troublesome to maintain in service. The idea of buying some of the trailer cars posed problems because of the special gangways and couplings used as well as the types of braking, lighting and heating the air-conditioned cars, which were certainly not compatible with steam traction. All the 'Blue Pullmans' were scrapped, and it is a matter of regret that the National collection did not decide to keep an example of these important Pullman cars for posterity.

With the increasing use of steam-hauled rail tours and other charter trains, the remaining BR Pullman cars often found weekend private charter use in addition to their ordinary duties. The second-class Pullmans were all withdrawn from Pullman service at the end of the 1960s by BR and used instead as ordinary *first*-class carriages, because their standards of seating and design was the equivalent of normal first-class stock (an interesting comment upon the superiority of Pullman design). These ex-Pullmans were painted in the standard blue and grey main line livery, and worked in such services as Motorail trains and the London Liverpool Street-Harwich boat trains, often used as dining cars. The *Hadrian Bar* saw further service from 1969 onwards as the *Nightcap Bar* on the Euston-Glasgow sleeper services (which also featured *Pegasus* the ex-'Golden Arrow' *Trianon Bar*). Both cars, when running as the *Nightcap Bars*, no longer carried the name Pullman on the sides, once they were repainted in standard BR blue and grey livery. Both are preserved.

British Rail was engaged on the development of the Advanced Passenger Train (APT) and on the prototype High Speed Train (HST) at the time the 'Blue Pullmans' were withdrawn. The APT, with gas turbine engines was still very experimental and regarded as a long-term solution to new passenger train design, whereas the diesel-electric HST was to be the stop-gap front line InterCity passenger train for the late 1970s and early 1980s, using the new standard Mk 3 air-conditioned carriages, and with an ability to run at 125mph. The standards of comfort offered by the HST was equivalent to the 'Blue Pullmans' but without a supplementary fare, and often without full waiter service; the catering being by Travellers-Fare, and increasingly with a buffet type of service. By 1976 full scale production of the HST was under way, and the new trains soon proved highly successful, and became so popular with the

213

213
The day every railway enthusiast will remember — 3 October 1971 — when BR lifted the ban on steam locomotives running on its tracks. The beautifully restored GWR 'King' class 4-6-0 No 6000 *King George V* (complete with the brass bell presented to it in America when it was exhibited there in 1927) and the 'Bulmers Cider Pullman Train' are seen on this historic day. The Pullman cars were repainted in Bulmers' house colours of green, cream and red, but the Pullman style was retained. The leading car in this superb picture is *Aquila*; the location is the entrance to the Severn Tunnel. What a magnificent sight it made! *R. O. Coffin*

public that there were problems of overcrowding at peak periods. With the introduction of HSTs on the East Coast Route in May 1978 the remaining ER Pullman services was withdrawn, leaving only the LMR 'Manchester Pullman' running, on the West Coast main line. It was then envisaged that this too would be displaced, by the introduction of new electrically-powered APTs, in service by 1984/85.

Meanwhile, nostalgia for Pullmans and all that they represented was steadily growing, and the preserved examples were gaining many admirers. In August 1974, the Kent & East Sussex Railway (reopened that year by the Tenterden Railway Co) introduced what was to become the 'Wealden Pullman', at first as an un-named experimental 'wine and dine' special train; in order to test public reaction. The Pullman car *Barbara*, one of the former Hastings line narrow-bodied Pullmans was the star attraction in this enterprising initiative, which was the first train operated by private railway in Britain since the war to offer a full catering service to the passengers. Other private lines have since followed their lead, but the 'Wealden Pullman' is a remarkable success story, to which I will return later in this chapter.

Largely thanks to the cinema and television industry, and to writings of the late Agatha Christie, Pullman cars and Wagons-Lits were gaining a glamour and prestige in the early 1970s such as had never been known during postwar days. The elegant marquetry and decor, the sumptuous furnishings and comfortable chairs and compartments were a film-maker's delight, and it was the film of Agatha Christie's classic *Murder on the Orient Express*, using beautifully restored Wagons-Lits and Pullmans on the Continent that really set the ball rolling. When what was announced as the last trip of the famous 'Orient Express' took place in May 1977, it seemed that the whole world knew about it, and mourned its passing.* In reality, the service had run out its last years as a rather second rate affair without a dining car, and mostly used by Turkish workers, peasants and students. Gone were the splendours depicted in the film of the Agatha Christie story, but the name alone

214
The narrow-bodied Hastings line Pullman car *Barbara*, as lovingly restored by the Kent & East Sussex Railway for use in their highly successful 'Wealden Pullman' wine-and-dine specials. A second 'Hastings' car *Theodora* is also on the Kent & East Sussex, but not in service. It was used as a static buffet and tea bar at Tenderden Town station, but the long-term intention is to have it running again in full Pullman splendour, together with *Barbara*. *Brian Stephenson*

215
Once James Sherwood had embarked upon his Venice Simplon-Orient-Express venture, the net was cast far and wide to find suitable old British Pullman cars for restoration to full working order. In France were two former 'Golden Arrow' cars — *Phoenix* and *Carina* — in use as static restaurants at the Hotel Mercure, near Lyon. These were brought back to England by road, for restoration at Steamtown, Carnforth, Lancashire. Whilst still in France *Phoenix* managed to get firmly wedged under a railway bridge, and the tyres on the transporter had to be deflated in order to free it. *Sea Containers Ltd (VS-O-E)*

214

215

*In fact the 'Orient Express' did *not* cease to run in 1977. What happened was that the through Paris-Istanbul service, latterly provided by the 'Direct Orient Express' routed via Dijon, Lausanne, Domodossola, Milan, Venice, Belgrade, etc was withdrawn. The 'Orient Express' itself, which originally provided the link, still runs between Paris and Bucharest, following the 1883 route via Strasbourg, Munich, Vienna and Budapest. Neither this train for many years past, nor, when they ran, the 'Direct Orient Express' or the latterday 'Simplon Orient Express' (from which the VS-O-E borrows its name) could lay claim to *de luxe* status. The VS-O-E train no longer follows the Simplon route, and the link to Istanbul is now made by a ship, owned by Sea Containers Ltd and named *Orient Express*, from Venice where the VS-O-E train terminates.

was famous enough to attract publicity world-wide at its demise. In the autumn of 1977 Sotheby's organised a sale of Wagons-Lits rolling stock at Monte Carlo, the five cars being on display behind the auctioneer, and the goods station where the auction was held was crammed with people, and television crews and press men. The event received enormous world-wide coverage, and the Wagons-Lits Co received enormous prices for the five cars, one of which had been used in the making of the film *Murder on the Orient Express*. If the train itself had ceased to run, the magic associated with its name nevertheless lived on and with new strengths.

Amongst the bidders at the Monte Carlo auction was Mr James B. Sherwood, President of the Sea Containers Group, and he purchased two of the Wagons-Lits for £41,600 and £23,400, having lost two earlier bids to the King of Morocco whose agent was buying for the royal train. The two 1929-built sleepers (one of English construction) were to form the nucleus of an idea which was beginning to take shape in James Sherwood's mind, namely to restore the famous 'Orient Express' back to life in all its finery, as a privately run *de luxe* train linking London, Paris and Venice, where Sherwood owns the Cipriani Hotel. For two years the two Wagons-Lits were stored at the Sea Containers depot at Bordeaux, whilst Sherwood investigated the possibilities of his ambitious idea, both in England and on the Continent. He was convinced it could be done, although certain ideas such as using steam locomotives in England, and taking the British Pullman cars across the Channel by train-ferry to join the Continental train, did not prove acceptable to the railway authorities. Instead, he opted for an all-Pullman train on the English side, and a mixed Wagons-Lits sleeping car and Pullman car train on the Continent. There was one big problem — where to get all the necessary additional cars required. Two old Wagons-Lits certainly would not suffice, and in any case they were too large to run in Britain.

It took James B. Sherwood and his Venice Simplon-Orient-Express Ltd (VS-O-E) organisation four and a half years from the date of the Monte Carlo sale, to locate the necessary rolling stock, learn how to restore it, negotiate the routes to be taken, engage the staff, and promote the enterprise. It cost £11 million to restore the British Pullman train and the 'Orient Express' Continental cars. The result was

216
At Steamtown a team of skilled engineers and craftsmen worked extremely hard to restore the assembled fleet of Pullmans, which had to be put back into full main line running order to satisfy BR specifications, as well as restored to their original Pullman finery. Each car was literally stripped to the bones and then put back together again. Here a carpenter is seen making a new oval toilet window for the car *Ione*. All the cars which originally had rectangular toilet or kitchen windows — 1951/52 build — have now received the famous oval type instead.
Sea Containers Ltd (VS-O-E)

217
The Steam Locomotive Operators Association (SLOA) purchased some of the 1960 East Coast Pullmans, which had become surplus to British Rail requirements, in order to have a modern train with both air and vacuum brakes, which would be available for charter use with either steam, diesel or electric haulage. On 13 February 1982 preserved Bulleid 'West Country' Pacific No 34092 *City of Wells* heads a southbound 'Cumbrian Mountain Pullman' service near Cotehill. Due to the fact that no 1960 Pullman brake vehicles were built, standard Mk 1 brake seconds are used, as seen here next to the locomotive. *R. Lumley*

indeed magnificent, and a splendid book has been written by Shirley Sherwood, the wife of James B. Sherwood, entitled *Venice Simplon-Orient-Express, the Return of the World's Most Celebrated Train* (published by Weidenfeld & Nicolson in 1983) which tells the full story of this supreme example of private enterprise, so entirely in the tradition of George Mortimer Pullman and Georges Nagelmackers. The 'Orient Express' features again in the next chapter, dealing with present-day Pullmans in Britain, and for the moment I merely want to outline how Sherwood managed to obtain the necessary cars for what is now called the 'British Pullman Train', although running under the VS-O-E banner. What VS-O-E was attempting to do was quite different from the private restoration and limited running of Pullmans that had so far been achieved. They had to find and restore vintage Pullman cars that would be permitted by BR to run on their tracks as a scheduled service, day-in, day-out. The condition of the bodywork and framing, the brakes, the electric lighting system, the bogies and wheelsets all had to match up to the required safety standards of a modern BR main line railway train. VS-O-E Ltd purchased a number of Pullmans from various sources, in order to make-up the British train. Whereas they wanted the train to be as authentic as possible, by careful restoration of the Pullmans to their original pristine condition and interior design, they also had to face an enormous programme of engineering work and essential repairs, into the bargain, in order to comply with the rigorous BR safety standards of the present day.

At this time BR did not want to part with the working Pullmans that it still possessed — the most modern examples — and the private preservation movement had already secured most of what was intact and available for restoration: although some cars rested in virtually scrap-heap condition, for need of funds to restore them. VS-O-E Ltd had no choice but to search among the examples already preserved or standing idle, and to negotiate the purchase of suitable cars, if their owners would agree. The net was cast far and wide, so that, for example, two suitable British cars were found in France, in use as roadside diners at the Hôtel Mercure, near Lyon as mentioned in Chapter 19. These former 'Golden Arrow' cars, *Carina* and *Phoenix*, were purchased and brought back by road and sea to England, in the course of which *Phoenix* got jammed under a railway overbridge in France, and the tyres on the low-loader road trailer had to be let down in order to release it! Other cars were purchased from private railway enthusiasts such as Mr David Lowther, a schoolmaster at Eton College, who had rescued the 'Brighton Belle' car *Audrey*, and Tony Robinson of Doncaster, who had similarly saved the car *Zena*. It was a long and painstaking search, and in some cases the cars were afterwards found to be unsuitable, because BR decided that it would not allow some wooden-bodied cars to run, although the cars *Ibis* and *Minerva* have steel-plated wooden bodies. BR also demanded that the wooden framed but steel panelled 1951 cars be treated for fire resistance on all the wooden framework; which VS-O-E undertook at Carnforth. Electric heating had to be installed in all the cars and all wiring renewed, and all the glass had to be changed to specially toughened safety glass, including the glass in the mirrors, and the unique oval toilet windows (where the existing antique glass was laminated to toughened glass to comply with safety regulations).

Most remarkable was the superb restoration of the cars' interiors, because over the years many of the lovely marquetry panels had suffered damage, or had even disappeared. In the car *Audrey* it was also discovered that splinters of glass and shrapnel had been embedded in the marquetry landscape scenes by the German bomb attack on Victoria station in 1940, related in Chapter 15. VS-O-E Ltd was able to locate the Dunn family of Chelmsford in Essex, whose occupation was marquetry and whose father had made the marquetry panels for at least 70 railway cars, including many prewar Pullmans. The Dunns did a magnificent job of restoration and renewal, sometimes using the original drawings which had survived in the family possession. No detail, large or small, was overlooked in the restoration of the Pullmans, and the finishing touch was the application of the full umber and cream livery by craftsman Bob Timmins to each car, with a wonderful finish effected by varnish layers, in time honoured Pullman style. The then Editor of *Modern Railways*, Charles Long, said to the Author, after he had inspected the restored cars, that in his opinion they were in better condition than at any previous time in their careers since first built. (Charles Long is himself a former Pullman Car Co employee, and a staunch believer in the Pullman tradition.) The Pullman cars used for the VS-O-E train in Britain were all restored at Steamtown, Carnforth in a specially-built workshop. The 1951/52 cars had to be stripped of blue asbestos insulation before work could commence, of which more anon. Details of further cars, acquired but not yet in service, are given in Table 2.

By the late 1970s the preservation movement, which had commenced life as a collection of various isolated initiatives, had become a highly organised affair, and one very much respected by British Rail. The Steam Locomotive Owners Association (SLOA) foresaw a future problem in the fact that BR was abandoning the use of the vacuum brake for rolling stock, and all modern vehicles were air-braked only. To continue to run steam specials, with locomotives fitted for vacuum brake train working only would soon pose operating difficulties. The answer was found in the 1960 Metro-Cammell East Coast replacement Pullmans, which were being withdrawn by BR in 1979/80, having been displaced by HSTs. In 1981 SLOA bought seven of the second-class parlour cars and the *Hadrian Bar*, so that it was assured of having vacuum-braked stock available for future main line steam running. In fact these Pullmans are now dual-braked (air and vacuum) and they are also dual-heated (steam or electricity) so that they can be worked by the latest diesel or electric locomotives, and their Commonwealth bogies permit a 100mph maximum speed. The saloon layout, and wide windows (the Pullman windows are 5ft wide, whereas on their contemporary standard BR Mk 1 carriages the windows were of only 4ft width) make them ideal for touring purposes. We will return to this train in the next chapter.

Some of the Metro-Cammell East Coast cars went for scrap, but quite a number have been rescued, in addition to the SLOA set, although in some cases the Commonwealth bogies have been retained by BR, for use under their recently refurbished Kent Coast SR electric trains; so that alternative types of bogie have had to be provided if the cars are required to run again. One such example is the car *Amethyst*, owned by Mr Bruce Wilson and located at Boat of Garten railway station, Inverness-shire. *Amethyst* now runs on BR Mk 1 heavy duty bogies, and is used in special 'wine and dine' trains on the Strathspey Railway. Known by the rather subtle name of the 'Boat Train Diner', which features very splendid menus whilst travelling amidst superb scenery, the Pullman is steam-hauled into the bargain. Another of the Metro-Cammell cars, No 340 is now a static extension, or annex, to the Avon Causeway Hotel, which is itself a former railway station building, at Hurn, near Christchurch. The Avon Causeway Hotel is a free house and has become something of a living railway museum and shrine to steam because of the enthusiasm of the proprietors, Mr and Mrs J. N. Palmer. The Pullman car looks completely at home in its setting — unlike some of the other Pullmans that have been used as pub restaurants, or roadside diners — and No 340 has been restored to umber and cream livery. It is often used for meetings and dinners on special occasions, as well as serving as an extension to the bar, where children are allowed in, with their parents. Hurn station was built in 1862 by the London & South Western Railway, and closed by the SR in 1935. The lines were removed in 1937, but the Palmers have placed a length of track alongside the existing platform, to exhibit the Pullman car properly.

Sister Metro-Cammell Car No 337 is also saved from the scrap dealers, but does not rest in such happy surroundings for the moment, as John Jeffrey of the Nene Valley International Steam Railway explained, in a letter to the Author in November 1983:

'This vehicle arrived on the Railway in 1981 and was immediately put into store, where it remains. It was withdrawn from Eastern Region service in 1978 and despatched to Salisbury for conversion into an exhibition coach. In the event this work was not carried out and No 337 was then moved to store at

218
For financial and operational purposes, control of the SLOA Pullmans and associated Mk 1 vehicles passed to the specially created 'Pullman Rail' organisation. Meanwhile a protracted and expensive operation took place to replace the dangerous blue asbestos insulation originally fitted to the 1960 Pullmans and repaint them in authentic umber and cream Pullman livery. Car No 353 is seen in March 1985 after full restoration. *C. J. Tuffs*

219
The Avon Causeway Hotel, at Hurn near Christchurch, has former East Coast Metro-Cammell second class kitchen car No 340 as an annexe used for social events, dinner parties, and as a bar extension in which children are allowed with their parents. The hotel owners have built-up a collection of railway relics, whilst the hotel building itself started life as a railway station, closed by the SR as long ago as 1935. Hence the Pullman car looks very much at home alongside the old platform. *Mrs N. Palmer*

Basingstoke from whence it was purchased by one of our members.

'Presently No 337 still carries its original Pullman umber and cream livery although this is now in a very poor state. The vehicle is usually stored inside Wansford tunnel and emerges only occasionally, during shunting operations. There are no immediate plans to commence restoration of 337, which is basically complete. We are currently building a carriage and wagon restoration workshop, which we hope to complete early in 1984 but the works programme for this new facility is fully booked for at least three years ahead. Unless, therefore, the owner commences the restoration himself (and there is no indication that this is his intention), it seems likely that No 337 will remain in store for the foreseeable future.'

Another working Pullman train set is in the offing, this time on the Severn Valley Railway, where six of the former East Coast replacement stock all-steel 1960 Metro-Cammell cars are based; privately owned by Mr Michael Bailiss, who wrote to the author in November 1983 giving the following details, which once again highlight the tremendous amount of work involved in private preservation, and subsequent return to service, of Pullman cars that had ended their days on BR destined for the scrap dealer:

'I own six Metro-Cammell 1960/61-built cars. Three Kitchen Cars and three Parlour Cars. All are as built for Pullman, including the Commonwealth bogies, apart from the 1967 're-modelling' by BR to increase the seating capacities, and in BR reversed fleet colours of blue and grey. All are first class cars.

'They were originally built for the "Queen of Scots", "Master Cutler" etc, all on the Eastern Region until the end of the Pullmans on ER when all three Kitchen Cars found use for a short while on London Midland Region Kensington-Scotland Motorail services as Restaurant Cars.

'The names & numbers with types are as follows:
Snipe E319 Kitchen 26 seats
Raven E317 Kitchen 26 seats
Finch E313 Kitchen 26 seats
Amber E324 Parlour 29 seats
Pearl E329 Parlour 29 seats
Topaz E331 see note

'Note: *Topaz* was built as a Parlour Car and served as such on BR. Prior to purchase it suffered much internal vandalism and items such as seating and fittings "removed". It is probable, that, subject to finances becoming available, *Topaz* will be used either as a Bar Car or Cocktail Car, in which case it may seat only 7/8 persons. The Kitchen Cars are all in basic working order and one is at Arley Station being externally repainted to Pullman colours. Perhaps two more will follow at both Arley and Bridgnorth over the next 3/4 months. One Kitchen Car is at the moment at Carlisle having been borrowed by BR to examine the dual braking and dual heating systems with a view to fitting the dual braking (in fact the air braking) to the SLOA Cars. This will be returned to the SVR when it is required. *Pearl* is in good internal and external condition except for the electrics wanting attention. The same applies to *Amber* except that the situation is slightly worse.'

Mr Bailiss went on to outline some possible future uses for the restored cars on the Severn Valley Railway, which included:

(1) 'Wine and dine' (similar to the 'Wealden Pullman').
(2) 'Dine and Theatre' — in which an amateur group of Players would stage a presentation on a pleasant summer's night, whilst the Pullman-conveyed audience watch from the station platform, beforehand or afterwards enjoying a dining car sevice similar to the K&ESR 'Wealden Pullman'.
(3) 'Educational' — made possible by the fact that all the cars are fitted with public address system (PA). A tutor could use this to lecture to students as they travelled up the line, advising on matters of geology, geography, land-use and history.

Until such a time as all the cars are restored and can be run as a complete train, individual Pullmans will be added to some ordinary Severn Valley Railway trains, as they become available.

Here I would remind the reader that some time has elapsed between these reports and the publication of the book. There may have been quite significant changes in some cases. Quite a lot of the preserved Pullmans have changed ownership, or been moved around the country, in recent years, and some will probably continue to do so until all find suitable roles and resting places.

In common with other modern carriages, many Pullmans, built since the last war, including the 1951/52 'Festival of Britain' cars, were insulated with blue asbestos; now forbidden on health grounds. The blue asbestos was, in principle, isolated within the car's panelling and bodywork, so that in everyday life the public does not come into contact with it. However, if an accident occurs, or if a hole is created in the protective layers, problems could arise. On BR, in collaboration with the railway trade unions, a major programme has been underway for some years to rid the network completly of all rolling stock contaminated with blue asbestos insulation by the end of 1987, partly by scrapping the older vehicles concerned, and partly by a process of removal of the asbestos in carriages which still have years of service left in them. It is an expensive process of removal, involving strict precautions for the workmen involved. It has unfortunately affected some of the preserved Pullmans (if required to run on BR), and has added considerable costs to their restoration. It has also prevented the purchase of some modern cars, which have had to be scrapped by a special process of incineration, which takes away the asbestos dust, in the meanwhile reducing the car body to a burnt-out shell.

On a much brighter note, the Kent & East Sussex Railway's 'Wealden Pullman', already referred to, was the venue for specially invited guests, including former members of the Pullman Car Co, on 8 August 1981, for a special run to commemorate the 150th anniversary of the birth of George Mortimer Pullman. The Kent & East Sussex Railway had steadily built up the patronage of their Pullman train, which is entirely run and staffed by volunteers, to achieve a remarkable level, whereby in 1980 they had provided 34 train services, all featuring splendid Pullman-style menus for a total of over 2,000 diners, in pre-booked accommodation. The same year saw the introduction of Pullman uniforms for the staff printed Pullman-style menu covers, and improvements in presentation. What had started life as an enthusiastic but amateur venture was now becoming established as a highly-regarded professional level of service, entirely in the traditions of George Mortimer Pullman, and the Pullman Car Co. The K&ESR then made history by introducing a Pullman car of their own creation. The K&ESR acquired a 20-year old BR unclassed restaurant car, No E1955, which had been withdrawn from traffic in 1978. The idea was to augment the accommodation on the 'Wealden Pullman', and to have it ready for the 150th anniversary celebrations, just mentioned. The work included repanelling the passenger area in varnished timber, re-upholstering the seating, and fitting carpets. The spacious kitchen was rebuilt, and all associated plumbing and gas and electricity brought up to scratch. It took the K&ESR, with volunteers, over a year to achieve. When completed, and in full Pullman umber and cream livery, the 'new' Pullman was ceremonially named *Diana*, because 1981 was the Royal Wedding year of the Prince of Wales to Lady Diana Spencer.

Preserved Pullman cars featured at the 'Rocket 150' parades at Rainhill in 1980, in the spectacular line-up of locomotives and trains of the past and the present that was arranged by BR to commemorate the 150th anniversary of the opening of the famous Liverpool & Manchester Railway. To haul the three cars — *Topaz* (I), *Audrey* and *Eagle* (all in variants of the umber and cream livery) — the preserved rebuilt 'Merchant Navy' class Pacific No 35028 *Clan Line* was suitably adorned with the regalia always carried by locomotives hauling the 'Golden Arrow'. The Keighley & Worth Valley Railway also has a replica set of these adornments, which are often carried by 'West Country' Pacific No 34092 *City of Wells*, when working the K&WVR 'Golden Arrow' specials. The cars *Topaz* and *Eagle* are from the National Railway Museum collection, and they had earlier featured, together with the Metro-Cammell 1960 car *Emerald*, also an NRM exhibit, in the special Travellers-Fare '100 Years of Refreshment' train which journeyed extensively around Britain in 1979. It was run to celebrate the centenary of railway dining car services in Britain, the first one, in 1879, being the Pullman car *Prince of Wales*, as related in Chapter 8. The former 'Brighton Belle' car *Audrey* is now one of the Venice Simplon-Orient-Express British Pullman train vehicles, once again in full revenue-earning use.

By the start of the 1980s there were considerably more Pullmans basically intact and in various private collections, or officially preserved, than were left on the BR stockbook, and the number of preserved cars passed as suitable for main line BR running was steadily growing; in particular with the entry into public service of the VS-O-E Pullmans, in 1982. The Pullman revival was really getting up steam!

220
The dedicated volunteer team, drawn from various walks of life, which staffs the Kent & East Sussex Railway's 'Wealden Pullman' — a service of such high quality that bookings have to be made far in advance. Such enterprise befits the name Pullman and demonstrates the qualities of the British railway enthusiast movement today, a highly professional presentation, down to the smallest details. The background is formed by the Pullman car *Barbara*, star attraction of the train. *Donald H. Wilson*

220

TABLE THREE:

VS-O-E Ltd Pullman Cars

British Pullman Cars used by VS-O-E Ltd
Showing Previous Owners

Note:
The VS-O-E organisation has also acquired the following cars: 'No 36' (sch No 194/built 1926); 'No 64' (219/1928); 'No 76' (229/1928); *Agatha* (239/1928); 'No 83' (261/1930); *Vera* (284/1932); *Aquila* (305/1951); and *Carina* (307/1951).

	When Built:	*Where Built:*	*Interior Designer:*	*Type Originally:*	*Type for VS-O-E*	*Acquired from:*	*Decoration:*
Audrey	1932	Metropolitan CW&F	G. F. Milne	Kitchen 20 seats	Kitchen 20 seats	David Lowther	Marquetry landscape panels; Art Deco strip lights
Cygnus	1951	Birmingham RC&W	Waring & Gillow	Parlour 26 seats	Parlour 26 seats	North Yorkshire Moors Railway	Mahogany panels, mirrors, old prints
Ibis	1925	Birmingham RC&W	Unknown	Kitchen 20 seats	Kitchen 20 seats	Standard Gauge Steam Trust, Birmingham	Greek dancing girls marquetry
Ione	1928	Metropolitan CW&F	Maple (?)	Kitchen 20 seats	Kitchen 20 seats	P. B. Whitehouse	Burr wood panels, Victorian frieze
Minerva (II)	1927	Midland RC&W	Unknown	Kitchen 22 seats	Parlour/Brake 26 seats	Lytham Creek Railway Museum, Lancashire	Edwardian-type marquetry
Perseus	1951	Birmingham RC&W	Waring & Gillow	Parlour 26 seats	Parlour 26 seats	North Yorkshire Moors Railway	Yellow wood panels, old prints
Phoenix	1952 (chassis '27)	Metro-Cammell	Waring & Gillow 1981	Parlour 26 seats	Parlour 26 seats	Hôtel-Restaurant Mercure, Lyon	Oval frames of marquetry flowers
Zena	1928	Metropolitan CW&F	Maple (?)	Parlour 24 seats	Parlour 24 seats	Terry Robinson	Art Deco marquetry
Lucille	1928	Metropolitan CW&F	Maple (?)	Parlour 24 seats	Parlour 24 seats	SE Steam Centre	Marquetry

221
Three preserved Pullman cars featured in the lavish 'Rocket 150' cavalcade at Rainhill in 1980. The train was hauled by rebuilt 'Merchant Navy' class Pacific No 35028 *Clan Line*, carrying the 'Golden Arrow' headboard and insignia. It is seen here leaving Bold Colliery on 26 May 1980, with the cars *Topaz* (I) (next to the engine), *Audrey* and *Eagle* in tow, the three cars representing some 50 years of Pullman progress. *G. Roose*

221

CHAPTER 21

The Pullmans back in Vogue

A red-letter day in British Pullman history, was 25 May 1982 when Mr James B. Sherwood gave his inaugural speech to present the 'Venice Simplon-Orient-Express' on its maiden run. The immaculate rake of Pullman cars standing at Platform 8 of Victoria station, ready to take invited guests on the British stage of the service to Venice, vied with the immaculate attire of the milling throng of guests and with the smartly turned out uniformed Pullman staff. Cameras and pressmen were everywhere, representing the whole world. There was already some 15,000 passenger bookings, despite the Falklands crisis which then occupied British attention, and the future of the train was assured. James Sherwood reminded his guests that one hundred years earlier almost to the day, the Pullman car *Jupiter* had left from the same station, for Dover (see Chapter 9) and he said:

'Unfortunately *Jupiter* was not a financial success because the prospective passengers did not want to pay the supplement on the fare to ride in her . . . I trust we will have no repetition of that sort of thing.'

He need not have worried because by the end of the 1984 season a total of 90,000 people had travelled on VS-O-E services, including special charter trains. Sherwood had proved beyond all shadow of doubt that there was a new need for leisurely, luxurious forms of travel, and he chose Pullmans, and Wagons-Lits, to provide it. The public image of Pullman was restored by Sherwood's enterprise and faith, to become almost a cult, setting new fashions for a generation that had thought the train to be a thing of the past, and whose lifestyle had been built around the motor car and the aeroplane since childhood. Many had never set foot in a Pullman car before, and it was a revelation to them to see how their parents and grandparents had travelled — Pullman style! The 'Orient Express' is a true masterpiece, with no detail overlooked, and the standards of service, menus, and passenger-care are exactly as in the great days of George Mortimer Pullman and Lord Dalziel.

The year 1982 had also witnessed the completion of the last of the fleet of High Speed Train units (HST) for British Rail, and the withdrawal from scheduled passenger service of the prototype electrically powered Advanced Passenger Trains (APT-P) due to persistent problems.

If all had gone according to plan, the APTs would be running in squadron-service on the West Coast main line from London (Euston) to the Midlands, North West and Glasgow as I sit and write these words. Instead, due to technical problems during development, the original idea of a fleet of high speed APT trains, able to tilt on curves to reduce passenger discomfort, but without reducing speed, is now officially abandoned. Meanwhile, the LMR electrified West Coast route services will have to soldier on with the locomotives and carriages that should by now have been displaced by APTs. As a result, whereas the East Coast route (now being electrified), the Midland lines, and Western Region services plus others, are worked by the very successful 125mph diesel-electric HSTs, the highest speed permissible on the LMR West Coast route is only 110mph, using selected

222

222
James B. Sherwood, President of the Sea Containers Group, and originator of the 'Venice Simplon Orient-Express', at Victoria station on 25 May 1982, giving his speech before the inaugural run of the 'Venice Simplon-Orient-Express'. Standing on his right is Mrs Sherwood (who has since written a brilliant book around the train's history; see Bibliography) and the gleaming Pullman car behind is *Audrey*, once a 'Brighton Belle' car, and now restored to pristine condition as part of the British Pullman train of VS-O-E. *Sea Containers Ltd (VS-O-E)*

223

223
In the VS-O-E British Pullman train no effort is too great to create a sense of calm well-being, and to some extent of deliberate nostalgia. If the 'Orient Express' has a somewhat theatrical atmosphere that may be described as deliberate, because the train is a high class voyage of leisure and pleasure — a holiday upon wheels akin to a four star hotel and restaurant. Here is the interior of the car *Minerva*, with coffee and dessert being served by the white-gloved attendant. The marquetry panels create the finishing touch to an elegant atmosphere, truly one of travelling in style!
Sea Containers Ltd (VS-O-E)

224

224
At the start of the VS-O-E project, James Sherwood had hoped to use steam locomotives to haul the Pullmans from London (Victoria) to the Channel Ports, but BR declined the idea. Today the train is usually hauled by SR electro-diesel locomotives of Class 73, as seen here approaching Tonbridge on 12 October 1982, with No 73132 in charge of the 12.44 Victoria-Folkestone Harbour VS-O-E working.
Colin J. Marsden

Mk 3 rolling stock and Class 87 electric locomotive haulage. Many WCML InterCity services are still limited to the 100mph level first made possible with electric traction in the 1960s.

The InterCity Sector (Sector management was created within BR by Sir Peter Parker during his chairmanship) was painfully aware of the unfulfilled promise of the APT by the early 1980s, and an interim strategy had to be adopted in order to bring up to scratch the facilities offered to the passenger, in particular on the West Coast main line. This was necessary to combat the growing threat from greatly improved long-distance motorway coaches, internal airline services, and the omnipresent private or company-owned motor car. Any great increase in speed was ruled out by the nature of the route, with its many curves and by the standards of the track. What was possible, pending the arrival of new trains, was an improvement in comfort and service to the passenger. Some of the Mk 3 carriages were coming due for heavy overhaul after about 10 years of intensive main line service, and a scheme was prepared to give these an internal refurbishment, with new APT-style seating, plus other improvements, in order to keep them in the best possible shape. Consideration was also given to ordering a new build of improved Mk 3s, as an interim solution to the inevitable delay before a Mk 4 design could be ready. In the midst of all this, the solitary 'Manchester Pullman' suddenly became the focus of much attention, still running regularly with its Mk 2 era Pullman cars, and still popular with its 'regulars' despite the supplementary fare charged for what was in many respects outmoded accommodation, even when compared to the earliest Mk 3s. The 'Manchester Pullman' cars were scheduled for withdrawal by 1985/86 and when they went, the Pullman marque would no longer exist on BR, except on privately-owned cars operating over BR tracks, in particular the SLOA Pullman train, and the VS-O-E Pullmans of the 'Orient Express', because the InterCity Sector had the idea of introducing 'Executive' trains, with a number of improvements in comfort and service for first class passengers in particular, using the refurbished Mk 3 carriages and a limited build of new ones, to be called the Mk 3b.

Meanwhile the 'Manchester Pullman' was getting a bit shabby, and with some years of service still ahead of the vehicles, the then InterCity Sector Director Mr Cyril Bleasdale decided to give the ever-popular train a facelift to keep it in good shape until the new Mk 3b stock could be built for the 'Executive' services that were being introduced. Rumours that the Pullmans were going to be withdrawn had already fanned the initial fires of protest amongst their devotees, and this, plus the evident popularity of the private Pullman car trains, raised doubts as to the merit of abandoning the Pullman image altogether. As a result, Mr Bleasdale made a bold move. He had the Mk 2 Pullman cars, by then all some 18 years old, overhauled and repainted in the striking new red, white and grey colour scheme which had originally been specially devised for the APT fleet. Moreover, he put back the Pullman coat of arms on the sides of each car, and he gave every car a name, in true Pullman tradition. The interiors were give a thorough facelift, with new furnishing materials, and the 'Manchester Pullman' was the subject of a special publicity exercise, with posters and other features prominently displaying the Pullman image. In October 1983 the revamped Pullman train was relaunched, and the passenger receipts started to rise immediately, until by March 1984 they were no less than 20% over those of the same period the previous year. A special 'Pullman Club' was formed, with membership limited to travellers on the

225

225
In October 1983, the by then 18 year-old, Mk 2 Pullmans of the 'Manchester Pullman' — the sole surviving BR service still carrying the famous Pullman name — were revamped with a striking new red and grey livery and were for the first time given individual car names. This was on the initiative of Cyril Bleasdale, then Director of the InterCity Sector. The same basic livery has since been adopted as the standard InterCity livery. In no time 'Manchester Pullman' receipts began to improve, aided by some good publicity and the improved appearance of the cars themselves. Being vacuum-braked only, the Mk 2 Pullmans could not be hauled by Class 87 electric locomotives, which are fitted with air brakes only. Class 86/2 No 86258 *Talyllyn* heads the 16.42 Manchester (Piccadilly)-Euston 'Manchester Pullman' on 14 May 1984. *Paul D. Shannon*

train; which still ran twice daily each way between London, Euston and Manchester Piccadilly. Without question, Pullmans were back in vogue, and Cyril Bleasdale was now convinced that they had a future role to play in his InterCity services; as we shall see later.

In the previous chapter I have related the problems of blue asbestos insulation, and the necessity to rid any car of it, if it was found to be in a potentially dangerous state. In late 1983 the SLOA Pullman car train, which was in regular use for charter running on BR tracks, revealed traces of blue asbestos dust entering the interiors during a routine inspection by BR staff at Carlisle Upperby, where the train was stabled and maintained under an annual contract with BR (gaining for it the popular nickname of the 'Carlisle cars'). SLOA had already spent over £50,000, mostly raised by railtour revenue and donations from enthusiasts, on restoration of the cars. It did not possess the necessary funds to pay for the removal of the asbestos over the next winter period. Happily a railway enthusiast, the well known businessman The Hon William MacAlpine (who had earlier saved the locomotive *Flying Scotsman*, as recounted in Chapter 19), came to the rescue by purchasing all the cars from SLOA, and arranging for the work to be done at his workshops at the Steamtown Railway Museum at Carnforth, which had already been involved in the restoration work on both the SLOA train and the 'Orient Express' cars. Each Pullman car was enveloped in a huge plastic cocoon and the asbestos sucked by machine out of the crevices. Once 'clean' the new insulation, manufactured by Fibreglass Ltd, replaced it. It took a team of four people six days for each car, to install the slabs of fibre glass below the floor and in the walls.

The SLOA Pullman car train has now been relaunched under the name Pullman Rail, and the objective is to provide an up-market charter train, with a choice of motive power: steam, diesel or electric (thanks to the fitting of dual braking) — although steam traction can only be used on routes agreed with BR. In a splendid publicity brochure entitled 'Privacy and Panache with Pullman Rail' the umber and cream livery with the Pullman coat of arms is used as the symbol of the new service. Amongst clients who have already used the services of Pullman Rail are British Telecom, who took potential customers to East Anglia to look at the latest developments in communications technology, and the Ministry of Defence who took a party of overseas military attachés to an Army proving ground to see the latest Challenger tank put through its paces. Pullman Rail also agree to the private hire of just one Pullman car, or more, and for the customer to make arrangements direct with BR for the route and locomotive; although Pullman Rail will undertake all such arrangements as part of the package. The complete Pullman Rail train can carry 348 passengers, complete with bar, kitchen and restaurant. If required, sleeping cars can be attached. The Pullman Rail emphasis is upon privacy and style. British Rail management for their part are enthusiastic supporters of this private enterprise, which is helping to restore luxury and elegance to railway journeys all around Britain.

The British Pullman Train of VS-O-E is also available for charter when not needed for scheduled services, and VS-O-E have themselves created an imaginative programme of day tours using the cars as an essential part of the attraction, with champagne and 'Buck's fizz', cream teas, luncheons and suppers, in full 'Orient Express' style served during the day. The destinations chosen for each excursion are among the most prestigious and beautiful of Britain's heritage, including Beaulieu, Bath and Bristol and Canterbury, and where necessary, luxury road coaches connect with the Pullman train to reach destinations such as Hever Castle and Leeds Castle. The VS-O-E Pullmans have also become a favourite way — in true Pullman tradition — to travel to race meetings, such as the Cheltenham Gold Cup, the Grand National, Ascot Diamond Day and Newbury Races. It is possible to charter the train in the various periods when the 'Orient Express' is not running, when the British Pullmans are not on day tours or race specials organised by VS-O-E Ltd. The Pullmans are intensively used, to justify the expense of their upkeep, although there are of course times when they have to be taken out of service for essential maintenance and repair. The Pullman Car Co had an established tradition for the revarnishing and repainting of their cars, whereby every three years they used to be revarnished, and every nine years, they were repainted. VS-O-E Ltd has respected this tradition, and in the winter of 1984, Bob Timmins, who painted the cars at Carnforth in 1981, carefully rubbed down and revarnished their exteriors; so that they commenced the 1985 season in tip-top condition.

The Bulmer Cider Pullman train which, as I have shown in Chapter 19 spearheaded the return of preserved steam — and Pullmans — to British Rail, is unfortunately not running on BR at present because, although in running order, the cars are not registered for use on BR lines. All five cars require an official 'lift' for bogie examination and brake overhaul before BR will accept them. (As this book was going to press it was learnt that these cars had been acquired by the VS-O-E organisation. Whether they were intended as spares for the VS-O-E Pullman train, or presaged some new venture, remained unclear, however.)

Although VS-O-E and Pullman Rail, and the revamped BR 'Manchester Pullman' stole much of the limelight during the years 1982-85, a great deal of further hard work was also put into Pullman cars by preservation groups and private owners over the same period. Some old cars, although 'saved' for posterity in theory still lingered on in dreadful condition, awaiting funds and volunteers for their restoration. Work started on restoring other cars, but this is in itself a prolonged activity, mostly undertaken at weekends and in holiday periods, and sometimes working out-of-doors and in all weather conditions. A classic example of recent Pullman car restoration, entitled 'Barbara Gets a Facelift', written by Philip Shaw was published in the summer 1984 edition of the *Tenterden Terrier*, the journal of the Tenterden Railway Company Limited; proprietor of the Kent & East Sussex Railway. I am grateful to all concerned for permission to reprint this feature, which gives a fascinating insight into the enthusiasm and sheer dedicated energy that has rescued so many fine Pullman cars for posterity.

'Pride of place in the Kent & East Sussex rolling stock collection goes to Pullman cars *Barbara* and *Theodora* — the last survivors of six narrow bodied vehicles, built in 1926 to run between London and Hastings. Both were condemned in December 1963, but rescued en route to South Wales for scrapping and delivered to Robertsbridge in September of the following year. Since 1974, *Barbara* has given sterling service as the bar car on "Wealden Pullman" trains, whilst *Theodora* has remained as a buffet at Tenterden, pending the erection of a purpose built catering block. Minor body repairs have been made from time to time and *Barbara* was last stripped and repainted in 1977, but it was not until this year, with the commissioning of the new carriage shed, that a long awaited major overhaul could be undertaken. Furthermore, as the vehicle is in constant use during the operating period, it was essential that the work should be finished in around four months, in time for the commencement of Pullman services at Easter.

'Under the guidance of Carriage & Wagon Department Manager, Paul Ramsden, stripping out commenced in early January. The frame, entirely of hardwood construction, was found to be in sound condition, but whereas the upper cladding panels had split, necessitating replacement with marine ply, the original matchboarding below the waistline was found to be in good order, requiring only minor attention and filling. Moulding and cills, where rotted, were replaced with mahogany timbers and the roof re-canvassed and given four coats of roofing compound. All brass fittings were removed from inside and outside the car, polished and re-laquered where appropriate.

'Before external painting could commence, research, mainly through the examination of old photographs, was necessary in order to restore the car to an authentic livery, as *Barbara* has appeared in many guises during her long life. After the war, she had her name removed and replaced, first by the legend "refreshment car", then "buffet car" and, as a final indignity, repainted in Southern green in 1958 and lettered "buffet". In the 1984 repaint, which is as far as possible in the 1930s style, three coats of coach enamel were applied, followed by two coats of varnish; lining out and lettering by professional sign writer, Brian Hart, took five days. Internally, the car has not changed significantly since its last refit in 1958, when the bar was lengthened and seating reduced. *Barbara* originally entered service as a first class kitchen car seating 20 passengers, which was later rearranged to accommodate 12 first and 11 third class; the change to a bar counter and 17 unclassified seats, stand seats and stools took place in 1946. The present accommodation is for eight seated passengers in Pullman armchairs (not original to the car) together with several stand seats; a bar pantry and kitchen make up the service accommodation, although meals as such are not normally served in this part of the train.

'All of the polished interior woodwork was completely stripped, treated with oxalic acid to bring out the grain and then given several coats of varnish. The exquisite panelling, a feature of all early Pullman cars, was removed and sent to Messrs A. Dunn & Son of Chelmsford, Essex, who, it is almost certain, carried out the original marquetry work in 1926. Mr Robert Dunn, Senior, recalls that the family firm started in 1898 and was responsible for the panelling in some 70/80 Pullman cars from 1910 — the last contract being carried out for the Brighton Belle in 1932. The main floral panels, of which there are 14, took an average of 20-30 hours each to restore. There was a great deal of unseen damage — wet rot to the backs which had to be stripped off, re-glued where sound and then relaid with waterproof plywood. The front veneer of Cuban sawn cut mahogany had to be carefully lifted, re-glued and pressed by hand clamps to get flat; then all the dirt, old varnish and polish, almost 1/32in thick in places, taken off with a mild paint stripper; next a very light rub down with sandpaper to bring back the original colours almost as good as new. Finally, one of the most difficult problems — to effect repairs invisibly and match the woods from stocks, many of which had been held for up to 80 years; Cuban mahogany, Honduras mahogany, boxwood, kingwood, rosewood, maple, sycamore, holly and sycamore dyed colours were all used and not only did Dunns restore the existing panels, but they replaced a missing one with an exact reproduction. Re-assembly and polishing of the panels was then carried out by members of the K&ESR team and at the same time, the opportunity was taken to instal a concealed burglar alarm and public address system.

'*Barbara* is no ordinary coach; she has style and elegance for those who demand the highest standards of cuisine and attention. She has earned her keep many times over in the "Wealden Pullman" trains; how long will it be before sister car *Theodora*, rejoins her rightful place in the Railway's premier service?'

There is a third old Pullman car on the Kent & East Sussex Railway, which arrived there in 1980, after 40 years of neglect, standing in all weathers, in use as a departmental vehicle in railway yards near Sheffield. It is the grand old twelve-wheeled, wooden-bodied car *Cambria*, built by Clayton in 1920 for the 'Hook Continental' London-Harwich service of the Great Eastern Railway. It is in a deplorable state, but given time and funds it will be restored to its 1924 condition as a brake parlour first, as used on the LNER East Coast main line, and then it will take its place in the 'Wealden Pullman' formation. Around Britain there are still a number of similar future restorations, stored away pending time and money. An acute example is the Car No 54, also built by Claytons, in 1923, and which was eventually purchased by James Sherwood's organisation, Sea Containers Ltd, after being in the hands of private owners since withdrawal in 1963. It was intended for the 'Orient Express' and was in a partially dismantled state during overhaul at Carnforth when BR decreed that only steel-panelled cars would be acceptable for main line running. The overhaul was abandoned, the car still in pieces, and now it has been sold in that state, awaiting removal by a private buyer associated with the Bluebell Railway, who also have the car *Fingall* awaiting restoration, and the former 'Brighton Belle' car *Bertha* which is owned by Mr Raymond Martin. In a letter to the Author in January 1984 Mr Martin explained that *Bertha* had been under restoration for 12 years — 'these things take time' — and that it will be used for 'wine and dine' specials on the Bluebell, once complete. It highlights the very different time scale involved for a small private preservation project, when compared to the amazing achievements of the 'Orient Express' and Pullman Rail organisations in less than half the time that *Bertha* has been under restoration.

226

226
The VS-O-E Pullman train was in the vanguard of the revival of Pullman charter travel on British Rail. When the 'Orient Express' proper is not scheduled to run, the cars themselves are made available for a variety of day trips to tourist attractions, or for charter to race meetings and other special events. The former SLOA Pullmans of the Pullman Rail organisation are also available for charter work, and BR InterCity itself is in the midst of similar developments on its own behalf, using the remaining Mk 2 Pullmans (see next chapter). Pictured here is the VS-O-E Pullman train on a 'Devon Belle' charter special, passing New Malden en route from Waterloo to Kingswear on 2 April 1983, headed by Class 33 locomotive No 33056. *Colin J. Marsden*

CHAPTER 22

InterCity Pullman Returns in Style

The vast carriage and wagon building and repair complex of workshops of British Rail Engineering Ltd (BREL) at Litchurch Lane, Derby, are just a short walk from the former site of the Pullman Palace Car Co's workshop, where George Mortimer Pullman's American-built cars had been assembled and repaired in the years 1874-83. His wooden-bodied cars had arrived in pieces in packing cases, and were carefully put back together, and given the finishing touches by hand. At BREL in the first months of 1985 a new production line of all-steel Pullman cars was underway, using the most modern techniques of assembly. Traditional craftsmanship still plays its role in the production of new railway rolling stock, in particular in the finishing touches, but the process of construction has undergone a revolution since the war, particularly in the use of pre-fabricated modular components; such as form the basis of the Mk 3b Pullman cars which have now been introduced by British Rail's InterCity Sector.

The Mk 3 carriage design, developed in the 1970s, has an impeccable service record as the front rank main line passenger carriage on BR, including its use in the High Speed Train (HST) sets. It gives a superlative ride at speeds up to 125mph (and more) and the passenger accommodation — all in open saloons — is comfortable, and has proved to be generally acceptable to the travelling public. The body design, which is of stressed steel, in uniform modular dimensions, is easily adaptable to suit first or second class seating or, if necessary buffet or other types of catering cars. The window apertures are the same for all, and in the case of first class cars they are spaced to coincide with the layout of seating and tables. This latter layout, which is a parlour car in Pullman terminology, has been adopted for the new Mk 3b Pullmans, with 2+1 seating using reclining seats of a design developed initially for the APT.

The Mk 3b Pullmans, also feature improved air conditioning, a new diffused interior lighting system, *plus* individual reading lights set in the luggage racks, and the décor has been improved with bulkhead carpeting, featuring abstract motifs. Compared to Pullman cars of the past, there is none of the traditional marquetry, the individual table lamps, or the private coupé-style compartments. In every other respect the new vehicles match up to the fine standards of the past. The tables are covered by fine quality white cloth when in service, the tableware is of first class design and manufacture, and the menus are devised and presented in true Pullman style. The exterior of the cars is finished in the current InterCity livery — which is described as a band of dark grey at window height, with light grey to the guttering above

227

227
The latest British Rail Mk 3b all-steel InterCity Pullman cars, photographed at the BREL Derby Litchurch Lane workshops in 1985, when just completed, and not yet individually named. Tokens of faith in the future of Pullman services on BR, these splendid new fully air-conditioned vehicles, capable of 125mph smooth running, were built just down the road from the site where George Mortimer Pullman's British workshop had been busy reassembling his Detroit-built wooden Pullmans for the Midland Railway more than a century ago. One retrograde feature of the Mk 3bs is that the passenger entrance doors open *outwards*, instead of inwards, due to the desire to retain standard BR parts in their production, but, in all other respects they are direct developments of the 'Blue Pullman' and Mk 2 Pullman designs of the 1960s. *BR*

228

and from below the waist. One broad white, and one broad red stripe run along the entire length of the sides just below the windows — a livery repeated on the electric locomotives of Classes 86 and 87 which haul the new Pullmans.

In the previous chapter I have related how the success of the revamped 'Manchester Pullman' convinced the then InterCity Sector Director, Mr Cyril Bleasdale, that retention of the Pullman marque (a name to which the BRB holds the British rights) had definite advantages. The outcome of that 1983 essay has now taken shape in the form of the InterCity Pullman services — regarded as the flagships of the fleet. The first three services went into operation on 13 May 1985, with the introduction of the summer timetable, between London and Manchester, Liverpool, Leeds and Bradford; followed by a fourth in the autumn, providing a London-Newcastle service. The trains were named: the 'Manchester Pullman'; the 'Merseyside Pullman'; the 'Yorkshire Pullman' and the 'Tees-Tyne Pullman', respectively. A further service, the 'Lancashire Pullman' (between London Euston and Blackpool) was introduced with the start of the May 1986 timetable, to be followed a year later by the 'Master Cutler' (now St Pancras-Sheffield) and the 'Birmingham Pullman'. The last is something of an oddity in that it runs in the southbound direction only — on a breakfast service from New Street to Euston! While the Manchester, Liverpool, Lancashire and Birmingham services are all locomotive-hauled formations, the other three are maintained by refurbished High Speed Train (HST) sets.

The introduction of the four services in 1985 involved an investment of around £5 million in new Pullman cars for the West Coast services, of the new Mk 3b design. The other two services, running on the East Coast feature existing Mk 3 cars suitably refurbished, with new style seats, within existing High Speed Train (HST) formations. All the Pullman services have full table service and special menus.

In fact the new Mk 3b Pullmans are only a part of a £12 million order for 60 new carriages and Pullman cars. The Mk 3bs represent 22 trimmed out as luxury Pullmans, and 16 as ordinary first class carriages; with three more first class carriages having guard and baggage accommodation as well. The remaining 19 new vehicles are second class carriages, of improved interior décor and comfort, and are used to increase seating capacity on the busy HST InterCity 125 services (which continue to suffer from overcrowding at peak periods). At the same time BREL is committed to a seven-year programme commenced in 1984, to refurbish 1,265 existing InterCity Mk 2f and Mk 3 carriages, at a cost of some £16 million. This work includes provision of new seats of APT style in the first class and modified seats with new moquettes in the second class. All will carry the new InterCity livery, and will complement the Pullmans and the repainted locomotives in this respect. The first 28 refurbished locomotive-hauled Mk 3a carriages were back in service at the same time as the first of the new Pullman cars. Clearly the InterCity Sector means business in its efforts to turn express passenger train operations from loss to profit, as has been requested by the Conservative government. Spearheads of the effort are the Pullman cars, with an accent on comfort and stylish service, entirely in Pullman tradition.

The new Pullman services also feature at least one car per train which is equipped with a coin-box radio-telephone cabin from which passengers can make national and international calls; a scheme which has been extended to more than 130 InterCity trains (including the Pullmans) on the main trunk routes. This is a service developed in conjunction with British Telecom. Significantly — because InterCity passengers value the privacy provided during a train journey — *incoming* calls will not be accepted.

The new services to Manchester and Liverpool, which commenced on 13 May 1985 displaced the 19-year-old 'Manchester Pullman' cars. The new trains are not exclusively formed of first class Pullman cars, and there are now some refurbished

228
Interior of a Mk 3b Pullman car, one of 22 new vehicles produced for BR InterCity services in 1985, all of which carry the Pullman coat of arms and names. The subtle lighting is a pleasing new development, complete with small individual reading lights inset into the luggage racks, to augment the diffused ceiling lights, but the traditional Pullman table lamps are no longer considered necessary. New seats, of a design initially developed for the Advanced Passenger Train (APT) project, have been further refined in style and comfort, and feature a control mechanism for reclining the back. The decor is in subtle shades of warm greys and matching decorative abstract motif fabrics cover the end bulkhead walls. *BR*

229
In conjunction with British Telecom, the new Pullman services feature 'Trainphones', coin-operated public telephones which work whilst the train is travelling at speeds up to 125mph. The kiosk is located at one end of certain of the Pullman cars. Calls can be made to phones anywhere in the United Kingdom, or to any of the 161 countries that can be dialled direct from Britain. *BR*

229

second class carriages as well, for use without a supplementary fare; and the idea is to provide at-seat catering throughout the train, including the second class. The full Pullman first class seat service (initially commanded a £2 supplementary fare (the same for all the new services introduced). Whereas the old 'Manchester Pullman' only made four runs daily (two in each direction), the new Manchester service was launched with *10* Pullman trains per day, including one each way calling at Stoke-on-Trent. The new 'Merseyside Pullman' service provided eight trains each day between London, Euston and Liverpool (Lime Street) and vice versa. All the Pullman cars on the Manchester and Merseyside routes featured the names of famous people associated with Manchester, Liverpool and Stoke-on-Trent (see Table 4b for details) and some of the names were chosen by the public in local competitions.

For the initial three new Pullman services, Travellers Fare, who were responsible for the catering, created no less than 57 new jobs — 26 in London, 16 in Manchester, 10 in Liverpool and five in Leeds — for stewards and stewardesses, in order to ensure a high level of personal at-seat service, in keeping with Pullman tradition. Smart new uniforms were designed to give a finishing touch to the services, and the Pullman coat of arms proudly featured once again as a lapel badge — gone forever, one sincerely hopes, are the stupid references to it as 'a badge of shame'. Nobody can afford such intolerant attitudes between railway unions and management in this present highly competitive climate, when InterCity is fighting to win.

Prior to the launch of the new InterCity Pullmans, the first Pullman lounge was opened, at King's Cross station on 22 March 1985 by Lord Brabazon, the Government transport spokesman in the House of Lords. A further six Pullman lounges were subsequently opened during 1985/86 at Euston, Liverpool Lime Street, Manchester Piccadilly, Edinburgh, Newcastle and Leeds. The King's Cross lounge, on platform 8, is typical. It has seating for 25 people. Light refreshments are served throughout the day with full bar service during licensing hours. Complimentary tea and coffee are always available. Facilities in the lounge include Teletext, Oracle and Prestel television information terminals, portable payphones and a photocopier in addition to hotel reservation and car-hire facilities. Entry to the lounge is restricted to passengers holding full fare or Executive rail tickets who are also American Express card holders. According to BR, this entry requirement is discretionary and those fulfilling the conditions should be able to take guests into the lounge with them. The Pullman lounge at King's Cross is open from 07.00 to 23.00 Monday to Friday. The aim is to provide Pullman style personal service *before* passengers board their train, considerably reducing the discomforts of any unforeseen delay.

The key to the new InterCity Pullmans is the emphasis upon personal service, at the seat, for every Pullman passenger. To get some idea of the scale of this initiative, the 1985 'Manchester Pullman' service, each weekday, offered 2,090 Pullman seats compared with only 816 on the trains it replaced, and each train on the 'Merseyside Pullman' (the first Pullman cars between London and Liverpool since the early 1970s) provided 161 seats, eight times a day; in other words 1,288 Pullman seats per day, making a total of 3,318. To add sources of potential Pullman patronage, all the 'Manchester Pullman' services called at Stockport in both directions, and one of the services was routed via Macclesfield and Stoke-on-Trent each morning to London, and each evening from London. The other eight 'Manchester Pullmans' all called at Wilmslow, whilst the Liverpool 'Merseyside Pullmans' called at Runcorn in each direction. Six Manchester and six Merseyside services were scheduled to call at Watford Junction — InterCity's important railhead station within easy reach of suburbs north and west of London. To begin with the usual Pullman practice of charging a supplementary (£2) fare was followed but in 1986 this was abolished. Travellers Fare*

*InterCity Catering took over Travellers Fare's on train main line catering duties in May 1986.

The 'Merseyside Pullman' and the 'Manchester Pullman' from 13 May 1985 (weekdays only)

UP SERVICES

Liverpool Lime Street	0710			1055		1600		1915	
Runcorn	0727			1112		1617		1932	
Manchester Piccadilly		0715	0750		1200		1640		2000
Stockport		0723	0758		1208		1648		2008
Wilmslow			0805		1216		1655		2016
Macclesfield		0736							
Stoke-on-Trent		0758							
Watford Junction	0925	0930				1815	1849	2145	2229
London Euston	0946	0954	1017	1345	1444	1840	1911	2206	2250

DOWN SERVICES

London Euston	0750	0800	1115	1130	1600	1750	1925	2020	
Watford Junction	0806	0816		1146	1616	1631	1807		
Stoke-on-Trent							1941		
Macclesfield							2001		
Wilmslow		1013	1333			1829			2247
Stockport		1024	1345			1840	2019		2259
Manchester Piccadilly		1033	1353			1849	2028		2309
Runcorn	1004			1349	1812			2142	
Liverpool Lime Street	1025			1411	1833			2204	

230

produced a series of new Pullman menus, offering a wider range of dishes, including the 'Great British Breakfast' which has become famous for its grill tray, and there is a wider choice of starters at breakfast time. Various *a la carte* dishes give alternatives to a varied *table d'hote* menu during the day, complete with carefully chosen wines as well as cocktails and aperitifs. At every seat, each Pullman passenger is now offered a complimentary fruit juice, or cup of tea or coffee before the end of the journey. It is this high level of passenger care which is so completely in the Pullman tradition, even if the new cars themselves no longer carry such hallmarks as oval lavatory windows, marquetry panelling and individual table lamps (all of which, one has to admit would seem a little out-of-place, rather like a fresh-faced young girl wearing too much make-up).

By the autumn of 1985 it was clear that the Pullmans were back in style! The official BR newspaper *Rail News* carried this report in the October issue:

'The Pullman comeback on BR is proving a runaway success.

'New luxury first class services introduced in May between Leeds, Manchester, Liverpool and London have brought businessmen flocking from their cars and the airlines during the first 12 weeks of operation. And the trend is continuing as plans to extend and strengthen Pullman services go into action.

'With customers spending between £70 and £80 a head for an executive package that includes meal vouchers, first class Pullman travel, reserved seats, car parking and car hire discounts on top of personal on-train service, the Pullmans have already proved that people are willing to pay more for quality.

'As well as booming business, timekeeping, especially of the Leeds to King's Cross "Yorkshire Pullman" has been remarkable by any standards, while the high-quality service given by specially-recruited Travellers Fare staff has earned special praise from InterCity managers.

'Already, "Yorkshire Pullman" crews have twice broken all previous records for restaurant car takings in one day, making it BR's biggest earner.

'The most outstanding and easily-measured success has been with the "Yorkshire Pullman". New levels of service over the Leeds Executive service it replaced have sent first class business on the southbound train soaring by 20 per cent and it has doubled on some return trips.

' "We've improved the product for first class customers by giving them a better deal and personal service which is bringing more people on to the trains," said Terry Wilkinson, Eastern's InterCity customer services and marketing manager.

'The increase in business on London Midland Region, which runs five Pullmans daily each way between Manchester and Euston and four between Liverpool and Euston, is not so easy to measure. That is because the entire West Coast main line timetable was recast with their introduction, when Pullmans became the premier first class services.

230

New uniforms, in Pullman tradition, have been designed for the InterCity Pullman staff, as seen here where the chief steward (or conductor in true Pullman parlance) waits on the platform with his seat reservation chart. Note the old Dalziel company coat of arms on his lapel! It is expected that a new house-style will be evolved for the Pullman services by the BR Design Panel. *BR*

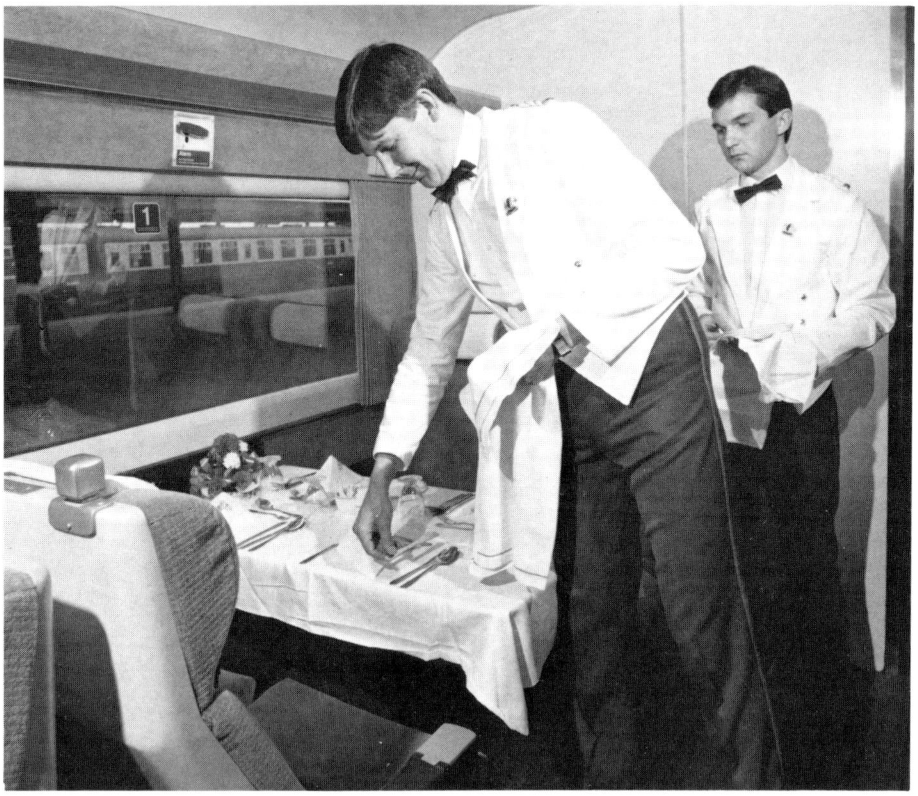

231

'Loadings on one easily measured service, the 16.40 Manchester-Euston, are up by a staggering 80 per cent compared with last year's old 16.42 Pullman.

'Generally, the best increases range from 15 per cent on the 07.10 Liverpool to Euston to nearly 39 per cent on the 16.15 Euston to Manchester. The 07.15 Pullman from Manchester via Stoke is carrying 38½ per cent more first class passengers than two previous trains — the 07.00 and 07.40 — put together with up to 180 Pullman passengers a trip in four new Mk III first class coaches.

' "Travellers Fare staff have been specially chosen and trained to a very high standard, " said Alan Bayliss, London Midland's InterCity sleepers and catering assistant. "They are cultivating customers and getting to know regulars, giving a high degree of personal service."

'On the "Yorkshire Pullman" a personal trolley service is also giving a major boost to catering business in the second class section of the train.

' "Over the first 12 weeks the 'Yorkshire Pullman' was early into King's Cross on all but three occasions and then the worst delay was only five minutes, " said Terry Wilkinson.

'That the timekeeping is so good, says Mr Wilkinson, is due in no small part to staff at Leeds Neville Hill depot where the InterCity 125 sets used for the service are maintained.

' "Drivers also treat the Pullmans as something special."

'Pullman services on both regions are now being extended to win even more business and the famous Newcastle to King's Cross "Tees-Tyne Pullman" — which last ran in 1976 — has already made a comeback with timetable alterations on 30 September.

'The "Yorkshire Pullman" had been expanded with an extra first class coach in place of second class to take on more business attracted by an extra stop at Wakefield.

'On London Midland, InterCity men are examining a number of options to extend and improve quality and profitability of Pullman services, including a possible new service from the North West.'

232

231
For the East Coast main line the introduction of InterCity Pullman services was achieved by refurbishing the interiors of selected Mk 3a HST first class carriages to Pullman standards. The 125mph HST fleet will continue to operate all the InterCity services from King's Cross until the main line electrification to Scotland is completed; thus it was logical to use them to form the new Pullman package. On all the new Pullman services, new crockery and cutlery has been introduced. In this picture the final touches are being added to the tables by Pullman staff of the 'Tees-Tyne Pullman' service, introduced in September 1985, using HTSs. *BR*

232
It was on 2 May 1985 that British Rail unveiled the new InterCity Pullman services, when press and travel trade representatives were invited on special demonstration trips to Liverpool. A West Coast train with new Mk 3b Pullmans was run from London (Euston) hauled by Class 87 electric locomotive No 87008 *City of Liverpool* (on the right) repainted in the new livery. The 'Yorkshire Pullman' was represented by a HST set (on the left) with refurbished Pullman-style accommodation, also painted in the new livery. This ran specially for the day across from Leeds to Liverpool via Manchester, and by a clever piece of stage management on BR's part, the two new Pullman trains ran into and out of Lime Street station, Liverpool, side-by-side at carefully co-ordinated speed. The two trains are seen here at Lime Street after arrival together; with their respective train crews posed for the camera. *BR*

There was more to come! The same issue of *Rail News* had indeed reported the return of the 'Tees-Tyne Pullman' as due to commence on 30 September 1985, which they reminded readers, had not run since the introduction of the HST fleet, which had displaced the remaining East Coast Pullmans, and subsequently the famous 'Deltic' locomotives. This was under the

InterCity 125

banner headline: 'Famous Pullman back in Service', and read as follows:

'After an absence of nine years, the Newcastle to King's Cross "Tees-Tyne Pullman" is back. Formed of a refurbished InterCity 125 it replaces the "Newcastle Executive", leaving Newcastle at 06.47, calling Durham, Darlington, York and Newark, arriving King's Cross at 09.59.

'The return service leaves King's Cross at 17.30, calling Doncaster, York, Darlington and Durham, arriving Newcastle 20.42.

'The highly-successful "Yorkshire Pullman", reintroduced last May, is to make extra stops at Wakefield Westgate.'

What they could *not* report however, because the event took place whilst *Rail News* was running off the presses, was that even before public introduction, the new 'Tees-Tyne Pullman' had set up a new start-to-stop diesel world speed record!

Exactly 50 years to the day since the famous pre-service demonstration record-breaking run of the new LNER London-Newcastle 'Silver Jubilee' train, when the new streamlined Class A4 Pacific *Silver Link*, designed by Sir Nigel Gresley had topped 112mph, and drawn worldwide attention the the train, BR InterCity laid on a similar pre-service demonstration for the 'Tees-Tyne Pullman'. As a mater of interest it was also 160 years to the very day since the opening of the Stockton and Darlington Railway.

On Friday, 27 September 1985, three days prior to entry into public service, the revived 'Tees-Tyne Pullman', formed by a specially-prepared HST set with invited guests, members of the press, and representatives of the *Guinness Book of Records*, sped from Newcastle to London (King's Cross) in 2hr 19min, at an average overall speed of 115.4mph; touching 144mph on one stretch near Essendine, where in 1938 the Gresley Class A4 Pacific *Mallard* had established the world speed record for steam traction of 126mph. The final 100 miles into London were covered at an astonishing average speed of 120.4mph, and arrival at King's Cross was half an hour inside the previous 2hr 51min fastest time set by a HST in 1983 and 10min ahead of the pre-set record schedule for the run! Throughout the journey the pasengers were wined and dined in true Pullman style, and of interest is the fact that the Travellers Fare crew included eight entirely new recruits, chosen from over 400 applicants to join the Pullman service, and who had started to train only eight weeks beforehand.

233
An executive-liveried HST set forms the 'Yorkshire Pullman' promotional train on a York-Leeds-Liverpool and return working. Photographed at Paddock, Huddersfield, on 2 May 1985 during the outward journey.
L. A. Nixon

233

InterCity Pullman means business, and it deserves to win. Indeed, spurred on by its own success and by the success of the privately-run VS-O-E and Pullman Rail leisure-based charter operations over BR tracks, British Rail InterCity has decided to retain 10 of the withdrawn Mk 2 'Manchester Pullman' vehicles, and to try to sell the remainder, rather than scrap them. The 10 BR cars are PFK Nos 504 and 506; PFP Nos 546/8-553 and PFB No 586. These vehicles are free from blue asbestos insulation, and have been refurbished for further life as a special charter train set, restricted to VIP-type work only. Air brakes are fitted, and because names clashed with those of some of the new Mk 3b Pullmans (in the cases of 506/46/49/50/1/3 which were the same as new 11079/81/2/92/3/5) the old cars have been renamed. The other cars from the old Mk 2 'Manchester Pullman' sets had asbestos insulation and have been withdrawn. They are to be stripped of spares to keep the 10 in traffic for the next decade — unless a buyer comes forward with enough capital to pay for the removal of the asbestos.

This new role as 'flagships' of the BR charter business for the 10 'Manchester Pullman' cars will be restricted, to begin with, to the West Coast main line electrified area, because the train heating system can only be powered by an ac electric locomotive. However, with the completion of the 'missing link', North London electrification in October 1988, the Pullman cars will be able to work into East Anglia, including Cambridge, Ipswich and Norwich with electric locomotive haulage. More exciting still is the 1989 completion date for King's Cross-Leeds electrification and then finally all the way through to Scotland. In this context the use of the name Pullman will vie with the VS-O-E and Pullman Rail privately-run charter trains, both of which carry the Pullman name and coat of arms with the blessing of BR!

Indeed, we should not overlook, in conclusion, those other ambassadors of the name Pullman on Britain's railways today; all of whom have played their part in bringing about the 'Pullman Revival'. Whether we talk of the 'Wealden Pullman' — giving elegant dinners in the vintage car *Barbara* as the steam-hauled train pauses in the moonlight upon a quiet stretch of line amidst Kentish watermeadows, so that no vibration from the old bogies beneath the car can upset the pleasant evening — or of the refined 'Venice Simplon-Orient-Express', whose British Pullman cars are one of the ingredients of many a once-in-a-lifetime romantic holiday — or of the Pullman Rail charter trains, giving privacy and panache to groups of travellers, sometimes from far afield — we are undoubtedly talking of travelling in style.

It is as a tribute to *all* who have worked so hard to retain, or regain, the pride and essential magic of the very name Pullman, that this book has been written.

Oh — and I nearly forgot. Included prominently on the InterCity Pullman menu, for the 'Great British Breakfast' are those 'marvellous, juicy and succulent' grilled kippers, once threatened with withdrawal by British Rail on economy grounds — along with the Pullman trains themselves!

Postscript: Finally, I must beg the reader's indulgence when it comes to relating *contemporary* Pullman affairs. Producing a book of this magnitude involves many months of work by author, printer and publisher. In the meantime the story goes on. For example, the 'Bournemouth Belle' has been reintroduced by VS-O-E Ltd and further important purchases of old Pullmans have taken place, and more are rumoured. Perhaps — in a way — it is this continuation of the 'Pullman Revival' that makes the story so worthwhile!

234

234
Fifty years to the day since Sir Nigel Gresley's locomotive *Silver Link* topped 112mph with the inaugural 'Silver Jubilee' streamlined express train a new start-to-stop diesel world speed record was created by a HST, to launch the new 'Tees-Tyne Pullman' service on Friday, 27 September 1985. The Pullman train ran from Newcastle to London in just 2hr 9min, at an average overall speed of 115.4mph and with a top speed of 144mph which was achieved on the famous descent of Stoke Bank, where the Gresley LNER Class A4 Pacific *Mallard* had achieved the world speed record for a steam locomotive, of 126mph on 3 July 1938. The East Coast Pullmans are also breaking records of another kind, because twice in the summer of 1985 the 'Yorkshire Pullman' Travellers Fare crews broke all records for restaurant car takings for a single day — making the train BR's biggest earner. As a BR press release put it: 'The revived Pullmans are on the right track'. Long may they remain so! On 9 October 1985 the up 'Tees-Tyne Pullman' approaches Stevenage. *David Percival*

TABLE FOUR:

Pullman Cars, 1966 to the Present Day

Pullman Cars for British Rail InterCity Services 1966 to the Present Day

Despite the sometimes uncomfortable ride provided by the 'Blue Pullman' diesel-electrics, the 'Midland Pullman' service introduced in 1960 between London, St Pancras and Manchester had proved to be popular with businessmen in particular. Hence, when the Euston-Manchester main line was electrified, giving an alternative and quicker journey time between the two cities, the London Midland Region opted for new Pullman trains (also to serve Liverpool) working over the newly-electrified route, to replace the 'Blue Pullmans'. For the new services, vacuum braked locomotive-hauled cars, suitable for 100mph running and fully air-conditioned, were provided. The integral steel bodyshell was basically that of the then new BR Mk 2 carriage, with the excellent B4 and B5 (heavy duty) bogies used beneath, to provide a ride infinitely superior to the 'Midland Pullman'. To conform to Pullman tradition, inward-opening bodyside doors were fitted at each end, whilst the general style and décor of the interiors followed the example created by the 'Blue Pullman' design, although the style of seating was improved. The cars were finished in the reverse BR blue and grey livery adopted for Pullmans, and carried no names. All 29 cars were built as first class, with 2+1 seating.

In due course the Liverpool service (which ran as part of an ordinary train with second class stock attached) was withdrawn, but the Manchester train carried on, to become the last BR Pullman still running, as related in the text. In October 1983 the surviving Pullman cars of the 'Manchester Pullman' service were relaunched, following a programme of heavy overhaul, and refurbishing. The livery was changed to the APT-style of light and dark grey with bold stripes of red and white at waist level, and the surviving cars were named in true Pullman tradition (each name being that of an individual associated with the area served by the train). No Pullman schedule numbers were allocated by BR. When first introduced there were 29 cars, five of which were normally 'spare' vehicles.

BR Mk 2 Pullman Cars. All built at BR Derby Works 1966

BR Nos:	*Name (after 1983):*	*BR Class/Pullman Description:*	*Type of Bogie:*
M500	*Sir Richard Arkwright*	PFK (18 seats)/Kitchen	B5 (heavy duty)
M501	*Sir Charles Hallé*	PFK (18 seats)/Kitchen	B5 (heavy duty)
M502	*Sir Joseph Whitworth*	PFK (18 seats)/Kitchen	B5 (heavy duty)
M503	*John Owens*	PFK (18 seats)/Kitchen	B5 (heavy duty)
M504*	*Robert Owen*	PFK (18 seats)/Kitchen	B5 (heavy duty)
M505	—	PFK (18 seats)/Kitchen	B5 (heavy duty)
M506*	*Francis Egerton*	PFK (18 seats)/Kitchen	B5 (heavy duty)
M507	—	PFK (18 seats)/Kitchen	B5 (heavy duty)
M540	*James Fraser*	PFP (36 seats)/Parlour	B4 (standard)
M541	—	PFP (36 seats)/Parlour	B4 (standard)
M542	—	PFP (36 seats)/Parlour	B4 (standard)
M543	*Sir William Fairbairn*	PFP (36 seats)/Parlour	B4 (standard)
M544	*Emmeline Pankhurst*	PFP (36 seats)/Parlour	B4 (standard)
M545	*John Dalton*	PFP (36 seats)/Parlour	B4 (standard)
M546*	*Elizabeth Gaskell*	PFP (36 seats)/Parlour	B4 (standard)
M547	—	PFP (36 seats)/Parlour	B4 (standard)
M548*	*Francis Burnett*	PFP (36 seats)/Parlour	B4 (standard)
M549*	*James Joule*	PFP (36 seats)/Parlour	B4 (standard)
M550*	*Ernest Rutherford*	PFP (36 seats)/Parlour	B4 (standard)
M551*	*L. S. Lowry*	PFP (36 seats)/Parlour	B4 (standard)
M552*	*Ford Madox Brown*	PFP (36 seats)/Parlour	B4 (standard)
M553*	*Sir John Barbirolli*	PFP (36 seats)/Parlour	B4 (standard)
M580	—	PFB (30 seats) Parlour Brake	B4 (standard)
M581	*Sir Humphrey Chetham*	PFB (30 seats) Parlour Brake	B4 (standard)
M582	*Daniel Adamson*	PFB (30 seats) Parlour Brake	B4 (standard)
M583	*Peter Mark Roget*	PFB (30 seats) Parlour Brake	B4 (standard)
M584	*Charles Scott*	PFB (30 seats) Parlour Brake	B4 (standard)
M585	—	PFB (30 seats) Parlour Brake	B4 (standard)
M586*	*John Rylands*	PFB (30 seats) Parlour Brake	B4 (standard)

Note: These cars were withdrawn from the 'Manchester Pullman' service in 1985, when new BR Mk 3b Pullman cars became available. The 1966 cars were originally fitted with vacuum brakes only and were incompatible with later air-conditioned Mk 2 and Mk 3 stock, fitted with air brakes, and could not be worked by air-braked locomotives. Nevertheless, 10 of these vehicles have been retained for special charter work (see Chapter 22). The cars concerned are marked with an asterisk (*) in the list above. Because many of their names clashed with those used on the newer Mk 3b cars however, they have been renamed (see list on right).

The unnamed cars in the list above were all withdrawn before 1983.

BR Mk 2 Pullman Cars. New names

BR No:	*Name (from 1987):*
IC 504	*Ullswater*
IC 506	*Windermere*
IC 546	*Coniston Water*
IC 548	*Grasmere*
IC 549	*Bassenthwaite*
IC 550	*Rydal Water*
IC 551	*Buttermere*
IC 552	*Ennerdale Water*
IC 553	*Crummock Water*
IC 586	*Derwent Water*

The set as a whole is known as the 'Lakeland Pullman'.

In May 1985 the Mk 3b Pullman cars started to enter service. They are all 48 seat first class parlour cars (PFP) and kitchen facilities are provided in other vehicles included in the trains concerned. Each car has two lavatories; some cars also have a phone kiosk. BR BT10 bogies are fitted, capable of 125mph running, and the vehicles are air-braked only. They are electrically heated, and equipped with public address.

Originally, specific cars were allocated to specific services as indicated in the table (right), and their names were chosen accordingly. But no more cars have been named since those shown and, as this book went to press, there appeared to be renewed doubts about the 'Pullman' branding and car naming policy, since some sets, such as that used on the 'Birmingham Pullman', are also used to form *non*-Pullman workings.

For the East Coast main line 'Yorkshire Pullman' and 'Tees-Tyne Pullman' and the Midland 'Master Cutler' services existing HST sets had their Mk 3a first class accommodation refurbished to Pullman trim. Because the HSTs also work ordinary non-supplement InterCity services, individual car names were not allocated although a suggestion has been made that the kitchen cars could carry names in Pullman tradition. To denote a Pullman working a special plaque carrying the Pullman coat-of-arms and the train name is inserted in a display window which is part of the toilet window at each end of each car. The livery is exactly the same as for West Coast Mk 3b Pullmans except for the lack of names and central coat-of-arms on the bodysides. No specific car details are thought necessary here, because the HSTs are interchangeable, although a determined effort is made to keep the Pullman sets in tip-top condition.

BR Mk 3b Pullman Cars. All built at BREL Derby 1985

BR Nos:	*Names:*	*BR Class/Pullman Description:*	*Type of Bogie:*
Liverpool service:			
11073	*William Ewart Gladstone*	PFP (48 seats)/Parlour	BT10
11074	*Thomas Brassey*	PFP (48 seats)/Parlour	BT10
11075	—	PFP (48 seats)/Parlour	BT10
11076	*John Lennon**	PFP (48 seats)/Parlour	BT10
11083	*Kitty Wilkinson*	PFP (48 seats)/Parlour	BT10
11084	*William Roscoe*	PFP (48 seats)/Parlour	BT10
11088	*Sir John Brunner*	PFP (48 seats)/Parlour	BT10
11089	*George Stubbs*	PFP (48 seats)/Parlour	BT10
Manchester service:			
11077	*Sir Richard Arkwright*	PFP (48 seats)/Parlour	BT10
11078	*John Owens*	PFP (48 seats)/Parlour	BT10
11079	*Francis Egerton*	PFP (48 seats)/Parlour	BT10
11080	*Emmeline Pankhurst*	PFP (48 seats)/Parlour	BT10
11081	*Elizabeth Gaskell*	PFP (48 seats)/Parlour	BT10
11082	*James Joule*	PFP (48 seats)/Parlour	BT10
11085	*Sir John Barbirolli*	PFP (48 seats)/Parlour	BT10
11087	*Sir William Fairbairn*	PFP (48 seats)/Parlour	BT10
11090	—	PFP (48 seats)/Parlour	BT10
11092	*Ernest Rutherford*	PFP (48 seats)/Parlour	BT10
11093	*L. S. Lowry*	PFP (48 seats)/Parlour	BT10
Manchester via Stoke service:			
11086	*Henry Doulton*	PFP (48 seats)/Parlour	BT10
11091	*Sir Stanley Matthews†*	PFP (48 seats)/Parlour	BT10
11094	*Arnold Bennett*	PFP (48 seats)/Parlour	BT10

* Chosen in a competition in Liverpool.
† Chosen in a competition in Stoke.

235

235
Arrival of the record breaking inaugural run of the 'Tees Tyne Pullman' at King's Cross on 27 September 1985. A time of 2hr 19min 37sec for the 268-mile journey between Newcastle and London resulted in an average speed of 115.5mph, a world record for a diesel-powered train. Photograph of head board and trailing power car No 43158. *Anthony Smith*

Appendices

1 Pullman Car Classification

The following details, extracted from official sources by George Behrend, first appeared in his book *Pullman in Europe* (Ian Allan, 1962) and are reproduced here by kind permission of the author:

Pullman cars, like ordinary carriages, had route restrictions allotted to them, if necessary, by the owners of the lines over which they ran. Prior to the grouping of the railways in 1923, the South Eastern & Chatham arranged for the Pullman cars then in service on their lines to receive type letters, which indicated certain operating features, described below, and indicated route restrictions. This system was peculiar to the SECR, and cars identical in general dimensions but in use on other railways at the time did not receive these type letters, which were as follows:

Type 'A': Four-wheel bogie, all-timber cars, with Pullman-pattern gangways, automatic couplings and retractable buffers.

These, the first Pullman cars to be introduced on to the SECR in 1910, were followed in 1914 by some of

Type 'B': Six-wheel bogie, all-timber cars, 6ft longer than Type 'A', but otherwise generally similar with Pullman gangways, automatic couplings, and retractable buffers. Type 'A' could therefore couple to type 'B' and through passage was possible. After the war these were joined by

Type 'C': Four-wheel bogie cars with timber bodies but steel bogies, rebuilt as Pullman cars from the SER Hastings Car Train Cars supplied by Gilbert. Bellows gangways, screw couplings and side buffers were fitted.

These were the smallest Pullman cars and, until 1926, the only ones which were able to run between Tunbridge Wells and Hastings; all others were restricted from this line. They would couple to ordinary carriages (when fitted with bellows gangways through passage was possible), but not to Types 'A' or 'B'

Type 'D': Four-wheel bogie cars with timber bodies on steel under-frames, rebuilt as Pullmans from the eight SER 'Folkestone Vestibuled Limited' cars supplied by Metropolitan C&W Co Ltd, and one by Jackson Sharp. These cars had special pattern gangways, screw couplings and side buffers.

Until about 1925, these cars could not be coupled to *any* Pullman car except another Type 'D', through gangway passage being possible between them. In addition, they could not be coupled to any railway rolling stock possessing a gangway. About 1925 the special gangways were replaced by Pullman gangways. Type 'D' could then be coupled to Types 'A', 'B' or 'F' (below) but not to any carriage possessing bellows gangways, including Pullman cars Types 'C' or 'E'.

Type 'E': Four-wheel bogie cars with timber bodies, mounted on separate steel underframes. The underframes were ex-LNWR ambulance train coaches built in 1918. Bellows gangways, screw couplings and side buffers were fitted.

These cars would not couple to other Pullman cars having a Pullman gangway. They would couple to Pullman cars Type 'C' or any railway vehicle without gangway, or with a bellows gangway, when through passage was possible.

In 1923, the newly-formed Southern Railway introduced four cars of a new type, generally similar in dimension to Type 'A', but one end vestibule was fitted with brake control and small tip-up seat for the guard. These are designated 'Guard Parlours' in this book to distinguish them from Brake Parlour Cars having a luggage compartment. As the SECR system was perforce continued for some years after the Grouping these became

Type 'F': Pullman gangway, automatic coupling, retractable buffers. They would couple to Types 'A', 'B' and after 1925 to 'D' as well.

Naturally the lot of the Pullman operator was made more difficult by not being able to place two Pullman cars next to each other in the same train if one had a Pullman gangway, and the other a bellows gangway.

On the LBSCR (Central Section of the SR from 1923 onwards) the following types of Pullman were in traffic:

Pre-1908 cars, some of which had screw couplings and side buffers. Six-wheel bogie all-timber cars.

Four-wheel bogie, timber-bodied cars mounted on railway underframes built by various railways for ambulance trains. Some were on ex-LNWR underframes, generally similar to the 'E' type cars on the SECR (Eastern Section of the SR). Three were built on ex-GWR underframes, and five more on ex-LYR underframes.

Finally in 1923 a new design of standard Pullman car had made its debut. It comprised at first a timber body on steel underframe, but later cars were of all-steel construction. None of these were given type letters before 1932 when it was decided to re-classify the type letter system to include all the cars likely to run on the Southern. The intention appears to have been to include all the cars wherever they were located in Britain, but in practice the type letter was only actually painted on cars running on the SR and not on every one of these, but only on the redesignated 'F', 'G' and some of the 'H' and 'J' classes. From this time it became customary to call locomotive-hauled Pullmans 'Steam' cars although those on the Metropolitan had been electrically hauled over part of their route for over 20 years! The new designation of type letters for 'steam' cars was as follows:

Pullman gangways unless otherwise stated

Type 'F': wooden bogie* The previous Type 'F'.

Type 'G': wooden bogie* The previous Type 'A'.

Type 'H': wooden bogie. The previous Type 'B' plus many twelve-wheeled cars not previously given a type letter.

Type 'J': steel bogie. The previous Type 'E' plus the cars from the SR Central Section on ex-LNWR underframes. Bellows gangways, screw couplings.

Type 'K': steel bogie. The standard Pullman of 1923.

Type 'T': steel bogie. Cars from the SR Central Section with ex-GWR underframes; bellows gangway.

Type 'S': 'Supernumerary'. Some of these cars became non-supplement buffet cars with bar counter. All were formerly Type 'H'.

Type 'U': Gresley bogie, etc, introduced 1951.

The 1960 Pullman cars with standard BR bodyshells, all steel construction, curved body sides:

Type 'B': First class kitchen cars.

Type 'C': Second class kitchen cars.

Type 'D': First class parlour cars.

Type 'E': Second class parlour cars.

Type 'F': *The Hadrian Bar* Bar car (Sch No 354).

All the 1960 cars have Commonwealth bogies.

In 1932, the following were excluded from the type letter scheme: cars of designs earlier than 1910, due for withdrawal soon afterwards; cars built on ex-LYR underframes; cars on the LMSR in Scotland, sold to that company in 1933; cars on the LNER of 1920 design, some of which were later included in Type 'H'; and cars on the Metropolitan Railway (which had no gangways and were specially designed).

The 'Electric' Car Types were as follows:

Type 'A': composite kitchen cars 1st-3rd class for '6-Pul' and '6-Cit' sets.

Type 'B' : First class kitchen cars, for 'Brighton Belle'.

Type 'C': Third class parlour cars, for 'Brighton Belle'.

Type 'D': Third class motor brake parlour cars, for 'Brighton Belle'.

*In the late 1950s, wooden bogies on the 'F' and 'G' class cars were replaced on all cars by steel bogies from scrapped SR 'NOL' two-car electric sets, except on *Montana* and *Emerald*.

2 Pullman Terminology

Over the years the Pullman Car Co built up its own language, and as a result definition of terms is very necessary, because the railway and Pullman meanings of the same words are often different.

Terminology of Cars

Buffet Car
Prior to 1932, a Pullman car with some means of providing refreshment to supplement-paying passengers in Pullman cars was called a buffet car (symbol B). From 1932, the symbol K for kitchen car has been used exclusively for such Pullman cars. The term 'buffet car' means in this book either a railway-owned, but Pullman-staffed railway buffet car usually without a bar, or a Pullman car, sometimes called 'refreshment car' or (when fitted with a bar) 'buffet bar' serving light refreshments, usually accessible without paying a supplement.

Bar Car
A Pullman car in an all-Pullman supplementary fare train, incorporating a bar with stools.

Bar Counter
As above, usually non-supplement. Term used prior to 1932 to distinguish a railway-pattern buffet car from a Pullman kitchen car.

Kitchen Car (Symbol K)
A Pullman car with supplementary fare Pullman chairs or seats and fitted for serving main meals. To avoid confusion, this term and symbol K are now used for all such cars, symbol B not being used nowadays for such cars.

Supply Car
A Pullman car without any seats, serving meals to passengers accommodated in railway vehicles fitted with tables. Equivalent to a railway 'kitchen car.'

Restaurant Car
A railway vehicle serving main meals, without supplement, staffed and victualled by the Pullman Car Company. (ROV)

Dining Car (Symbol D)
A Pullman-owned restaurant car as above with seating normally not requiring a supplement for use, but not always. Early dining cars were sometimes called hotel cars in contemporary reports.

Hotel Car
A Pullman car with sleeping berths, and a kitchen. This type was never used in Britain.

Pantry Car
A railway-owned, Pullman-staffed, refreshment vehicle without any cooking facilities. (ROV)

Parlour Car (Symbol P)
A Pullman car with swivel or armchairs, and sometimes end coupés, but without kitchen, brake control, bar, or sleeping berths.

Brake Parlour Car (Symbol Bke P)
A parlour car, one end of which is fitted with brake control guard's accommodation and van space for luggage, with opening doors.

Brake Parlour Car (Symbol Gd P)
A parlour car with brake control and tip-up seat in one end vestibule, the doors of which are lettered 'Guard'.

Motor Brake Parlour Car (Symbol M Bke P)
An 'electric' Bke P with motorman's compartment for driving multiple-unit 'Brighton Belle' Pullman trains.

'Steam' Car
A locomotive-hauled Pullman car.

'Electric' Car
A Pullman car running in an electric multiple-unit train.

'Diesel' Car
A Pullman car running in an all-Pullman multiple-unit diesel driven train.

Schedule No
Pullman accounting number.

Running Number
Number displayed on the sides of second and third class cars.

Dunkerque Train
The 'Night Ferry'.

Ostend or Calais Service
London-Dover boat train. 'Calais' also refers to London-Folkestone boat trains.

Ocean Liner
A London-Southampton special boat train.

Additional terms used by the British Pullman Car Co

Coupé
Name given to a compartment containing Pullman chairs (four unless otherwise stated).

Pullman couplage
One Parlour car marshalled next to the kitchen end of a Kitchen car.

Hassocks
Footstools (six per first class car).

Runners
Covers placed over carpets to protect them when necessary.

Commissary sheet
Catering and Supplement documentation.

Attendants
Pullman train crew (changed to *Stewards* by BR).

Conductor
Pullman employee in charge of train.

Chef
The person who does the cooking.

Supplement
Supplementary fare for Pullman service.

Acknowledgements to George Behrend and Julian Morel for information contained above.

236
'Britannia' class 4-6-2 No 70004 *William Shakespeare* blasts its way out of Victoria at the head of the 'Golden Arrow' in May 1952. *BR*

236

Acknowledgements

In the Introduction I have recorded my personal discovery of the family feeling that seems to surround all things, and all people, concerning Pullmans. To write this book I had first of all to make contact with a considerable number of people who had something to contribute to the story, and many hours were spent writing letters, or telephoning, and in some cases personally meeting those involved. To each and every one I am most deeply indebted for their generosity in giving their time, and in many cases personal photographs and historical ephemera to add interest. The list below is as complete as I can make it, and I hope nobody has been inadvertently overlooked. My sincere gratitude therefore to : Charles Long (who read the script), George Behrend (a constant help), Julian Morel and Doug Lindsay (two Pullman devotees) and also: Alec Swain, H. C. Casserley, John Edgington, James Cousins, Cyril Bleasdale, Ian Allan, Colin Marsden, Raymond Martin, Graham Warburton, Brian Radford, Roman Schmidt, Mrs P. E. Thornhill, Mrs D. Turner, S. C. Pritchard, Mark Toynbee, K. Roberts, Michael Bailiss, R. Lambert, John Jeffrey, J. R. Hymas, E. T. Taylor, E. C. Haydon, Mrs N. Palmer, Geoffrey Cox, Sonja Scott, M. J. Draper, P. D. Short, Alex Murray, Peter Mills, David Malden, D. E. Reynolds, J. B. Howes, B. J. Holden, Brian Perren, L. N. S. Wooler, Alan Butcher, G. J. Smith, Andrew Fox, John V. Simpson, Sam Morris, Douglas Ferreira, D. Wigley, Heather Batchelor and David Benson, Derrick and Shirley Amos, M. Jöel Robuchon, Arthur Dubin, and Prof Liston E. Leyendecker.

The following companies or Societies, etc, are also thanked for their help:

Sea Containers Ltd/Venice Simplon-Orient-Express.
The Bluebell Railway Preservation Society.
The Ravenglass & Eskdale Railway Co Ltd.
British Rail InterCity, and the Press Officers of the London Midland, Western Eastern and Southern Regions.
Travellers Fare.
North Norfolk Railway plc.
British Railways Staff Association (Western Region).
Bentley Simpson Partnership.
The Kent & East Sussex Railway.
City Industrial Limited.
Pritchard Patent Product Co Ltd, (PECO)
Modelrama, Beer, Devon.
The Southern Steam Trust.
Severn Valley Railway (Holdings) plc.
Western Athletic Clubs Inc USA.
Avon Causeway Hotel, Hurn.
The Dart Valley Light Railway plc.
Steam in Hereford Limited.
Colne Valley Railway Co Ltd.
Nene Valley International Steam Railway.
Strathspey Railway Association.
The Keighley & Worth Valley Railway Preservation Society.
National Railway Museum.
Midland Railway Trust Ltd.

Bibliography

The books that the Author has mainly consulted during the preparation of the manuscript are mentioned in the narrative, where appropriate, and the various quotations therein already have their source indicated. Nonetheless, the reader in search of more information about Pullmans, in particular, may find this list helpful:

Pullman, Julian Morel, David & Charles, 1983.
Pullman in Europe, George Behrend, Ian Allan, 1962.
Pullman and Perfection, F. Burtt & W. Beckerlegge, Ian Allan, 1948.
The Brighton Belle, Nicholas Owen, Southern Electric Group, 1981.
Mansions on Wheels, Lucius Beebe, Howell-North Press, California, 1959.
Grand European Expresses, George Behrend, Allen & Unwin, 1962.
History of Trains de Luxe, George Behrend, Jersey Artists/TPC, 1982 (published in USA as *Luxury Trains*, Vendome Press, New York, 1982).
The London Brighton & South Coast Railway, C. Hamilton Ellis, Ian Allan, 1971.
The World's Rail Way, J. G. Pangborn, Bramhall House, New York, 1894.
The Railroad Passenger Car, A. Mencken, John Hopkins Press, Baltimore, 1957.
The American Pullman Cars of the Midland Railway, J. B. Radford, Ian Allan, 1984.
Venice Simplon Orient-Express, Shirley Sherwood, Weidonfeld & Nicolson, 1983.
Railway Carriages in the British Isles from 1830 to 1914, C. Hamilton Ellis, George Allen & Unwin, 1965.
The Restaurant Car, Geoffrey Kichenside, David & Charles, 1979.
Midland Railway Carriages, R. E. Lacy and George Dow, Wild Swan Publications, 2 volumes 1984.
Midland Carriages — An Illustrated Review, David Jenkinson and Bob Essery, Oxford Publishing Co, 1984.
The American Railroad Passenger Car, John H. White Jun, John Hopkins University Press, 2 volumes 1978.
Le Carrozze Pullman, Renzo Perret, Edizioni Elledi, Turin, 1982 (also published in French as *Les Voitures Pullman*, Les Editions du Cabri, 1983).
Les Voitures Pullman 4001/4030 de la CIWL, J.-P. Lepage, published by the author at Meun, France, 1985.
Wagon Lits — Das Exquisite Reisen, Dr Fritz Stöckl, Josef Otto Slezak, Vienna, 1984.
Night Ferry, George Behrend & Gary Buchanan, Jersey Artists Ltd, 1985.
Britain's Railway Liveries, Ernest F. Carter, Harold Starke Ltd, 1963.
Rail Centres: Brighton, B. K. Cooper, Ian Allan, 1981.
Pictorial Encyclopedia of Railways, C. Hamilton Ellis, Hamlyn, 1968.
Some Classic Trains, Arthur D. Dubin, Kalmbach Pub Co 1964.

Relevant books by the same Author:

Railway Design Since 1830 (2 volumes), Brian Haresnape, Ian Allan, 1968.
BR 1948-1983 A Journey by Design (second edition), Brian Haresnape, Ian Allan, 1983.
Design for Steam 1830-1960, Brian Haresnape, Ian Allan, 1981.
Sealink, Brian Haresnape, Ian Allan, 1982.
British Rail Fleet Survey: 5, High Speed Trains, Brian Haresnape, Ian Allan, 1983.

Many references to Pullmans can be found in the following journals
Railway World, Railway Magazine, Modern Railways, Railway Gazette, The Locomotive, Tenterden Terrier, Railway Observer, Rail News, Model Railway Constructor and The Railway & Travel Monthly.

237
Once upon a time the pride of George Mortimer Pullman's Detroit workshops, an old Midland Railway car ends its days as an abandoned grounded body at Hellifield in 1967.
R. K. Graham